Microplastic

Andreas Fath

Microplastic

Distribution, Avoidance, Usage

 Springer

Andreas Fath
Fakultät Medical and Life Sciences
Hochschule Furtwangen
Villingen-Schwenningen, Germany

ISBN 978-3-662-69843-3   ISBN 978-3-662-69844-0   (eBook)
https://doi.org/10.1007/978-3-662-69844-0

This Springer imprint is published by the registered company Springer-Verlag GmbH, DE, part of Springer Nature.
The registered company address is: Heidelberger Platz 3, 14197 Berlin, Germany

If disposing of this product, please recycle the paper.

Preface

The rafting trip that Thomas Kipp (1st Chairman of the Rafting Association Schiltach) undertook with his self-built raft together with me and my son Enzo, Juri Jander (Master's thesis in the course SBE = Sustainable Bioprocess Engineering), Michael Kipp (son of Thomas) and Martina Baumgartner (Press Offenburger Tageblatt) on Saturday, April 29, 2017, from Biberach (Schwaibacher Bridge) to Gengenbach (Rafting Museum) in glorious sunshine, was a premiere and an unforgettable event of the most beautiful kind for everyone, except for the gentlemen Kipp (Rafting Association Schiltach) (Fig. 1). All participants could only understand the enthusiasm of Mark Twain from 1878, which is associated with the incomparable nature experience during a rafting trip (Twain 2014), even if there are many other more comfortable travel alternatives today, with one single restriction, which I will go into in more detail later.

The limitation is based on the spread of plastic waste along the river, which disrupts the superficial idyll as soon as the focus of the journey can only be directed towards it. The bushes, shrubs, and trees lining the riverbank reveal the plastic yield of a flood through their artificial adornment. As in a rake of a sewage treatment plant everything that the river carries with it gets stuck in the branches. When the water level drops again, the catch becomes visible to everyone, in stark contrast to the green nature.» What is frightening is not just the quantity, as every bush was successful with its filter arms, but the fact that the riverside vegetation only filters the river's edge currents and the main bulk of the plastic load, no longer visible, was transported with the faster flowing midstream into the Rhine and into the sea. This means, the visible plastic waste is only the tip of the iceberg and it is already unmistakable (Fig. 2).

The disposal of plastic waste in the river and its surroundings by some irresponsible and uninformed citizens was, viewed from the middle of the river, shockingly apparent, and it became clear that the growing massive plastic islands in the world's oceans are not exaggerations. Neither the young Dutchman Bojan Slat with his kilometers-long floating tube barriers nor a plastic-eating caterpillar will absolve us of the responsibility for our waters, because only proper disposal of our plastic waste can significantly reduce further increase. Also not visible is the microplastic load, which the water carries with it. A close look at the "plastic harvest" immediately shows what happens sooner or later with the plastic waste. It slowly disintegrates into ever smaller particles and seems to dissolve (Fig. 3).

Fig. 1 Rafting trip of the garbage collection action

Fig. 2 Plastic waste in the branches of the Kinzig

However, the problem is not solved, because what is missing from the plastic film has not dissolved at all, but is found as microplastic in our waters, where it can become a serious threat to marine habitats, but also to humans.

Fig. 3 Torn plastic bag

When I swam the Rhine in the record time of 28 days from source to mouth in the summer of 2014, I analyzed the 1231 kilometers of the river under various scientific questions together with a team of students and staff from different institutes. For the first time, as part of the science project "Rheines Wasser" (www. rheines-wasser.eu), the river was also examined for microplastic pollution along its entire length. From the Swiss Alps to the North Sea, the HFU research team filtered every 100 kilometers 1000 liters of the near-surface river water through an extremely fine metal sieve of a specially made portable filter pump for the project, and the filter residues obtained were meticulously evaluated with the support of the Alfred Wegener Institute on Helgoland. The results are presented and discussed in this book.

Even the Source of the Rhine is Polluted
In total, about eight tons of microplastic particles are carried by the surface water of the Rhine into the North Sea each year. This is only the proverbial tip of the iceberg. The actual burden of the Rhine with microplastics is likely to be many times higher. After all, we have only filtered the water to a depth of 15 centimeters. The majority of the microplastics, however, sink and are located in the lower layers of the river water or in the sediment, which has not yet been investigated.

In total, ten different types of plastic are found in the surface water of the Rhine. Particularly high were the proportions of polypropylene (PP), which is used, for example, for the production of cups, buckets and their plastic lids, and of Polyethylene (PE), from which plastic bags, tubes and other Packaging are produced. Together, these two types of plastic make up around 90% of the particles that the team has filtered out from the near-surface water. This is due to the fact that 15 centimeters below the water surface, one primarily finds those plastic particles that float on the water surface due to their low density—and PP and PE having lower density than water. In deeper water layers, you will probably increasingly find plastics with a higher density such as polyvinyl chloride or polyurethane.

The contamination of the Rhine water with tiny pieces of plastic begins already in Lake Toma at 2345 meters altitude in the Grisons Alps, which is commonly regarded as the source of the Rhine. This result initially surprised me, as sources of contamination, in this more or less untouched alpine landscape, were not immediately apparent.

The alarming results of the plastic pollution in the Rhine, about which I have reported in many lectures and presentations within the project "Rheines Wasser", have inspired me to write this book.

Consistently Too Low Pollutant Measurements?
The potential danger of water pollution by microplastics is primarily determined by certain properties of the small plastic particles that have adverse effects on humans and the environment. Foremost among these is the ability to attract organic pollutants—such as the highly toxic perfluorinated surfactants (PFT)—like a magnet. Microplastics can bind other pollutants, and transport them further in a more concentrated manner. However, since microplastics are usually filtered out of water samples to prevent sensitive analytical equipment from clogging, some of the pollutants are not detected by the standard water tests currently in use, such as those at the Rhine monitoring stations. Therefore, I assume that the contamination of waters with PFT and other organic pollutants, which are not completely eliminated in sewage treatment plants, is higher than the usual measurements revealed to us. Unless additional solid-phase extractions are performed, we get a false picture of the actual quality of our waters. The discrepancies can be quite significant depending on the type, quantity, and surface structure of the microplastic load and depending on the distribution equilibrium.

The fact that plastic particles act as carriers of pollutants, such as PFT, is particularly concerning because organisms living in the waters, as several studies have now shown, ingest microplastics. There, it not only enters the digestive organs, but it can also penetrate into the tissue and body cells—including the adsorbed pollutants. Another risk also comes from the additives such as plasticizers, flame retardants or dyes, which were originally intended to improve the properties of the plastic, but which, in the course of decomposition in water, separate from the plastic and are released into the environment. There is a possibility that these concerning ingredients can also be released from the plastic matrix by the gastric secretions during digestion in fish and also stored in the tissue. Section 2.6 is specifically dedicated to these additives.

Solutions must Address the Root Causes

To minimize the dangers posed by the contamination of waters with microplastics, it is necessary to address the causes of this contamination. A significant source of contamination are so-called Microbeads. These are plastic bodies in the micrometer range, which are mainly produced as additives for almost all types of personal care products—from exfoliants to sunscreens to toothpastes. There are now alternative solids that cosmetic manufacturers are increasingly switching their products to under appropriate pressure from environmental organizations like BUND (www.bund.net/mikroplastik). Here, both the consumer behavior and the willingness of the cosmetics industry to truly sustainable action are crucial to achieve improvements.

This also applies to the handling of plastic waste. Because microplastics also originate from the decay of macroplastics, that is, when larger plastic parts such as discarded PET beverage bottles or plastic bags are broken down into ever smaller components by physical, chemical, or biological means. I experienced firsthand that the Rhine is a huge plastic mill during my swimming marathon. Gravel, sand, and rocks carried by the Rhine are harder than plastics and grind down sunken plastic. Thus, a lot of microplastics are created by purely mechanical abrasion from macroplastics.

The incomplete or improper burning of plastics, the washing of synthetic textiles, sludge contaminated with plastic, sewage sludge or organic waste are further examples that favor microplastic contamination of waters. Often it is short-sighted action that causes us problems—for example, when recycling strategies are not thought through to the end. In the fermentation of expired fruit and vegetable products from supermarkets, for example, the products are often not carefully enough or not at all freed from their plastic packaging before they are shredded. This results in the packaging materials also being shredded and contaminating the fertilizer that is spread on the fields. These shredded plastic residues then end up in the groundwater with the next rainfall or in our rivers and lakes during heavy rain via the sewer system.

The main concern of this book is to create an awareness of the impact of microplastics on humans and nature. It explains the causal relationships between plastic waste, microplastics, and environmental chemicals. In addition, it provides the reader with a working guide to be able to examine microplastic particles and their ingredients themselves. An introduction to the laser diffraction spectroscopy and the infrared spectroscopy, enriched by experiences and practical examples from the industry in the field of plastic analytics, are part of the necessary tools. Sources and dangers of microplastics are presented and results are shown using the example of the most important inland water in Europe, the Rhine. The book is not only about exposing and denouncing grievances, but as a "hydrophilic" teacher, I see myself obliged to continue pointing out the "creeping danger", to continue investigating it intensively scientifically and on the other hand to recognize opportunities on the research side. Chap. 4 shows the reader a perspective on how the special properties of microplastics could be used for water purification.

In honorable duty, I would like to thank everyone who supported the "Rheines Wasser" and "TenneSwim" projects, especially the master's students Jonas Loritz, Juri Jander, and Darius Hummel, as well as Dr. Thorsten Hüffer (University of Vienna) and Dr. Habil, Nikolaus Nestle (BASF), who provided me with some as yet unpublished data. Special thanks also go to the master's student Philipp Walter Neek for drawing the structural formulas. Many thanks to Ms. Birte Bayer from the Biological Institute Helgoland of the AWI, who, under the direction of Dr. Gunnar Gerdts, together with Jonas Loritz, analyzed the Rhine samples for microplastics. In addition, I would like to thank my assistants Ms. Dipl. Ing. Helga Weinschrott and Lars Kaiser, who supervise the physical-chemical and analytical internship with me, from which some results have found their way into this book.

Dear readers, last but not least, I wish you an inspiring and exciting read. You can get additional information at www.rheines-wasser.eu, www.facebook.com/ RheinesWasser and www.tenneswim.org.

Heidelberg on the Neckar   Most sincerely
In May 2018   Prof. Dr. Andreas Fath

Reference

Twain, M. (2014). Collected Works: Journey Around the World; Journey Through Germany. Volume 5 of the Selected Works in twelve volumes by Mark Twain, edited by Karl-Heinz Schönfelder at Aufbau Verlag, Berlin. Translated from the American by Ana Maria Brock. © Aufbau Verlag GmbH Co. KG, Berlin 1963, 2008. Reprinted with kind permission.

Contents

Introduction: Microplastics—a Growing Threat to Humans and the Environment

Contents

The topic of "environmental pollution by humans", whether through CO_2 emissions or radioactive nuclear waste, and how to counteract it, is becoming increasingly important. In this context, the entry of plastics into the environment plays a very special role (Cressey 2016). Plastics can be found almost everywhere in the environment, even far away from human civilization, and they do not degrade within a human lifetime. Thus, we ourselves feel the consequences of this type of environmental pollution on our own bodies and the action motto of many citizens: "Out of sight, out of mind" literally falls at our feet on the beaches (Cressey 2016) or via the sea salt into our food (Jander 2017).

1.1   How Much and Where are Micro-, Meso- and Macroplastics Found?

Since 1976, plastic has been the most used material in the world (ifw-Hamburg 2017). Since its boom in the 1960s, an estimated 8.3 billion tons of plastic have been produced worldwide (Garms 2017). According to a study in the journal *Science*, up to 12.7 million tons of plastic entered the world's oceans in 2015. Due to the further increasing production of plastic, without an immediate drastic change in our consumption behavior and intelligent disposal or reuse, the peak of annual pollution has not yet been reached even in this century. Current forecasts

A. Fath, *Microplastic*, https://doi.org/10.1007/978-3-662-69844-0_1

predict that annual pollution will increase tenfold by 2025 (Hoornweg et al. 2013). It is irrelevant whether the plastic waste is introduced into nature as so-called macro- (>25 mm), mesoplastic—(5–25 mm) or microplastic (particles from 100 nm–5 mm) (Napper et al. 2015).

Microplastic, plastic in the micrometer range, is a threat to the entire food chain in inland waters and oceans (Eerkes-Medrano et al. 2015; Desforges et al. 2015; Güven et al. 2017) that can also reach humans. Microplastics can harm organisms either through toxic plastic ingredients like DEHP or Bisphenol A or due to their large surface act as a "magnet" for toxins and pollutants in the water, for example pharmaceutical residues, and thus smuggle these substances into the microplastic-absorbing organism as a Trojan horse. Polymer particles in the form of nano- and microplastics thus act as vectors for pollutants (Anderson et al. 2016; Fröhlich and Roblegg 2012; Li et al. 2016; Ziccardi et al. 2016).

Animals are harmed both by the consumption of macroplastic in the form of, for example, plastic bags, which disrupts body functions due to its indigest-ibility, and by the fact that microplastics can be deposited in the tissue of plants and animals (EFSA) 2016; Avio et al. 2015; Taylor et al. 2016). The occurrence of microplastics in the environment is increasing, as is the release of the chem-icals contained in the plastics (Teuten et al. 2009). So far, the focus of research has been on the investigation of waters (Lassen et al. 2015; Anderson et al. 2016; Li et al. 2016; Deng et al. 2017). However, as soon as microplastics are deposited in animal tissue, humans are also affected by microplastics through food intake (Rochman et al. 2015).

Primary microplastics are mainly found in cosmetic products, in the size range defined for microplastics. Secondary microplastics are formed by erosion of meso- and macroplastic under thermal, mechanical and photochemical influences.

While the American Congress under the Obama administration passed the "Microbead-Free Waters Act 2015" (https://www.congress.gov/bill/114th-con-gress/house-bill/1321/text), a law prohibiting the use of microplastics in cosmetic products from July 2017 in the USA, in Germany we rely on the insight of the producers. With recipe changes, they react on the one hand to the requirements of environmentally conscious customers, on the other hand, they want to shed the negative image that comes from a company presence on the BUND list as quickly as possible.

But even if we, legally forced or voluntarily, completely refrain from using microplastics in cosmetic products, the microplastic pollution of our waters will hardly change due to irresponsible disposal, because according to the results of the International Union for Conservation of Nature (IUCN), less than 2% of the microplastic amount in cosmetic products is responsible (Boucher and Friot 2017).

Rather, it can be assumed that the pollution of the oceans by plastic waste will continue to increase. According to a study by the Ellen MacArthur Foundation, by 2050 there will be more plastic by weight than fish in our oceans (World Economic Forum 2016). Despite improved waste management, the amount of plastic waste that is washed into the sea from the mainland each year is estimated to be 4.8–12.7 million tons (Jambeck et al. 2015). For the estimate, global data on waste quantity, population density, and economic status of 192 coastal countries were used. The respective infrastructure of waste management was also taken into account. The authors of the cited publication see their forecast of the continuously increasing amount of plastic waste in our oceans confirmed and state that without drastic improvements in waste disposal, this trend will continue.

Due to the resilience of plastic waste, the cumulative amount increases annually, even if the plastic waste slowly breaks down into smaller pieces and eventually into microplastics (if their diameter is less than 5 mm) (secondary microplastics), it still remains present, with all the negative aspects of the microscopic synthetic material mentioned in Chap. 2. The total plastic waste load that is introduced into the world's oceans already has a share of 15–31% microplastics (primary microplastics) (Boucher and Friot 2017). The main source for this is not cosmetic products, which as already mentioned only make up a share of 2%, but mainly car tires and synthetic clothing (Boucher and Friot 2017). Microplastic particles rub off on car tires made of vulcanized synthetic rubber, e.g. from NBR (acrylonitrile butadiene rubber), continuously while driving (Fig. 1.1). Each car driver can measure how much plastic is released into the environment each year by the decreasing tread depth of his tires per year. Tire wear therefore accounts for 28% of the total microplastic waste amount (Fig. 1.1) (Boucher and Friot 2017). The problem is not only the "rubber particles", but also their ingredients.

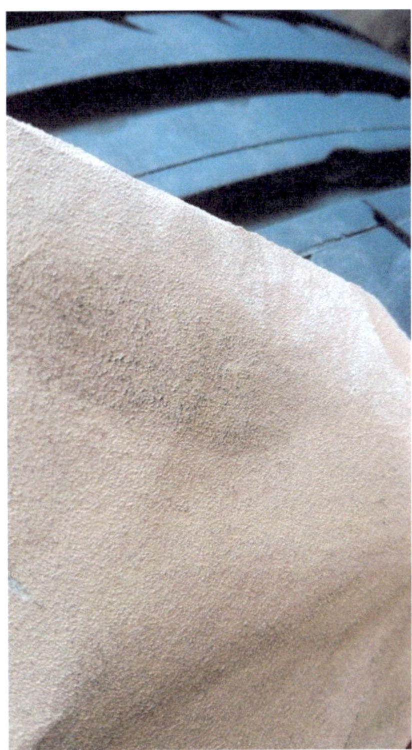

Fig. 1.1 Abrasion of a car tire with a sandpaper

In addition to the PAH-containing plasticizer oils, the plastic tire blackened with soot also contains carcinogenic polycyclic aromatic hydrocarbons (PAHs) (Federal Environment Agency 2016). The properties, detection and further occurrences of PAH are informed in Chap. 2.

The problem of the spread of tiny tire particles and their ingredients arises not only from abrasion, but also during the recycling of old tires, which are used, among other things, as shredded granulate as filling materials, e.g. on artificial turf sports fields (Fig. 1.2). On the plastic turf fields filled with rubber granulate, the granulate between the polyethylene artificial fibers ensures that the fibers are aligned accordingly and the low-maintenance field receives sufficient injury-preventing cushioning. This plastic is even less firmly attached to its place of use than the car tire and is exposed to all weather conditions. As a result, the granulate, which by definition also belongs to microplastics, is found in cleated shoes, jerseys, washing machines, households, parking lots, sewer channels and in the passing river (Fig. 1.2).

Carried by different paths of wind, rainwater, sewage or rivers, all microplastics end up in the sea as a collection basin, including the plastic granulate, which is spread as litter in animal stables together with or instead of straw (Environment Agency Austria n.d.). When the animal excrement is recycled in the biogas plant and the fermentation residues, which still contain the plastic granulate, are spread as fertilizer on the fields, the journey of the microplastics begins from there towards the sea.

Another source of contamination of rivers and seas with tiny plastic particles are synthetic fibers, which are released from the clothing during washing and enter bodies of water with the wastewater. These fibers account for about one third (35%) of the amount of microplastics in the world's oceans (Boucher and Friot

Fig. 1.2 Distribution of plastic granulate on an artificial turf football field

2017). No small problem, as these fibers add up to the equivalent of 15,000 plastic bags per 100,000 inhabitants, as researchers at the University of California in Santa Barbara have calculated (Hartline et al. 2016). But there is now a solution for this too: "Guppyfriend" (http://guppyfriend.com), a fabric bag for the washing machine, in which you put the clothing made of synthetic material before the washing process. This laundry bag made of a high-tech material acts like a filter and retains 99% of the broken fibers. The fact that a lot of fibers come together in one wash cycle is shown in Fig. 1.3. Here, only the swab of the sieve of a clothes dryer is visible. In the actual washing cycle before, an even larger amount of synthetic fibers is released and not retained by a sieve. In the washing drum, the aim during spinning is to separate the water from the clothing as quickly as possible, which would be prevented by a fine sieve. When washing a single fleece jacket, 1.7 grams of microfibers are released (Hartline et al. 2016; Fig. 1.3). In order to reduce the amount of synthetic fibers released into bodies of water, in addition to the laundry bag as an "intrinsic filter", we also have the option of replacing our synthetic clothing with textiles made from natural fibers such as cotton, jute, hemp, linen and silk.

Most polymers are rather non-polar, causing non-polar organic substances in water to accumulate on the surface through adsorption, and depending on the properties of the polymer, within the polymers through absorption. Due to the increased surface area relative to the mass and volume of the particle, microplastics can sorb larger amounts of pollutants than meso- and macroplastics. The uptake of polymer particles by animals and humans occurs rather arbitrarily with decreasing size, as the plastic particles can hardly or not at all be perceived

Fig. 1.3 Synthetic textile fibers. (Swab from the clothes dryer sieve)

(Fröhlich and Roblegg 2012). Due to physiological conditions, such as altered pH value and, for example, existing bile juices or similar in organisms, the desorption of organic substances from the polymer particles is favored. This establishes the bioavailability of the sorbed substances. Depending on the type and amount of substances, these can demonstrably negatively affect organisms (Bakir et al. 2014; Sleight et al. 2017).

The calculated amount of 4.8 to 12.7 million tons of plastic waste, which is transported annually from the coastal states into the sea (Jambeck et al. 2015), is contrasted by an amount of 236,000 tons of microplastic particles, which were determined in 2015 with the ejection of 11,854 nets. In this "combing" of the sea surface, only the Arctic Sea was excluded (van Sebille 2015). So the question arises, where the rest has remained, especially since the amount of plastic washed into the sea in previous years and that which is disposed of directly at sea, was not taken into account at all. Large amounts of plastic waste are collected in the great Pacific and other garbage vortices of the world's oceans (Moore et al. 2001). These amounts are also known. Millions of tons of so-called "Missing Plastic" are elsewhere: Part of it is certainly in the deep sea on the ocean floor, where plastics with a higher density than salt water sink or are pulled down by the attachment of marine organisms to the plastic particles floating on the surface (Woodall et al. 2014). While a portion of the plastic waste is washed back to our coasts, a considerable amount of microplastic accumulates in Arctic ice, which is orders of magnitude higher than in heavily contaminated surface water (Obbard et al. 2014). There are many indications of further microplastic sinks, such as a progressive fragmentation to nanoplastics, which is difficult to investigate and quantify. A difficult-to-estimate proportion of the plastic waste input into the oceans is absorbed by all possible marine habitats, starting with the smallest form of marine life, the plankton (Cole et al. 2013).

1.2 How Dangerous is Microplastic Really?

Many studies on the toxicity of microplastics for different species are based on too high concentrations, which have not yet been reached in waters. The direct influence of microplastics on the fertility was observed in 2016 in Pacific oysters, whose eggs and sperm showed poorer quality and consequently 41% fewer larvae could be produced than in the control group, which was in clean water without microplastic particles. The microplastic concentration to which the oysters were exposed corresponded to that in the sediment of their natural habitat (Sussarellu et al. 2016). A study in the same year showed that perch larvae preferred microplastics of comparable concentration to their natural food occurring in the aquatic environment, thereby significantly worsening their growth and chances of survival (Lönnstedt and Eklöv 2016).

1.3   The Microplastic Problem Starts at our Doorstep and Returns to the House

Not only the coastal states bear responsibility for the pollution of the oceans with plastic waste. Inland waters such as the Rhine also transport plastic waste and microplastics from riparian states into the sea (Fath 2016). Even the Black Forest contributes to the annual load into the Atlantic basin via the Kinzig, a tributary of the Rhine (Jander 2017).

In one cubic meter of Kinzig water, there are about 800 microplastic particles in the size range of 25 µm to 500 µm. Microplastic particles were also found in the stomach of a fish examined from the Kinzig, mainly polyethylene and polypropylene (Fig. 1.4), which it cannot distinguish from a natural food supply.

Compared to the number of microplastic particles in the Rhine (on average about 200/m^3) or the Yangtze (9000/m^3) (Wang et al. 2017), the 800 particles per cubic meter seem very high. However, the total load is crucial for the increasing pollution of the world's oceans with plastic waste. With a volume flow of the Kinzig of 5–10 m^3/s, 15–30 kg of microplastic particles between 10 µm and 500

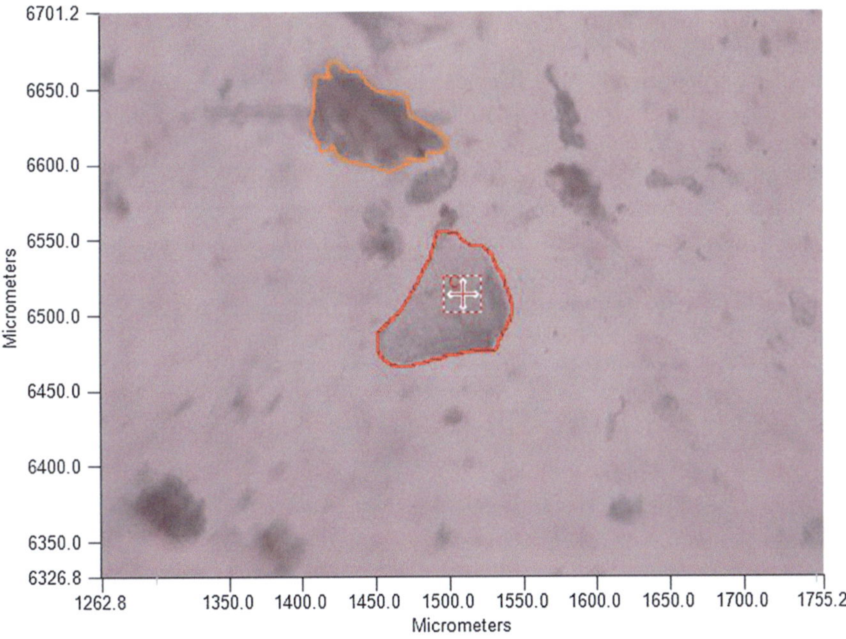

Fig. 1.4 Microplastics under the IR microscope. Polypropylene (PP, red) and polyethylene (PE, orange) from the stomach of the Kinzig fish

µm in diameter flow into the Rhine each year. In comparison, the Rhine transports about 8 tons of microplastics into the North Sea with an average volume flow of 2500 m^3. However, this is only the tip of the iceberg, as only the plastic particles near the surface are included in the calculation. So far, neither a depth profile has been created, nor do we know the sediment concentrations.

Due to the shallower depth and some rapids, plastics with a higher density than water, such as polyvinyl chloride, are measured in the Kinzig, which were detected in significantly lower proportions in the Rhine measurement, as they settle in the deeper river bed. Table 1.1 shows the percentage distribution of the different types of microplastics in 1 m^3 of Kinzig water in April 2017.

Studies on table salt, which is obtained in salt pans, show that some of our plastic waste finds its way back to us in the kitchen and our digestive tract as microplastics. We are essentially eating our own plastic waste. While no microplastics were detectable in rock salt from the mine, up to 18,400 particles per kilogram are found in Mediterranean salt (Jander 2017). Compared to the recently published values of only about 10 particles/m^3 in various commercially available salt products (Karami et al. 2017), the particle count in Mediterranean salt is very high. However, the values are not comparable. In the commercial salt products, only microplastic particles with a diameter larger than 145 µm were quantified. The particles that we (Jander 2017) measured had a diameter of 25–500 µm. This means that the number of smaller microplastic particles is orders of magnitude higher than that of larger particles.

In all samples, mainly polypropylene and polyethylene were found, which due to their lower density float on the water surface and thus predominantly ended up in the examined salt samples, which are obtained in salt pans (Jander 2017).

The detection of plastic particles in a salt sample over 30 years old (Fig. 1.5) from a salt pan in Formentera that was shut down in the 1980s shows that the problem of "microplastics in waters" did not just arise in recent years, but is probably as old as plastic production itself. Only the amount of microplastics has increased with rising production numbers. While the salt of the old sample contained about 6000 microplastic particles per kilogram, the number has tripled in a current salt sample from the same region (Jander 2017).

Table 1.1 Types of microplastics in Kinzig water. (Jander 2017)

PE (Polyethylene)	38.44%
PF (Phenoplasts)	23.12%
PP (Polypropylene)	7.66%
PS (Polystyrene)	15.45%
PSU (Polysulfone)	7.66%
PVC (Polyvinyl chloride)	7.66%

Fig. 1.5   Salt block

1.4   Countermeasures or Fight Against Plastic in the Water

Unfortunately, we can't do anything about the microplastics already distributed in our waters. It will gradually sink to the bottom of the oceans and embed itself in the sediment. However, we can already influence the thickness of this layer, which will be a testament to the plastic generation in a few centuries, by firstly no longer using primary microplastics in cosmetics and hygiene products and secondly reconsidering our plastic waste management and our plastic consumption by strictly following the three "r's": *reduce, reuse, recycle.* The photographed shopping cart in Fig. 1.6 with 51 plastic bags provides enough potential for this (Further tips for plastic reduction are listed in Chap. 4).

The consequences of a change in the composition of plastic waste entering the waters are visible in a short time even far from the entry point. Fulmars are effective biological indicators for these changes. A reduction of plastic granules in plastic waste since the 1980s led to a 75% reduction in fulmars as well as a 75% reduction in the subtropical North Atlantic plastic garbage vortex, while no trend was evident for consumer plastic products (van Franeker and Law 2015). Conversely, the example shows that improved global plastic waste disposal can lead to visible and tangible changes in the environment in a manageable period of time, even before our oil resources will be exhausted. But what to do about the gigantic amounts of macroplastic waste in the large ocean vortices? Boyan Slat, a young ambitious Dutchman, is cleaning up the oceans with kilometers-long U-shaped anchored floating barriers. They are not firmly anchored, but drift slightly slower than the ocean currents with them. In the process, the surface-near

Fig. 1.6   Shopping with 51 plastic bags

floating plastic waste collects in the nets attached under the barriers (The Ocean Cleanup 2018). Even though biologists are skeptical of this project, fearing that fish populations and plankton may be harmed, and even though engineers question the resilience of the barriers against the rough sea, this is so far the only approach to clean up before the plastic waste is no longer reachable. This approach must be supported. The barriers are not trawl nets and therefore no danger to healthy organisms, and the question of functionality under changing weather conditions will be answered in practice.

A completely different question arises in relation to the sorption properties of microplastics. Said properties are already being used in the form of passive samplers in analytics (Sect. 3.1; Müller 2017), in which special polymers are capable of binding pollutants in detectable concentrations and releasing them to a corresponding solvent. Could we not use these sorption properties to clean bodies of water, almost like a filter material? The fact that the sorption of pollutants to polymers is potentially possible has already been demonstrated (Muhandiki et al. 2008; Matsuzawa et al. 2010). If finely ground plastic waste could also be used for this purpose, after appropriate pretreatment for cleaning and roughening the surface, before it is burned anyway, it would become a valuable material again, in the sense of upcycling. The plastic waste would no longer be waste and would not be thoughtlessly discarded in the environment everywhere. This sounds like a utopian idea. Into which Sect. 3.2 provides an outlook (preliminary). On microplastic powder, which was produced from plastic waste using a cryogenic mill, as well as on commercially available plastic powder (sintering material for 3-D printers), the sorption of hormones, depending on the type of plastic and the surface, is investigated and quantified (Hummel 2017). It is determined by using different techniques whether the specific hormone is adsorbed or absorbed by plastic particles (Hummel 2017). This is a crucial question for the further development of microplastic as an effective water filter, to quantitatively remove hormones from wastewater and possibly even recover them through desorption with a suitable solvent and regenerate the filter material.

For the sorption experiments, the hormones 17α-ethinylestradiol (EE2), norethisterone (Nor) and estrone (E1) were used. They are also like DEHP, atrazine, imidacloprid and thiacloprid, to name just a few more representatives of this substance class, among the endocrine disruptors that occur in our wastewater. In sewage treatment plants, these trace substances are not completely degraded and therefore also found in rivers. Even in very low concentrations, hormones are capable of influencing the metabolism and reproduction of aquatic organisms (Manickum and John 2014).

References

Anderson, J. C., Park, B. J., & Palace, V. P. (2016). Microplastics in aquatic environments: Implications for Canadian ecosystems. *Environmental Pollution (Barking, Essex: 1987), 218*, 269–280.

Avio, C. G., Gorbi, S., & Regoli, F. (2015). Experimental development of a new protocol for extraction and characterization of microplastics in fish tissues: First observations in commercial species from Adriatic Sea. *Marine Environmental Research, 111*, 18–26.

Bakir, A., Rowland, S. J., & Thompson, R. C. (2014). Enhanced desorption of persistent organic pollutants from microplastics under simulated physiological conditions. *Environmental Pollution (Barking, Essex: 1987), 185*, 16–23.

Boucher, J., & Friot, D. (2017). *Primary microplastics in the oceans: Global marine and polar programme*. Switzerland: IUCN.

Cole, M., Lindeque, P., et al. (2013). Microplastic ingestion by zooplankton. *Environmental Science and Technology, 47*, 6646–6655. https://doi.org/10.1021/es400663f.

Cressey, D. (2016). The plastic Ocean. *Nature, 536*, 263–265.

Deng, Y., Zhang, Y., Lemos, B., & Ren, H. (2017). Tissue accumulation of microplastics in mice and biomarker responses suggest widespread health risks of exposure. *Scientific reports, 7*, 46687.

Desforges, J.-P. W., Galbraith, M., & Ross, P. S. (2015). Ingestion of microplastics by zooplankton in the Northeast Pacific ocean (eng). *Archives of Environmental Contamination and Toxicology, 69*(3), 320–330.

Eerkes-Medrano, D., Thompson, R. C., & Aldridge, D. C. (2015). Microplastics in freshwater systems: A review of the emerging threats, identification of knowledge gaps and prioritisation of research (eng). *Water Research, 75*, 63–82.

EFSA (European Food Safety Authority). (2016). Presence of microplastics and nanoplastics in food, with particular focus on seafood. *EFSA Journal, 14*(6), 1–30.

Fath, A. (2016). *Rheines Wasser*. München: Hanser.

Fröhlich, E., & Roblegg, E. (2012). Models for oral uptake of nanoparticles in consumer products. *Toxicology, 291*(1–3), 10–17.

Garms, A. (2017). Globale Statistik: Plastik-Welt. http://www.spiegel.de/wissenschaft/natur/plastik-menschen-haben-mehr-als-8-milliarden-tonnen-produziert-a-1158676.html. Zugegriffen: 23. Nov. 2017.

Güven, O., Gökdağ, K., Jovanović, B., & Kıdeyş, A. E. (2017). Microplastic litter composition of the Turkish territorial waters of the Mediterranean Sea, and its occurrence in the gastrointestinal tract of fish. *Environmental pollution (Barking, Essex: 1987), 223*, 286–294.

Hartline, N. L., Bruce, N. J., Karba, S. N., Ruff, E. O., Sonar, S. U., & Holden, P. A. (2016). Microfiber masses recovered from conventional machine washing of new or aged garments. *Environmental Science and Technology, 50*(21), 11532–11538. https://doi.org/10.1021/acs.est.6b03045.

Hoornweg, D., Bhada-Tata, P., & Kennedy, C. (2013). Comment waste. *Nature, 502,* 615–617.

Hummel, D. (2017). *Untersuchung der Sorption wässrig gelöster organischer Substanzen an Polymerpartikel.* Masterthesis, Nov. 2017, HFU.

ifw-Hamburg. (2017). Alles Plastik: Kunststoffe in Wissenschaft und Alltag. http://www.ifw-hamburg.de/daten/id_104/110715_albis_seiten.pdf. Zugegriffen: 23. Nov. 2017.

Jambeck, J. R., et al. (2015). Plastic waste inputs from land into the ocean. *Science, 347*(6223), 768–771. https://doi.org/10.1126/science.1260352.

Jander, J. (2017). *Mikroplastik in Flüssen und Lebensmitteln.* Masterthesis, HFU.

Karami, A., et al. (2017). The presence of microplastics in commercial salts from different countries (eng). *Scientific Reports, 7,* 46173.

Lassen, C., Hansen, S. F., Magnusson, K., Hartmann, N. B., Rehne Jensen, P., Nielsen, T. G., & Brinch, A. (2015). Microplastics Occurrence, effects and sources of releases to the environment in Denmark. www.eng.mst.dk.

Li, W. C., Tse, H. F., & Fok, L. (2016). Plastic waste in the marine environment: A review of sources, occurrence and effects. *The Science of The Total Environment, 566–567,* 333–349.

Lönnstedt, O. M., & Eklöv, P. (2016). Environmentally relevant concentrations of microplastic particles influence larval fish ecology. *Science, 352,* 1213–1216.

Manickum, T., & John, W. (2014). Occurrence, fate and environmental risk assessment of endocrine disrupting compounds at the wastewater treatment works in Pietermaritzburg (South Africa). *The Science of The Total Environment, 468–469,* 584–597.

Matsuzawa, Y., Kimura, Z.-I., Nishimura, Y., Shibayama, M., & Hiraishi, A. (2010). Removal of hydrophobic organic contaminants from aqueous solutions by sorption onto biodegradable polyesters. *Journal of Water Resource and Protection, 2*(3), 214–221.

Moore, C. J., Moore, S. L., Leecaster, M. K., & Weisberg, S. B. (2001). A comparison of plastic and plankton in the north Pacific central gyre. *Marine Pollution Bulletin, 42,* 1297–1300.

Muhandiki, V. S., Shimizu, Y., Adou, Y. A. F., & Matsui, S. (2008). Removal of hydrophobic micro-organic pollutants from municipal wastewater treatment plant effluents by sorption onto synthetic polymeric adsorbents: Upflow column experiments. *Environmental Technology, 29*(3), 351–361.

Müller. J. (2017). *Qualitativer Nachweis von Schadstoffen in Gewässern und deren Abbaupotenzial: Charakterisierung von Schadstoffen im Rhein, die mithilfe eines Passivsamplers während des Projektes „Rheines Wasser" nachgewiesen wurden.* Masterthesis, HFU.

Napper, I. E., Bakir, A., Rowland, S. J., & Thompson, R. C. (2015). Characterisation, quantity and sorptive properties of microplastics extracted from cosmetics. *Marine Pollution Bulletin, 99*(1–2), 178–185.

Obbard, R. W., Sadri, S., Wong, Y. Q., Khitun, A. A., Baker, I., & Thompson, R. C. (2014). Global warming releases microplastic legacy frozen in Arctic Sea ice. *Earth's Future, 2,* 315–320.

Rochman, C. M., Tahir, A., Williams, S. L., Baxa, D. V., Lam, R., Miller, J. T., et al. (2015). Anthropogenic debris in seafood: Plastic debris and fibers from textiles in fish and bivalves sold for human consumption. *Scientific reports, 5,* 14340.

Sleight, V. A., Bakir, A., Thompson, R. C., & Henry, T. B. (2017). Assessment of microplastic-sorbed contaminant bioavailability through analysis of biomarker gene expression in larval zebrafish. *Marine Pollution Bulletin, 116*(1–2), 291–297.

Sussarellu, R., et al. (2016). Oyster reproduction is affected by exposure to polystyrene microplastics. *Proceedings of the National Academy of Sciences of the United States of America, 113,* 2430–2435.

Taylor, M. L., Gwinnett, C., Robinson, L. F., & Woodall, L. C. (2016). Plastic microfibre ingestion by deep-sea organisms. *Scientific Reports, 6,* 33997.

Teuten, E. L., Saquing, J. M., Knappe, D. R. U., Barlaz, M. A., Jonsson, S., Björn, A., et al. (2009). Transport and release of chemicals from plastics to the environment and to wildlife. *Philosophical Transactions of the Royal Society of London. Series B, Biological Sciences, 364*(1526), 2027–2045.

The Ocean Cleanup. (2018). www.theoceancleanup.com.

Umweltbundesamt. (Hrsg.). (2016). Polyzyklische Aromatische Wasserstoffe – Umweltschädlich! Giftig! Unvermeidbar? https://www.umweltbundesamt.de/sites/default/files/medien/376/publikationen/polyzyklische_aomatische_kohlenwasserstoffe.pdf.

Umweltbundesamt Österreich (o. J.). Mündliche Mitteilung eines Mitarbeiters.

van Franeker, J. A., & Law, K. L. (2015). Seabirds, gyres and global trends in plastic pollution. *Environmental Pollution, 203,* 89–96.

van Sebille, E., et al. (2015). A global inventory of small floating plastic debris. *Environmental Research Letters, 10,* 124006.

Wang, W., Ndungu, A. W., Li, Z., & Wang, J. (2017). Microplastics pollution in inland freshwaters of China: A case study in urban surface waters of Wuhan, China. *The Science of The Total Environment, 575,* 1369–1374.

Woodall, L. C., et al. (2014). The deep sea is a major sink for microplastic debris. *Royal Society Open Science, 1,* 140317.

World Economic Forum (2016) The new plastic economy: Rethinking the future of plastics. http://www3.weforum.org/docs/WEF_The_New_Plastics_Economy.pdf.

Ziccardi, L. M., Edgington, A., Hentz, K., Kulacki, K. J., & Kane Driscoll, S. (2016). Microplastics as vectors for bioaccumulation of hydrophobic organic chemicals in the marine environment: A state-of-the-science review. *Environmental Toxicology and Chemistry, 35*(7), 1667–1676.

Microplastics

2

Contents

© The Author(s), under exclusive license to Springer-Verlag GmbH, DE, part of
Springer Nature 2024
A. Fath, *Microplastic*, https://doi.org/10.1007/978-3-662-69844-0_2

2.1 Definition, Origin, and Use

Plastics, colloquially also referred to as plastic (not to be confused with a sculpture or sculpture of a sculptor), are used extensively due to their durability. This results in problems with environmentally friendly disposal, with microplastics playing a particularly important role.

The prefix "micro" comes from the Greek *mikros,* which means "small". Today, we use "micro" to denote the one-millionth part of, for example, a meter. This puts us in the order of magnitude of μm. Microplastics are therefore small plastic particles or fibers.

Plastics, on the other hand, are semi- or fully synthetic macromolecular materials. For the production of semi-synthetic materials, natural polymers, so-called biopolymers, such as cellulose, are used, which are further processed into artificial silk by esterification. The fully synthetic and non-degradable plastics with a decomposition time of several thousand years are synthesized from so-called petrochemically produced monomers. Depending on the functionality of the monomers, the individual building blocks are either linked together by a radical polymerization, polycondensation or polyaddition to form linear or branched high-molecular chains with n > 1000 chain links. During polymerization, different monomers can also be used in a chain propagation reaction, resulting in a multitude of possible polymers. As a result, there is now a wide range of plastics with different properties for a variety of applications. Plastics are extremely durable, flexible, and easily moldable, they impress with easy processing and are above all a cheap starting material. For a variety of applications, plastics are indispensable today. They are used in the automotive industry, in medical technology, in electrical engineering, in building services, in the construction industry, in games and sports, and much more.

The largest share of produced plastics is used for packaging. About 39% of the plastics produced in Europe are used for packaging purposes (PlasticsEurope 2013). The demand is increasing: Thus, the global plastic production increased from half a million tons in 1950 to 288 million tons in 2012. In 2012 alone, global plastic production increased by 2.8% compared to the previous year. However, at the same time, there was a decrease in production in Europe by 3%, which still represents an enormous amount with 57 million tons per year (PlasticsEurope 2013). In a study by the Wuppertal Institute (special issue 12. Vol. 46 of the plastic newspaper *Blessing or Curse*), the use of plastics is predicted to increase by a further 28% by 2030 if no recyclate initiative is undertaken. With increasing demand, the resulting waste volume also increases. No longer used plastic products end up in waste dumps, are burned or recycled, or to a large extent improperly disposed of, leading to headlines like "Garbage vortex burdens North Pacific—millions of tons of plastic end up in the sea every year". The accumulation of plastics in the Atlantic is also unmistakable (Fig. 2.1). This accumulating mountain of plastic waste

Fig. 2.1 Yield after a flood phase in a 10 m wide bay on the Cantabrian Atlantic coast near Comillias. Macroplastic on the way to microplastic. The plastic yield after the subsequent flood was comparable. The abrasion of the plastic articles over the rock chunks during the wave movements leads to mechanical crushing

obviously cannot be completely disposed of. Recycling often yields too little profit, and so it is inevitable that a large part enters our environment through various channels and can cause devastating damage with as yet unforeseeable consequences for our environment. Once plastics are exposed to the external influences of nature, the plastic breaks down into ever smaller particles—we speak of microplastics, an invisible threat to our wastewater.

The problem of the accumulation of plastics as macro or microplastics in our environment has been partially recognized, so that developments for products made from compostable plastics or degradable plastics such as the lactic acid-based Polylacticacid (PLA) have started. The advantage lies in the renewable raw material and the recyclable polymer (https://www.thyssenkrupp.com/de/produkte/polylactide.html).

Researchers do not agree on the exact definition of microplastics. There is not yet a unified definition for this form of plastic. Thus, Moore et al. (2011) refer to plastic particles larger than 5 mm as macroplastics, and those smaller than 5 mm as microplastics. In the publication by Browne et al. (2010), plastic fragments smaller than 1 mm are referred to as microplastics. The term "mesoplastics" is used by Andrady (2011) to distinguish between plastic that is visible to the human eye and that which is only perceptible under a microscope.

Plastic particles are now defined by their size (diameter) (Table 2.1).

Primary and Secondary Microplastics

Microplastics are further divided into primary and secondary microplastics. Primary microplastics include the smallest plastic particles in the micrometer range, so-called "Microbeads". These plastic shapes are produced by the industry for further processing (Liebezeit and Dubaish 2012). The smallest plastic particles are used in the cosmetics industry. In care products such as shower gel, washing peels, make-up or even toothpaste, plastic is added. Many toothpaste manufacturers have now refrained from using microplastics in their products due to the "invisible danger". This partial success is certainly also due to the Federation for Environment and Nature in Germany (BUND), which lists companies and their products containing microplastics on its website (www.bund.net/mikroplastik) and thus exposes them as "environmental sinners". The product range is several pages long and includes face care, body care, foot care and hand care products as well as shampoos, shower gels, powders, makeup, concealers, blush, eyeshadow, mascara, eyeliner, eyebrow pencils, lipsticks, lip gloss, lip liners, sun creams, shaving foam

Table 2.1 Size classification of "plastic". (Andrady 2011; Cole et al. 2011; Ryan et al. 2009)

Particle size	Designation
>25 mm	Macroplastics
5–25 mm	Mesoplastics
1–5 mm	L-MPP (Large Microplastic Particle)
<1 mm	S-MPP (Small Microplastic Particle)

and deodorants (Fig. 2.2). If you, as a reader of this textbook, discover affected cosmetic articles that are not yet listed, you are welcome to send them to the BUND as an addition to the product list. You will learn how to recognize whether a product contains microplastics in the following sections.

Some companies are already making a switch and no longer use plastic particles in their products, instead using alternative solids with the same effects. The Fraunhofer Institute Umsicht is currently researching an alternative to microplastics. In this process, the scientists produce the smallest particles from bio-wax using a high-pressure method. The cold-ground particles correspond in shape and size to traditional microplastics. As a result, they can be used without concern in hygiene and care products (Fraunhofer Umsicht 2014). Other natural products, such as ground walnut shells or grape seeds, also fulfill the function of an abrasive effect in peelings. Furthermore, the finest plastic particles made of polyethylene are used for air pressure cleaning to remove dirt and rust. In medicine, small plastic particles are used as a vector for various active ingredients (Patel et al. 2009). Before going into mass production, microplastics in the form of polyamide powder are used in the production of prototypes, which are melted together into a three-dimensional component using a position-controlled laser beam.

Microplastics also include microfibers, which are released during washing from synthetic fleece textiles. In Fig. 2.3, the microscopically small fibers can be seen. Those that do not flow off with the washing machine's wastewater towards the sewage treatment plant remain stuck in the dryer's sieve and can thus be examined optically and IR-spectroscopically.

Fig. 2.2 Different microplastic particles

Fig. 2.3 Microfibers from clothing

Secondary microplastics are created by the decomposition of macroplastics. Larger plastic parts are broken down into ever smaller components by physical, chemical, or biological processes (Fig. 2.4).

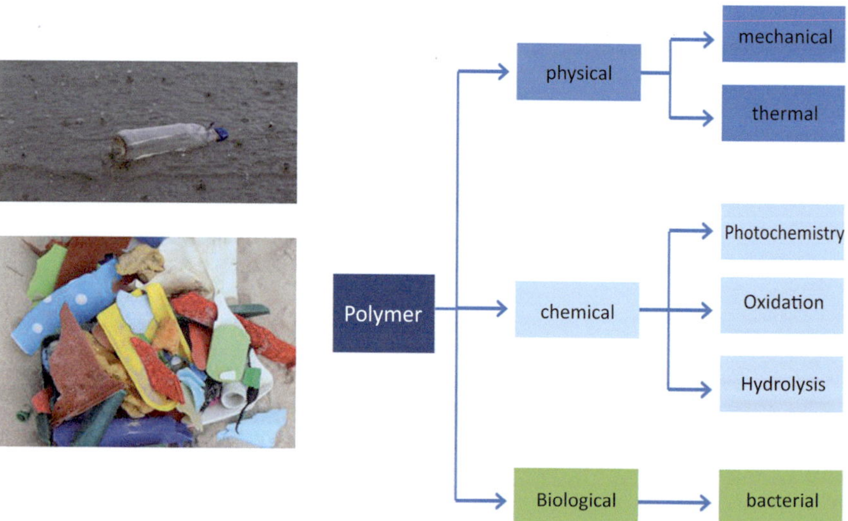

Fig. 2.4 Decomposition processes in the breakdown of macroplastics to microplastics

Fig. 2.4 shows the path by which macroscopic plastic items, such as a PET bottle, decompose into smaller particles in the environment. This can happen purely physically, for example in a major fire, where combustion occurs under high temperature. In addition to oxidation and complete combustion to carbon dioxide, the smoke from the fire cloud contains plastic particles that are hurled into the atmosphere. It cannot be ruled out that microplastic particles are hurled into the air attached to soot particles during fires. This could explain why microplastic particles are also found in waters far from civilization. In waste incineration plants, these particles are captured by soot filters.

The mechanical decomposition is easy to understand. Sand and rocks are harder than plastics, even if their glass fiber content can be up to 60%. When sand and rocks grind over plastics or vice versa, as with tides and strong wave movements or gravel in fast-flowing rivers like the Rhine, purely mechanical particulate abrasion occurs and macroplastics become microplastics. In rivers, this situation mostly arises from sunken plastic bottles that accumulate in large numbers at the weirs.

Chemical decomposition can occur through exposure to oxygen (oxidation), UV radiation, or through a reaction with water (hydrolysis). The aggressive oxygen attacks as a diradical, for example, unsaturated compounds like polybutadiene, the energy-rich UV radiation leads to the breaking of covalent bonds, and polyimides, polyesters, or polyethers can be hydrolyzed into shorter fragments depending on the pH value. All the mentioned reactions lead to a brittleness of the polymers. Over time, the plastics become brittle and, supported by mechanical action, fall apart into smaller pieces.

Instead of the decomposing attack of water, UV radiation and oxygen, bacteria can also occur.This bacterial decomposition of plastics can of course also be used in a controlled manner to reduce plastic waste. The Institute for Molecular Microbiology and Biotechnology at the Westphalian Wilhelms University of Münster deals with the microbial degradation of synthetic polymers (http://mibi1.unimuenster.de/Biologie.IMMB.Steinbuechel/Steinbuechel/Index.html). The working group led by Prof. Steinbüchel investigates the biological degradation of various synthetic polymers, primarily polyethylene glycol (PEG) and polypropylene glycol (PPG), but also polyvinyl alcohol (PVA) or polyacrylate (PA). The biological degradation of these water-soluble polymers is of particular importance, as they cannot be recycled and reused due to their use as non-ionic detergents (PEG) in detergents, but also as emulsifiers in cosmetics (PPG) or chelating agents (PA).

Sources of Microplastics and Entry Into Waters
Microplastics enter our rivers and lakes in various ways. Here, one can distinguish between direct and indirect paths into the water. Especially in industrial shipping, in fishing or also through the consequences of tourism, plastics enter the waters consciously or unconsciously in a direct way. Sources of microplastics are granules from plastic production, so it is possible that plastic granules already enter the environment accidentally during production or transport (Cole et al. 2011). This is evidenced by several granule finds on sea beaches (Claessens et al. 2011; Rios

et al. 2007). In the oceans, no longer used or torn fishing nets make up the largest share of found plastic waste (Andrady 2011). So-called "ghost nets" remain at the bottom of the waters and continue to catch fish after they have been sunk or lost (Moore 2008; Lopez and Mouat 2009). Indirect entry of microplastics into a body of water often occurs through the use of microplastic-contaminated hygiene and care products. Through domestic use of the mentioned products, plastic particles enter the sewage treatment plant via the wastewater. Since adequate filtration is currently not feasible in the sewage treatment plants, the plastic particles enter our rivers and lakes almost unhindered (HELCOM BASE Project 2014). Thus, textile fibers made of polyester enter the wastewater with each wash cycle. A study by Browne et al. (2011) showed that about 1900 plastic fibers are released per garment. But also plastic bags and plastic bottles are sources of microplastics, they either enter a body of water directly or the plastic decomposes and smaller particles seep into the groundwater and enter the water in this way. Thus, Zubris and Richards (2005) found smallest plastic fibers in sewage sludge. In agriculture, sewage sludge is often used as a fertilizer for the fields, and so agricultural products come into direct contact with microplastics, or rain leads to a seepage of the particles. Extreme weather conditions are also responsible for the entry of plastic particles. Heavy rainfall washes the particles into the waters via the sewer system. Storms and strong winds carry finest particles into near-surface layers of the earth's atmosphere, in this way microplastics can reach all conceivable places over kilometers (Liebezeit and Liebezeit 2014).

Not only does the sewage sludge contribute to plastic fertilizer on German fields. A report on the show *Kontraste* (ARD 2015) led to general confusion among environmentally conscious consumers. Whether eggshells, fruit and vegetable scraps, coffee grounds or wilted flowers, all of this belongs in the organic waste bins and can thus be recycled as compost. The organic waste becomes necessary cheap fertilizer for industrialized agriculture. Actually a sensible thing, but on closer inspection it was found that the organic waste bin is the cause of environmental pollution with microplastics. Too many plastics still end up in the compost. When the compost is processed into fertilizer for agriculture, it is shredded for homogenization and then spread. The plastic is shredded along with it and also ends up on the fields. Since farmers prefer to spread their fertilizer before a rain event so that it penetrates well into the soil with the rain, there is of course the risk that the light plastic particles will run off with the surface water, without passing a sewage treatment plant. Through this supposedly environmentally conscious action, microplastics are also distributed in our environment and exactly the opposite of sustainable action is practiced.

It goes without saying that this procedure endangers the groundwater and thus our food through the continuous leaching of additives from the plastics.

Another example where a recycling strategy was not thought through to the end is the fermentation of expired food, mainly fruit from supermarket chains. Since large amounts of biomass are produced here, the approach of producing bioethanol from it, for example, is ecologically and economically sensible. However, only if the fruit is unpacked before the shredding process. Since this costs time and money, unpacking is sometimes not done carefully or even omitted altogether. As a result, the packaging materials also go through the shredder and are shredded. Since the microorganisms that initiate the fermentation process cannot metabolize the microplastic particles, they remain in the fermentation residues. In addition to the plastic residues from our organic waste bin, these are then spread on the fields as fertilizer, including microplastics. This example also shows that there are additional, previously unknown entries of microplastic particles into our waters. The problem in this case is that the surface water does not flow into the sewage treatment plant, but directly into the rivers, and that the smallest microplastic particles can be carried into our groundwater together with the pesticides used.

2.2   Potential Dangers of Plastics and Microplastics

The effects of microplastics on rivers and lakes are still largely unexplored. Countless studies on marine ecosystems and their inhabitants, however, clearly show the potential consequences of microplastics. To approach the topic, it is important to first clarify how plastics are structured, how the diverse properties of plastic arise, and what dangers are associated with them.

Plastics are divided into thermo-, endo-, and duroplastics. Thermoplastics are linearly arranged carbon chains, consisting of thousands of consecutive monomers. This type of plastic becomes malleable and melts at elevated temperatures.

Thermoplastics are the most commonly used type of plastic. Since this type is little to not branched due to the weak physical bonds, this type of plastic degrades the fastest due to external influences. In addition to thermoplastics, there are cross-linked macromolecules, less cross-linked plastics are elastic, they are assigned to the elastomers. Highly cross-linked plastic is hard and resistant, called duroplast (Saechtling and Baur 2007).

To improve the properties of plastics, additives are added to them during production, such as plasticizers, dyes, UV stabilizers, flame retardants, and other inhibitors. Some of the additives are toxic. Phthalates are used, among other things, in food films or cosmetics. In addition, other pollutants such as Bisphenol A or Nonylphenol can be found in many plastic products (Liebezeit and Dubaish 2012). When plastic enters a body of water, decomposition begins: physical, chemical, and biological processes break down the plastic into ever smaller fragments. Since additives are not chemically bound to the plastic, they leach out or separate from the plastic during the degrading process and are then released into the environment. If this happens in a body of water, the organisms living in it can be harmed.

A far greater danger to aquatic inhabitants, however, is not the degrading process itself, but the end product. Microplastics can be ingested by various organisms. For example, fish often mistake the smallest plastic fragments for their food. Lusher et al. (2013) examined ten different species of fish from the English Channel. Microplastics were found in the gastrointestinal tract of 36.5% of the total 504 fish caught. It is also possible that the fish's food already contains microplastics. Cole et al. (2013) demonstrated the ingestion of the smallest plastic particles with a diameter of 1.7–30.6 μm by zooplankton, which typically represents the lowest trophic level in the food chain.

The ingestion of microplastics can in turn lead to blockage of the gastrointestinal tract or the animals may feel a "false" sense of satiety due to an accumulation of plastic. Not only fish, but also reptiles, birds, and mammals are at risk from the ingestion of microplastics. In a long-term study on seabirds in the Mediterranean, microplastics were found in the stomach of almost every animal between 2003 and 2010 (Codina-Garcia et al. 2013). It is known that 44% of all seabird species studied so far ingest plastic through their food. The number of fish, mammals, and birds that die each year from ingesting plastic is unknown, but it is estimated to be in the millions (Moore 2008). It is assumed that the ongoing fragmentation of plastic in bodies of water increases the likelihood of microplastic ingestion by organisms living there (Barnes et al. 2009). As the fragmentation of plastic parts increases, so does the likelihood that microplastics will enter the tissue of organisms. For example, studies on mussels *(Mytilus edulis),* showed that microplastic particles (80 μm in diameter) can enter cells and even cell organelles, leading to severe pathological changes in the organs (Moos 2010). Microplastic particles can in turn absorb pollutants such as additives and even heavy metals. As the surface area of the particles increases due to the ongoing degrading process, the adsorption of chemical pollutants increases (Barnes et al. 2009). Mato et al. (2001) examined plastic granules made of polyethylene, which were found on the coast of Japan. They showed high concentrations of PCB, DDE, and Nonylphenol, organic toxins with partly carcinogenic effects. In an experiment with comparable granules in seawater, he found that the concentration of pollutants within the plastic granules significantly increased (Mato et al. 2001).

When microplastics are ingested, there is also the risk that pollutants are released into the animals. Indeed, a study by Oehlmann et al. (2009) shows that ingested pollutants from plastic particles are transferred to organisms. Teuten et al. (2009) describe in an experiment on seabirds *(Calonectris leucomelas)* that after ingestion of contaminated microplastic particles, pollutants are sequentially released into the body. Oehlmann et al. (2009) investigated the effects of additives on the development and reproduction rate of marine fish, crustaceans, molluscs, and amphibians. It turned out that phthalates and bisphenol A can have a negative impact on some creatures, although not all examined species showed negative changes. In swordfish *(Xiphias gladius, L.),* intersexual characteristics were found in a quarter of the 162 examined fish (Metrio 2003). Female polar bears were also documented, which had additionally developed rudimentary male sexual organs. It is suspected that the animals' hormone system was disrupted by synthetically produced pesticides (Wiig et al. 1998). Since corresponding long-term studies are lacking, such effects on humans can only be suspected. There are fears that additives contained in plastic, phthalates or bisphenol A, can damage the hormone system and other biological mechanisms of the human body (Meeker et al. 2009). Figure 2.5 schematically shows how listed surface-active substances with high adsorption capacity and the corresponding polarity can attach to the surface of microplastic particles. Microplastic particles are accumulators for hydrophobic substances (Ziccardi et al. 2016)

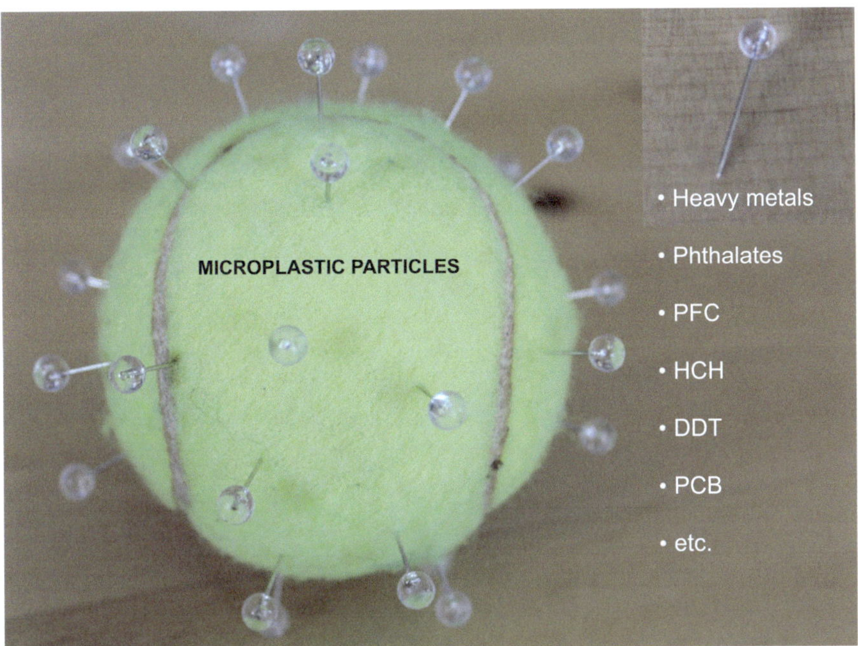

Fig. 2.5   Attachment of pollutants to microplastic particles

At the top of the food chain are human beings—through the consumption of fish and other marine and river dwellers, microplastics can be ingested by us. Microplastics therefore pose a great danger. Researchers have now even detected microplastics in various foods that were bought in store. For example, Liebezeit and Liebezeit (2014) examined German beers from different manufacturers in their studies on microplastics. They found microscopic plastic fragments in all 24 tested types of beer. The number of particles varied per type of beer between 5 and 79 per liter of beer. According to the authors, this is not yet a worrying amount; however, it shows that microplastics are ubiquitous in our environment. Not only our food is affected by microplastics, but also our drinking water is at risk. Plastic particles end up in landfills or as accompanying substances of fertilizer on our fields. There, photochemical and microbial decomposition processes start, which make the plastic brittle and break it down, so that an entry into our groundwater and drinking water are inevitable consequences. The microplastic load of humans, how macroplastic becomes microplastic and how it enters our food chain is vividly illustrated in the graphic of the Fraunhofer Institute Umsicht (Fig. 2.6).

The environmental problem "microplastics" encompasses a scale that will concern us far into the future. Especially for humans, further studies are needed to investigate the effects of microplastics. Consequences for human health are not yet known and can only be speculated, but countless studies on animals show alarming results (Fig. 2.6)!

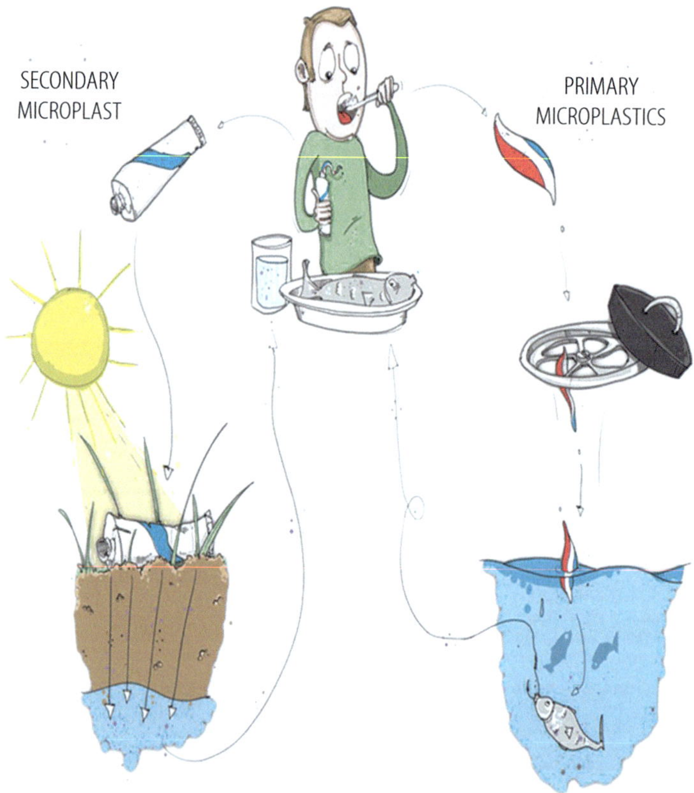

Fig. 2.6 Microplastics from macroplastic and the effects on humans

2.3 Investigation Methods of Microplastics

2.3.1 Particle Size Distribution Using Laser Diffraction Spectroscopy

Another method to investigate the particle size distribution of microscopically small plastic particles involves the analysis of diffraction spectra. Under an infrared microscope, it is of course also possible to count and measure the individual plastic fragments within a filter cake, but this only provides the distribution on the surface and not the integral distribution in the underlying layers in the sand matrix. With laser diffraction spectroscopy, there is a faster method to measure the particle size distribution in suspensions, emulsions, aerosols, or sprays. However, for the examination of the filter cake, a separation of sand and plastic particles by a density separation would be necessary first. The plastic suspended particles in the aqueous suspension could then be introduced directly into the measuring cell via ultrasonic-assisted wet dispersion or after a filtrationand drying via compressed air dry dispersion.

In many hygiene, cosmetic, and care products (BUND 2018), microplastics are used in the form of small pellets. The plastic granules serve a function as abrasives, fillers, film formers, molders, or as binders (Liebezeit and Dubaish 2012). As primary microplastics, the particles enter the sewage system and our waters through daily use. For the examination of various cosmetic products regarding the size distribution of the contained plastic particles, the optical method of laser diffraction is suitable. Here, important insights about the selection of filter fabrics for wastewater treatment or for the examination of waters for microplastics in general can be obtained. The following presents the basics of laser diffraction spectroscopy, briefly laser diffraction.

When light hits the interface of two different media, the interactions shown in Fig. 2.7 are possible in principle.

Information about the material is obtained through each type of interaction of the light with the medium. Absorption, where the intensity of the outgoing light beam decreases, provides information about the concentration within the irradiated material (photometry). Refraction also provides information about the purity (concentration) of a liquid sample (Abbe refractometer, degrees Oechsle = sugar content in must). The reflection is used for color measurement ($L_{a,b}$-value determination) of surfaces and for turbidity measurement of water samples. Diffraction provides information about the particle size, particle size distribution, and the shape of particles in a special optical arrangement, as can be seen in Fig. 2.9, both in scattering phenomena such as refraction, diffraction and reflection, as well as in absorption, there is a weakening of the incident radiation I_0. This effect is known as extinction. The Fraunhofer diffraction spectra take into account the diffraction component in the forward direction. This diffraction component strongly depends on the particle size (Fig. 2.8).

When high-energy monochromatic and coherent light, such as that produced by a helium-neon gas laser, hits particles of different sizes, the red light waves (633 nm) are elastically scattered in a characteristic manner depending on the size of the particles. The larger the particles, the stronger the scattering in the forward direction. For particles smaller than 100 nm, the scattering intensity is almost

Fig. 2.7 Interaction of light rays with a spherical particle as an optically denser medium

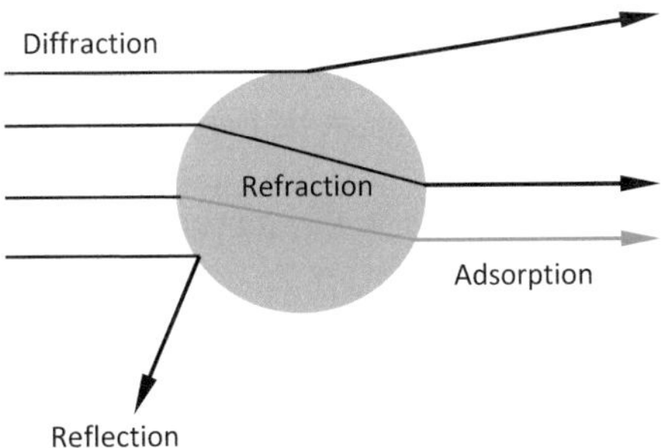

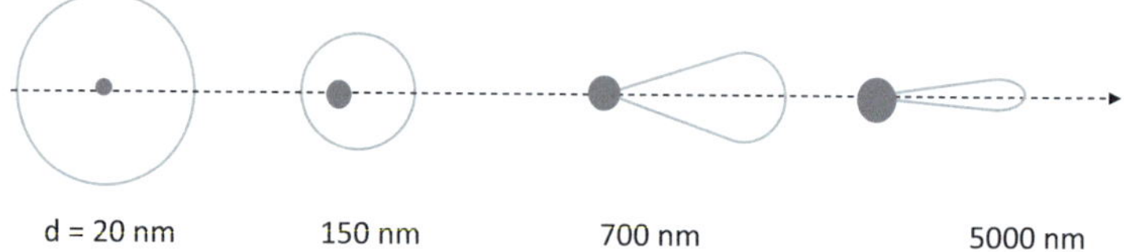

Fig. 2.8 Dependence of the laser diffraction in the forward direction on the diameter d of a spherical particle

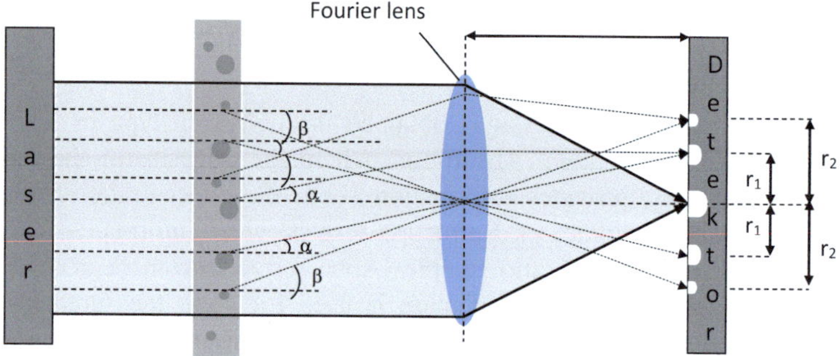

Fig. 2.9 Optical arrangement for generating diffraction spectra

identical in all directions. The scattered light of a helium-neon laser beam on particles, whose diameter is in the micrometer range, is bundled by a collecting lens (Fourier lens) in a laser diffraction measurement and imaged on a fixed detector depending on the scattering angle. Since the scattering angles β on small particles are larger than the scattering angles α on large particles, there is a light scattering intensity distribution on the detector, as shown in Fig. 2.9.

Monodisperse particles interact with the laser radiation and generate a convergent beam, at a certain distance r around the center of the bundled laser beam, which had no interaction with particles in the measuring cell. The use of a Fourier lens has the advantage that scattered light can also be detected at very large diffraction angles. This creates the possibility to measure very small particles as well. The Fourier collecting lens transfers an angle-dependent intensity distribution into a radius-dependent distribution ($r_2 > r_1$) around the central focal point of the laser beam. To achieve the highest possible accuracy, the particles to be measured must approximately lie in one plane, as the size of the scattering image in the convergent beam depends on the distance of the particles to the detector.

In the case of laser diffraction on spherical particles, a diffraction image is created on the detector, consisting of concentric bright and dark circles around the focal point. The intensities and distances depend on the particle size. Figure 2.10

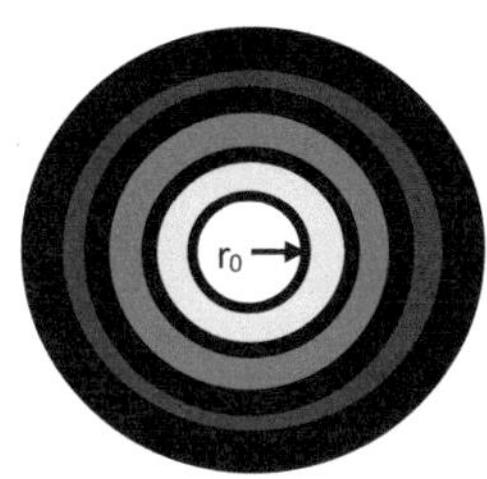
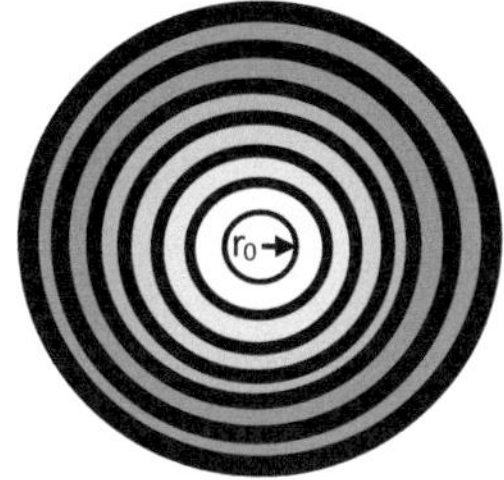

Fig. 2.10 Diffraction images of different sized particles. Left: small particles. Right: large particles

shows the diffraction images of two different particle sizes of spherical particles. Depending on the size of the particles, the distances of the bright and dark areas on the detector are different. Small particles create large ring distances (left image), while large particles lead to intensity distributions with closely adjacent rings (right image).

The diameter of the rings is larger the smaller the circular particles are. If you bring a multiple of these same-sized particles into the laser beam, the intensity of the diffraction image increases by the same factor, but the position of the rings remains the same.

The determination of the particle size distribution from the measured scattering spectra can either be calculated with the Fraunhofer or the MIE theory (Bohren and Huffmann 2008), depending on whether the model for opaque spheres or the model for translucent spheres is better suited for the spherical particles. When evaluating using the MIE theory, the complex refractive index (scattering and absorption) of the particles must be known. The laser scattering determines the particle volume. The result of a laser scattering measurement calculated from the intensity distribution indicates in the sum curve or cumulative passage curve $(Q3(X))$ how many percent of the particles in the total sample volume contain a certain particle size. The so-called histogram $(dQ3(x))$, which states how many percent of the sample volume are in particles of a certain size, is calculated from the first derivative of the sum curve. This also allows the total surface area of all particles to be calculated, an important size for estimating the amount of adsorbable substances.

The following examines the particle size distribution of selected cosmetic products (Loritz 2014). This includes both the procedure and the interpretation of the results. Products containing microplastics are listed on a product list for cosmetics and cleaning agents, which has been published by the Federation for Environment and Nature Conservation Germany (BUND 2018). Five products were selected from this product list: two toothpastes, an eyeshadow, a cleansing mousse, and a shower gel. The products are listed in Fig. 2.11.

The analysis of particle distribution is helpful for selecting a suitable filter fabric to capture microplastics from waters that originate from cosmetic products. After separating the plastic particles from the gel, the plastic can be quickly and

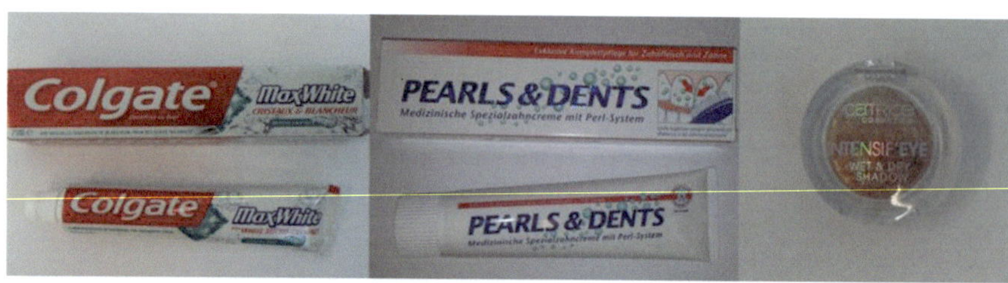

Fig. 2.11 Selected products containing microplastics. From top left to right: Two different toothpastes and an eyeshadow. Below: a cleansing mousse (left) and a shower gel with peeling effect

Table 2.2 Cosmetic products with various microplastic contents

Product	Usage	Type of plastic
Catrice Intensif Eye Wet & Dry Shadow	Optical function	Polyamide
Colgate Max White	Abrasive	Polypropylene
Melt Away Foaming Cleansing Mousse BeBe MORE	Binder	Acrylate-copolymer
Pearl and Dents Toothpaste	Abrasive	Ethylene vinyl acetate
Daily Wash peeling NIVEA	Peeling	Polyethylene

easily identified using IR-ATR spectroscopy. Table 2.2 shows a selection of the different types of plastic used in the examined care products.

Polypropylene, polyamide, and polyethylene are presented in detail in Sect. 2.5. Ethylene vinyl acetate (short EVAC) is a copolymer made from the monomers ethylene and vinyl acetate (Fig. 2.12). Depending on the proportion of the individual monomer types, a polymer with the depicted general formula is obtained.

The copolymer is produced using the same method as PE. The vinyl acetate component improves the elongation at break and ageing resistance compared to pure polyethylene and, like this, is mainly used for films, but also as EVAC

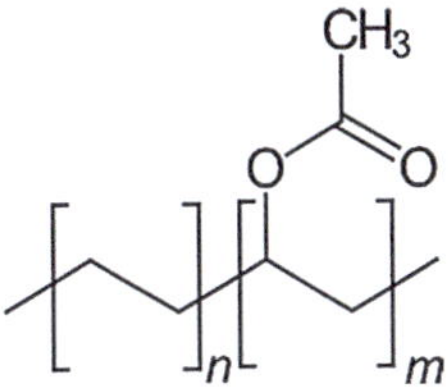

Fig. 2.12 Structural formula
of the copolymer EVAC

micropellets or PEVA (polyethylene vinyl acetate) in cosmetic peeling products, as the example in Table 2.2 shows.

The acrylate copolymer is found not only in the cleansing mousse but also in many other shower gels with a designated peeling effect. A look at the back of the product into the list of ingredients (Fig. 2.13) provides certainty about this.

In addition to the polymers listed in Table 2.2, polyethylene terephthalate (PET), polyester (PES), polyurethane, polyimide (PI), acrylate crosspolymers (ACS) and polyquaternium-7 (P-7) are also used in cosmetic products. Under the collective term ACS, copolymers of acrylic acid alkyl esters (C10–C30), acrylic acid and methacrylic acid are found, which are crosslinked via propenyl (allyl) modified sugars. Polyquaternium salts are a class of polymeric compounds produced by radical polymerization of acrylamide and dimethyldiallylammonium chloride. They are used in cosmetics and hair care products, and depending on their function, a number is added to the name. The use of the quaternary ammonium chloride monomer results in a polycation, which is used as a film former and antistatic in care products. The Federal Institute for Risk Assessment (BfR)

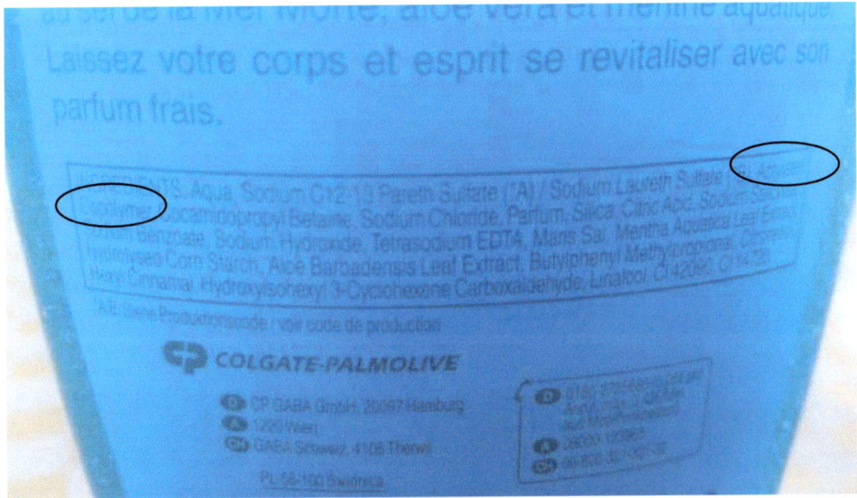

Fig. 2.13 List of ingredients listed on the back of the product, where the acrylate copolymer (circled) is listed

has identified polyquaternium-7 as the source of the carcinogenic substance polyarylamide (BfR 2003), as it can contain polyarylamide as a not fully converted polymer, which then ends up as a microplastic component in wastewater.

Even high-temperature-resistant and comparatively expensive plastics like polyimides (PI) are used as microplastics in cosmetics. Polyimides are obtained by a polycondensation reaction. The process goes through two stages. In the first stage, a tetracarboxylic acid dianhydride is converted with a diamine to a polyamidocarboxylic acid. Despite aromatic groups, the polyamidocarboxylic acid is still viscous enough to be processed into shape before the polycondensation to the solid polyimide is thermally initiated.

As a strength-giving aromatic anhydride, the aromatic benzene-1,2,4,5-tetracarboxylic acid-1,2:4,5-dianhydride (pyromellitic dianhydride) is often used, and a 4,4'-diaminodiphenyl ether is used as the diamine component, so that the structural feature shown in Fig. 2.14 is obtained.

By now, some cosmetic product manufacturers have already carried out a production change and, due to the described problem, no longer added microplastics or found a functional replacement through degradable alternative materials. Currently, BUND is not aware of any tooth cleaning that contains microplastics. This positive trend is also noticeable in shower gels. The topic "Microplastics as an environmental pollutant" has reached cosmetic customers to the extent that product marketers are already actively promoting the non-use of microplastics (Fig. 2.15). Instead of plastic particles, natural products like almond stone or olive kernel meal take over the "peeling". Also, granulations from the wax of the jojoba plant or from beeswax are capable of removing dirt and dry skin flakes.

Particle Size Analysis in Cosmetic Products

To carry out the particle size analysis, the products are suspended in various solvents. About 5 g of a product are taken and the soluble components are dissolved in distilled water. The resulting suspension, if it contains microplastic particles, is placed in the sample containers of the analysis device. The stirrer speed and pump power of the device, as well as the sonication time to break up agglomerations, can be selected. Several fractions of the suspension can be fed into the measuring cell one after the other, wet or dry, so that the sonication time adds up. If there are significant differences in the particle size distribution per run, this provides information about the agglomeration of the particles. The result of the measurement is a graphical representation of the cumulative distribution in percent and the average

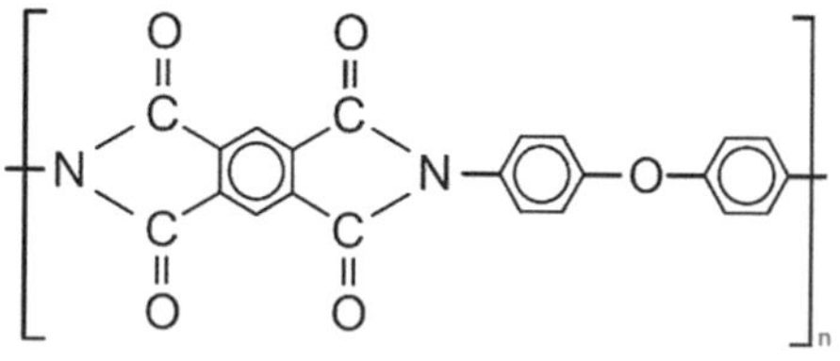

Fig. 2.14 Structural feature of the polyimide

Fig. 2.15 Peeling with jojoba wax instead of microplastics

particle diameter in micrometers. The distribution of the particles of a sample run across the different particle diameters can be read from a kind of Gaussian distribution curve, which represents the first derivative of the cumulative curve. The runs are repeated five times per product and solvent in the study presented here. Distilled water and methanol are used as solvents. If components of a product, other than microplastics, dissolve poorly or not at all in water, the experiment is repeated with methanol. A tested toothpaste had to be suspended in ethyl acetate as the only product. After filtering in a Büchner funnel with filter paper (10 µm pore size), the filter cake can be suspended in distilled water and measured again for comparison. If the results of individual runs with different solvents are compared, it can be determined which particle size can be attributed to the microplastics. A reference measurement with distilled water is carried out before each run. The diagrams in Fig. 2.16 show the result of the microplastic particle size distribution in various products.

The abscissa indicates the pore size in micrometers, the ordinate gives the cumulative distribution in percent. The maximum of the resulting derivative curve describes the value x_{50}. This indicates that 50% of the existing particles have a smaller diameter than the corresponding x-value. The optical concentration c_{opt} indicates concentration of a particle solution, it was between 4% and 25% in all experiments. The diagrams reveal that the size of the plastic particles used can vary greatly, within a range of <10 µm to >100 µm. While the plastic of the

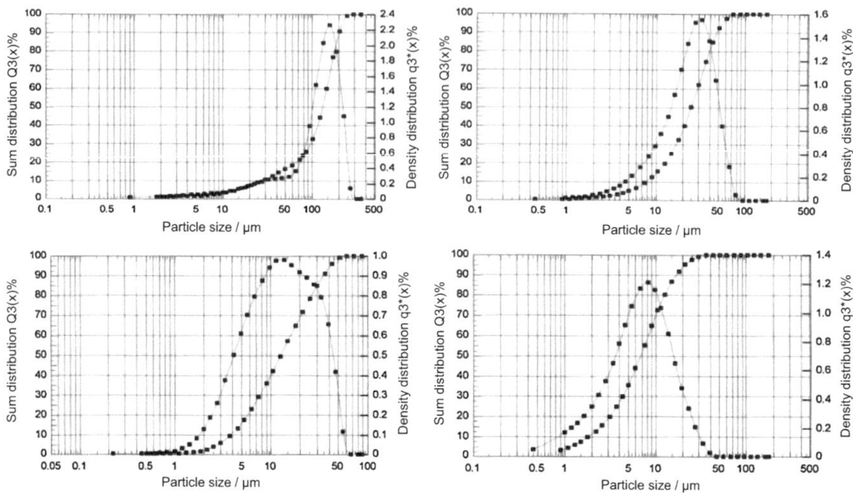

Fig. 2.16 Microplastic particle sizes in a wash-peeling (top left), an eyeshadow (top right), a toothpaste (bottom left) and a foaming cleansing mousse (bottom right)

foaming cleansing mousse (Fig. 2.16 bottom right) has an averaged particle diameter of 7.542 µm for 50% of all particles occurring in the suspension and the particles have the smallest particle diameter compared to the other tested products, the averaged particle diameter of the daily wash-peeling reaches the largest averaged particle diameter with 131.558 µm (Fig. 2.16 top left). For the two other products, the averaged particle diameters are in the middle range between 10 and 40 µm. To capture the microplastic particles in a river water filtration, whose origin can be traced back to cosmetic products, special stainless steel filter cartridges with a stainless steel mesh, whose pore size is 10 µm, are used.

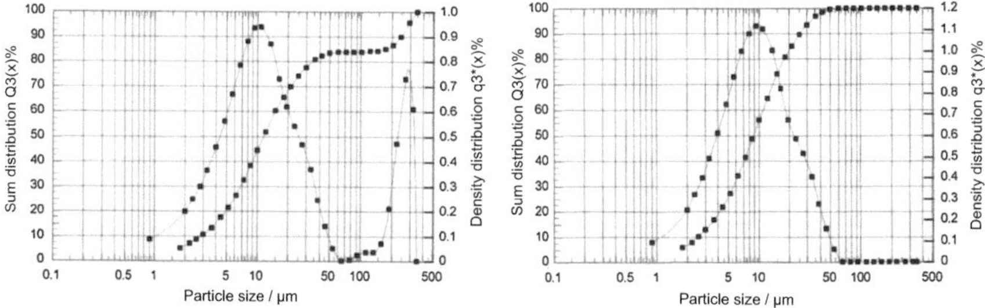

Fig. 2.17 Results of the particle size analysis of a toothpaste. The left diagram shows the result with methanol as the solvent. The right diagram shows the run with distilled water as the solvent after treatment with ethyl acetate

Another examined toothpaste shows the particle distribution depicted in Fig. 2.17. It is noticeable that in addition to the maximum at 10 μm particle diameter, there is a second maximum at about 380 μm. If the paste is dissolved in ethyl acetate, in which the polypropylene micro particles (Table 2.5) are insoluble, the second maximum disappears in a subsequent measurement of the suspension in the organic solvent.

The second particulate component soluble in ethyl acetate is the so-called CelluloseGum. This cellulose derivative is used as a "natural product" as a thickener or binder in pastes and creams (including ice cream) and is often found on the list of ingredients (Fig. 2.18).

Cellulose Gum is a viscous cellulose derivative, synthesized by etherification of varying proportions of the hydroxyl groups of the natural β-D-glucose units of cellulose. In this process, alkali cellulose reacts with chloroacetic acid to form the corresponding carboxymethyl celluloses (CMC). The solubility of the CMC depends on the number of carboxymethyl groups and the pH value. The anionic form is also used as a cation exchange material. Due to their high adjustable viscosity and their non-toxic properties, carboxymethyl celluloses are mainly used in the food industry as an additive with the number E466. CMCs are also used as microparticles in bone tissue engineering *(bone tissue engineering)* (Gaihre et al. 2016).

Fig. 2.18 Ingredients of a toothpaste

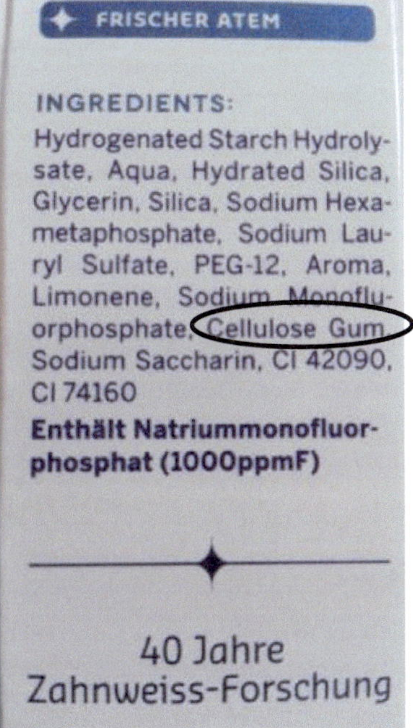

Table 2.3 Results of the particle size analysis of a peeling; optical concentration of the suspension c_{opt}: 32.5–22.29%; measurement range: 1.8–350 µm; solvent: distilled water; x_{10}: particle diameters corresponding to 10% of the cumulative passage-Q3(x); x_{50}: corresponding to 50%; x_{90}: corresponding to 90%

Measurement	x_{10} [µm]	x_{50} [µm]	x_{90} [µm]	Sonication [s]
1	2.65	9.02	22.68	120
2	2.34	8.10	19.70	240
3	2.07	7.36	17.88	360
4	1.89	6.84	16.92	480
5	1.75	6.39	15.96	600
Average	**2.14**	**7.542**	**18.628**	

The subsequent particle size analysis of two products provides information about the agglomeration behavior of the microplastic particles. While in the peeling product (Table 2.3) the particle size only reaches a constant value after 600 s, the particle size distribution of the plastic particles in the examined toothpaste is constant and independent of the ultrasonic irradiation time of the suspension (Table 2.4).

Table 2.4 Results of the particle size analysis of a toothpaste; optical concentration of the suspension c_{opt}: 20.9–21.8%; measurement range: 1.8–350 µm; solvent: distilled water

Measurement	x_{10} [µm]	x_{50} [µm]	x_{90} [µm]	Sonication [s]
1	3.91	12.70	35.43	120
2	3.91	12.72	35.46	240
3	3.91	12.71	35.43	360
4	3.92	12.74	35.55	480
5	3.91	12.71	35.45	600
Mean value	**3.912**	**12.716**	**35.464**	

Table 2.5 Resonance frequencies ($\tilde{\nu}$) of selected bonds depending on their dissociation energy and bond length

Bond	Bond length [pm][a]	Dissociation energy [kJ/mol][a]	$\tilde{\nu}$(cm^{-1}) (Hesse et al. 1987)
C–C	154	348	$\approx$1000
C=C	134	611	$\approx$1640
C≡C	120	837	$\approx$2200
C–H$_{aliphatic}$	110	411	<3000
C–H$_{aromatic}$	108.5	468	>3000

[a] Bond lengths and dissociation energies from textbooks of organic chemistry (e.g., Streitwieser et al. (1994) Organic Chemistry, Weinheim: Wiley-VCH)

Since 1950, 8.3 billion tons of plastic products have been produced and by 2050 it is expected to be 34 billion tons. So far, only 9% of this has been recycled worldwide and 12% incinerated (source: Roland Geyer/University of California 2017). The remaining 79% are either still in use or in landfills or in the environment. This means that without massive countermeasures, the amount of plastic waste in our oceans will continue to increase.

The fact that some bacteria and fungi are capable of degrading polyethylene (Yoshida et al. 2016; Yamada-Onodera et al. 2001) does not change this, as the degradation process is very slow. The wax moth *Galleria mellonella* works much faster. One hundred waxworms eat 92 mg of a plastic bag in 12 hours (Bombelli et al. 2017; Dickmann 1933). Waxworm larvae prefer to feed on beeswax, a composition of alkanes, alkenes, fatty acids and esters, in which the most common structural element is the ethylene grouping CH_2–CH_2, just like in polyethylene. Whether a specific enzyme is responsible for the metabolism of PE to ethylene glycol is the subject of current investigations. If this were the case, a wide range of biotechnological applications would be conceivable. Nevertheless, waxworm larvae, fungi and bacteria capable of degrading plastics will by no means absolve consumers of the responsibility to handle plastic waste more conscientiously.

Due to the resilience of plastic waste, the cumulative amount will increase annually, despite waxworm larvae. Even if the plastic waste slowly breaks down into smaller pieces and eventually into microplastics when their diameter is <5 mm (secondary microplastics), it still remains, with all the negative aspects of the microscopic synthetic material mentioned in Sect. 1.2. The total plastic waste load entering the world's oceans already consists of a proportion of 15–31% of microplastics (primary microplastics) (Boucher and Friot 2017), which joins the microplastics already formed from the 150 million tons of plastic waste (secondary microplastics). The artificial suspended particles are carried everywhere by ocean currents. They are found in sediments of the deep sea (Van Cauwenberghe et al. 2013) as well as in the Arctic (Bergmann et al. 2016). Swimming barriers are now being used to try to fish the plastic waste out of the seas (Spiegel online 2016).

For the already distributed microplastics, this measure comes too late. However, it is still a sensible preventive measure to reduce further microplastics. To achieve a significant reduction of large and small plastic waste in our oceans, prevention must start even earlier and address our handling of plastic waste. For this approach, it is necessary to know the sources of waste.

Another source of contamination of rivers and seas with microplastic particles are synthetic fibers, which are released from clothing during washing and enter bodies of water with the wastewater. These fibers account for about one third (35%) of the microplastic content in the world's oceans (Boucher and Friot 2017)—no small problem, as these fibers add up to the equivalent of 15,000 plastic bags per 100,000 inhabitants, as researchers from the University of California in Santa Barbara have calculated (Hartline et al. 2016). But there is now a solution for this too: "Guppyfriend" (http://guppyfriend.com), a fabric bag for the washing machine, in which you put synthetic material clothing before washing. This laundry bag made of a high-tech material acts like a filter and retains 99% of the broken fibers. The fact that a lot of fibers accumulate during a wash cycle is shown in Fig. 2.21. Here, only the smear of the sieve of a tumble dryer is visible. In the actual wash cycle before, an even larger amount of synthetic fibers is released and not retained by a sieve. In the washing drum, the aim during spinning is to separate the water from the clothes as quickly as possible, which would be prevented by a fine sieve. Washing a single fleece jacket releases 1.7 g of microfibers (Hartline et al. 2016; Fig. 2.21). In order to reduce the amount of synthetic fibers released into bodies of water, in addition to the laundry bag as an "intrinsic filter", we also have the option of replacing our synthetic clothing with textiles made from natural fibers such as cotton, jute, hemp, linen, silk, etc.

2.3.2 Introduction to IR Spectroscopy

The infrared spectroscopy is the most suitable analysis method for the identification of plastics of both macroscopic and microscopic dimensions. Not only the polymeric material, but also its additives can be identified and quantified with this simple and inexpensive spectroscopic method using infrared radiation. Since the quality of plastics is strongly influenced by additives and these are partly health and ecologically questionable substances (Table 2.11), quality control is indispensable. The analytical control extends over the entire supply chain of a plastic article, from the raw materials (monomers) to the granulate manufacturer, compounder, plastic processor, to the product manufacturer and the end customer. For the end product and its eco-balance, the whereabouts of the product after the end of its life cycle is becoming increasingly important in the sense of the sustainability strategy of many companies, and the question of recyclability is fortunately gaining in importance. Cost-effective and thus practical recycling is only possible if plastics can be separated by type and contain no or few of the same ingredients. The tool to check all the mentioned requirements for quality, harmlessness and recyclability of plastics is the IR spectroscopy. This also explains the necessity to introduce the theoretical basics of IR spectroscopy in a textbook/non-fiction book on the topic of "microplastics", despite a multitude of analytical textbooks that describe this method from a similar motivation (Günzler and Heise 2003; Hesse et al. 2016; Alsonso and Finn 2000; Atkins 2013).

With the help of infrared (IR) spectroscopy, unknown organic substances can be determined both qualitatively and quantitatively. Absorptions in the infrared range are measured. This is conveniently divided into the short-wave near infrared (wavelength NIR: 800 nm–2.5 µm), the medium infrared (wavelength MIR: 2.5–50 µm) and the long-wave far infrared (wavelength FIR: 50–10^3 µm). The infrared light is only absorbed if the substance to be examined is a dipole molecule. The dipole moment µ interacts with the electric vector of the light and absorption occurs (Naumer and Heller, 1997) when the dipole moment μ_1 at one extreme of the vibration is not equal to the dipole moment μ_2 at the other extreme of the vibration.

Therefore, only those vibrations in a molecule are active where $\Delta\mu \neq 0$. All other vibrations, where $\Delta\mu = 0$, are IR-inactive. The result of this selection rule is that all vibrations symmetrical to a center of symmetry in a molecule are IR-inactive. This applies, for example, to all biatomic gases such as N_2, O_2, H_2, Cl_2. Symmetrical vibrations can, however, be depicted in Raman spectroscopy as the polarizability changes during a symmetrical vibration. In Raman spectroscopy, a sample is irradiated with monochromatic laser light. The frequency differences to the incident light correspond to the energies characteristic of the material of rotations and vibrations of the molecules in the sample, in which the polarizability changes.

In this respect, both types of spectroscopy complement each other excellently, because all molecules that are IR-active are Raman-inactive and vice versa. The reason why IR spectroscopy is the better-known method for the identification of substances is that most functional groups of organic molecules do not have a center of symmetry and there are also asymmetric IR-active vibrations in symmetrical molecules that contribute to structure elucidation with IR spectroscopy.

The example of two different stretching vibrations (Fig. 2.19) (change of the bond length) in the CO_2 molecule shows how both types of spectroscopy complement each other due to their selection criteria. The asymmetric stretching vibration v_s, as the simplified graphic shows, is IR-inactive, but Raman-active, while the asymmetric stretching vibration v_{as} behaves exactly the opposite.

The position of the absorption band in the IR spectrum can be expressed on the abscissa in units of the wavelength λ of the absorbed light. However, the specification in units of the reciprocal wavelength, the so-called wave number $\tilde{v}$ (cm^{-1}), has prevailed; it increases on the abscissa from right to left. The numerical value indicates how many waves of the infrared radiation fit into one centimeter (Hesse et al. 2016).

The wave number $\tilde{v}$ is defined by:

$$\tilde{v} = \frac{1}{\lambda} \tag{2.1}$$

For the conversion of wave numbers into wavelengths, the following applies when using the common units:

$$\text{Wellenzahl } \tilde{v} \left(cm^{-1}\right) = \frac{10^4}{\textit{Wellenlänge } \lambda \; (\mu m)}$$

Wavenumbers $\tilde{v}$ have the advantage that they are directly proportional to the frequency v of the absorbed radiation and thus also to the energy or light intensity ΔE (Hesse and Meier 2016).

It applies:

$$v = \frac{\lambda \cdot c\, v}{\lambda} = \frac{c}{c} \cdot \tilde{v} \tag{2.2}$$

The ordinate scale of the IR spectrum uses the transmission or absorption, each given in percent.

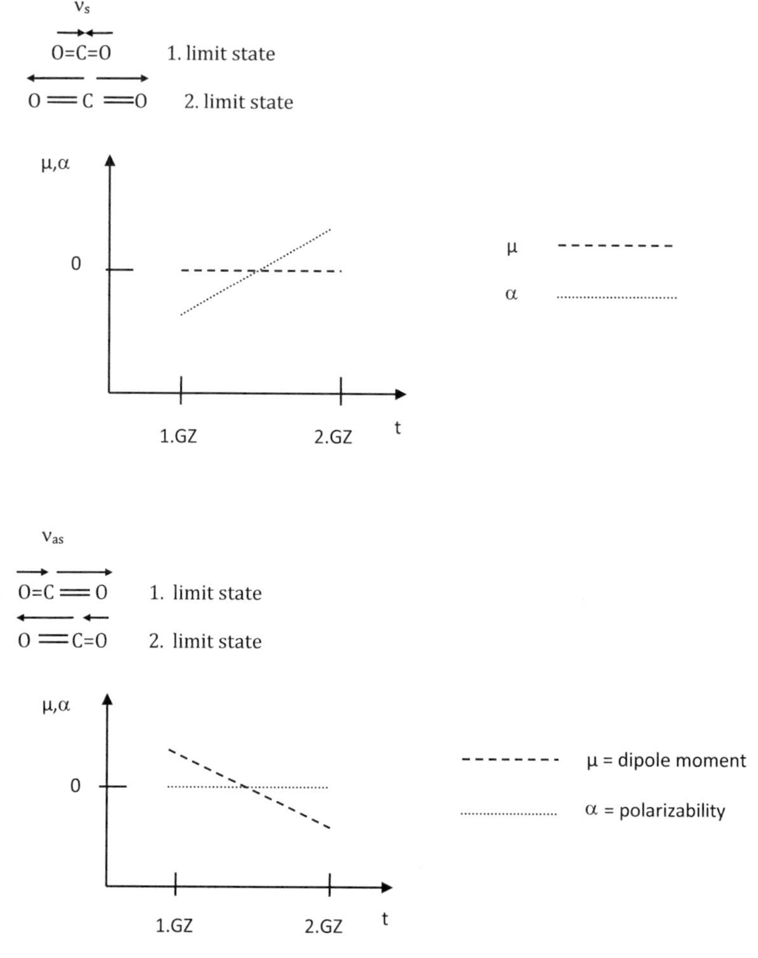

Fig. 2.19 Change of the dipole moment μ and the polarizability α during two different stretching vibrations in the CO_2 molecule

The transmission T of a sample is defined as (Spangenberg n.d.):

$$T(\lambda) = \frac{I(\lambda)}{I_0(\lambda)} \cdot 100\,\%$$ (2.3)

The light intensity experiences an exponential decrease when passing through the measuring cell (Sect. 2.3.2, "Quantitative Spectrum Evaluation").

For the energy or light intensity ΔE applies (Günzler and Heise 2003):

$$\Delta E = h \cdot v$$ (2.4)

Through the frequency, Eqns. 2.2 and 2.4 are linked to each other (Günzler and Heise 2003):

$$\Delta E = h \cdot v = \frac{h \cdot c}{\lambda} = h \cdot c \cdot \tilde{v} \tag{2.5}$$

The normal or medium (MIR) infrared range, in which molecular vibrations are excited, lies between the wavenumbers 4000–400 cm^{-1}. Overtone and combination vibrations are excited by the higher energetic infrared rays in the near infrared range (NIR) at 12,500–4000 cm^{-1}. NIR spectroscopy is mainly used to determine the water content, protein or fat content of food. Absorption of longer-wavelength infrared rays in the far infrared range (FIR) (Fig. 2.20) mainly leads to the rotational excitation of entire molecules.

The position and intensity of the absorption bands in the IR spectrum are extraordinarily substance-specific.

Therefore, the IR spectrum of a substance is a highly characteristic feature, similar to a human's fingerprint. The basis for the evaluation of absorption spectra is the quantum theory.

The occurrence of the absorption bands of various chemical compounds at their respective characteristic wave numbers in the spectrum is shown in Fig. 2.21.

Molecular Vibration Possibilities

To make the processes involved in the creation of an IR spectrum understandable, the model of the harmonic oscillator from classical mechanics can be used.

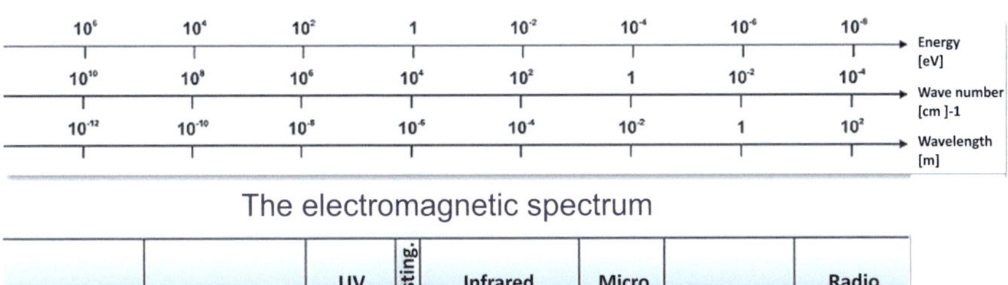

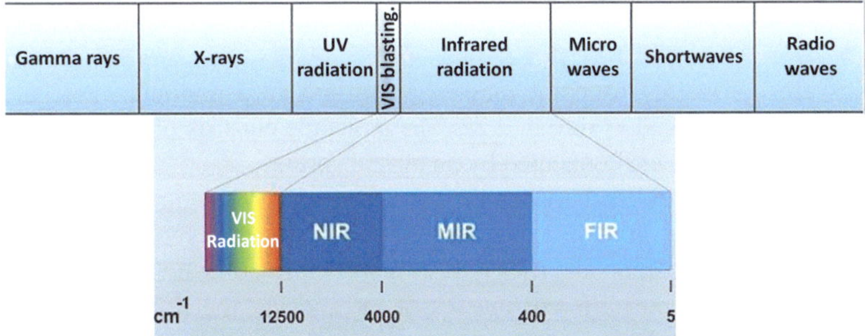

Fig. 2.20 Electromagnetic spectrum

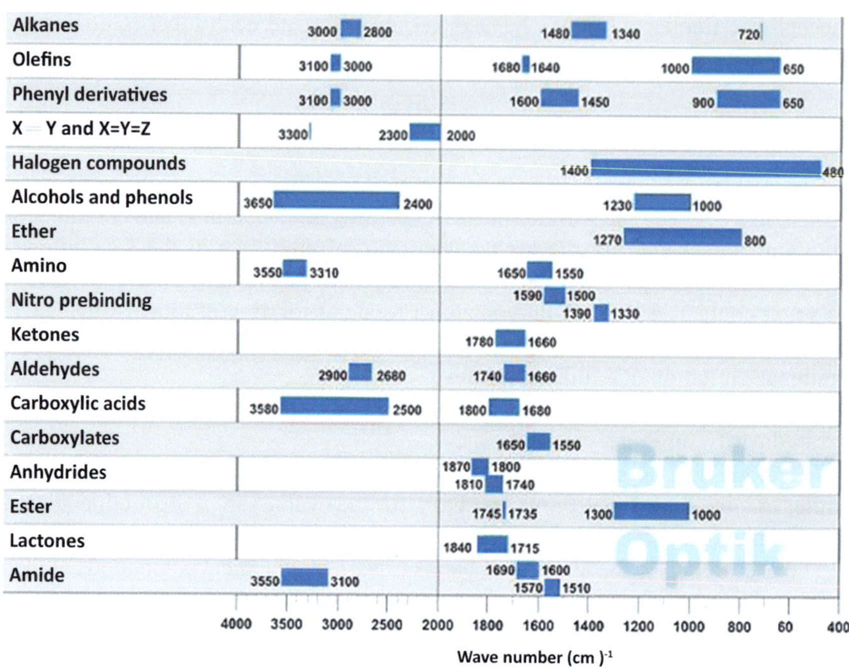

Fig. 2.21 Characteristic wave numbers of chemical compounds. (Bruker Optik 2008)

Fig. 2.22 Dumbbell model of a diatomic molecule. (Bruker Optik 2008)

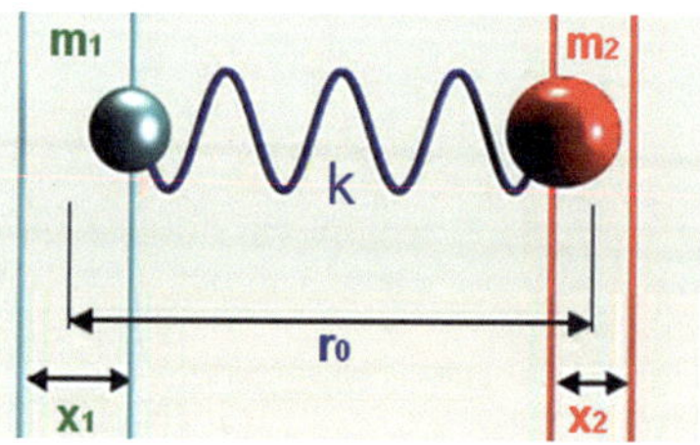

If atoms are considered as point masses, vibrations in a diatomic molecule (e.g., HCl) can be described as in Fig. 2.22. The molecule consists of the different masses m_1 and m_2, which are connected by an elastic spring.

If the equilibrium distance r_0 between the two masses is stretched by the amount $x_1 + x_2 = \Delta r$, the restoring force F is created. When released, the system oscillates around the equilibrium position and tries to restore it (Günzler and Heise 2003).

To take into account the different masses of m_1 and m_2, the reduced mass μ is introduced. The larger the mass of m_2 compared to m_1, the more the center of gravity s, around which the two masses m_1 and m_2 rotate, moves in the direction of m_2 (Naumer and Heller 1997).

$$\mu = \frac{m_1 \cdot m_2}{m_1 + m_2} = \frac{m_1}{\frac{m_1}{m_2} + 1} \tag{2.6}$$

According to Hooke's law, the restoring force F is proportional to the displacement Δr:

$$F = -k \cdot \Delta r \tag{2.7}$$

Since the force is opposite to the displacement, a negative sign appears.

In the mechanical model, the proportionality factor k corresponds to the spring constant. In the molecule, k is a measure of the bond strength between the atoms and is referred to as the force constant.

According to Newton's equation of motion, the following applies (Naumer and Heller 1997):

$$F = \mu \cdot a = \mu \cdot \frac{\partial^2 x}{\partial t^2} \tag{2.8}$$

When the forces from Eqns. 2.7 and 2.8 are equated, we obtain the basic equation of simple harmonic oscillation (Alonso and Finn 2000):

$$\frac{\partial^2 x}{\partial t^2} + \frac{k}{\mu} x = 0 \tag{2.9}$$

This differential equation has a real solution (Alonso and Finn 2000):

$$x_A = A_S \cdot \sin\left(t \cdot \sqrt{\frac{k}{\mu}}\right) \tag{2.10}$$

This equation describes the oscillation process of a diatomic molecule, where the mass point changes its location x periodically depending on the time t.

After the time for one period P, the mass point is again at the same location as one oscillation (Alonso and Finn 2000):

$$P = 2\pi \sqrt{\frac{\mu}{k}} \tag{2.11}$$

This results in the frequency ν_0 of the mass point:

$$\frac{1}{P} = \nu_0 = \frac{1}{2\pi} \sqrt{\frac{k}{\mu}} \tag{2.12}$$

This frequency is referred to as the classical natural frequency of a harmonic oscillator and describes the kinetic energy E_{kin} of the system.

The oscillation frequency ν increases with an increase in the force constant k, i.e., with a stronger bond (Bruker Optik 2008). This results in the ability to distinguish the type of bond between the carbon atoms in hydrocarbons, i.e., whether there is a single bond or a stronger multiple bond. Table 2.5 illustrates this fact by comparing the IR absorption of the type of bond. Hydrocarbons naturally contain C–H bonds, which absorb IR radiation in a certain frequency range. Here, the s-component of the σ-bond has a significant influence on the absorption frequency. The higher the s-component, the shorter and stronger the bond. This allows an aromatic C_{sp2}–H bond to be distinguished from an aliphatic C_{sp3}–H bond (Table 2.5).

Furthermore, the oscillation frequency v increases as the atomic masses decrease (Fig. 2.23). Consequently, the wavenumber increases, as it represents the reciprocal value of the wavelength (Bruker Optik 2008). This is also evident when comparing the absorption of a stretching vibration of a carbon atom, which is connected with different substituents. If a light hydrogen atom is connected to an sp^3-hybridized carbon atom, absorption occurs in the higher energy range at ≈2900 wavenumbers, while a heavy chlorine atom shifts the absorption of the stretching vibration to the lower frequency range at ≈700 wavenumbers. This makes it possible to recognize functional groups and substituents connected to a carbon skeleton by infrared spectroscopy and, for example, to distinguish PVC from PE.

The total energy of the oscillating system is composed of kinetic and potential energy. At the moment of maximum displacement, the potential energy reaches a maximum, E_{kin} is zero at this point.

When swinging through the equilibrium position, the kinetic energy reaches its maximum and E_{pot} is zero (Fig. 2.24). However, the sum of the two forms of energy remains constant (Günzler and Heise 2003).

The potential energy of the harmonic oscillator is calculated taking into account Eq. 2.12 (Alonso and Finn 2000; Atkins 2013):

$$E_{pot} = \int_0^x k \cdot x dx = \frac{1}{2} k \cdot x^2 = 2 \cdot \pi^2 \cdot \mu \cdot v_0^2 \cdot x^2 \tag{2.13}$$

This function results in a parabola, the vertex of which is located at the equilibrium position.

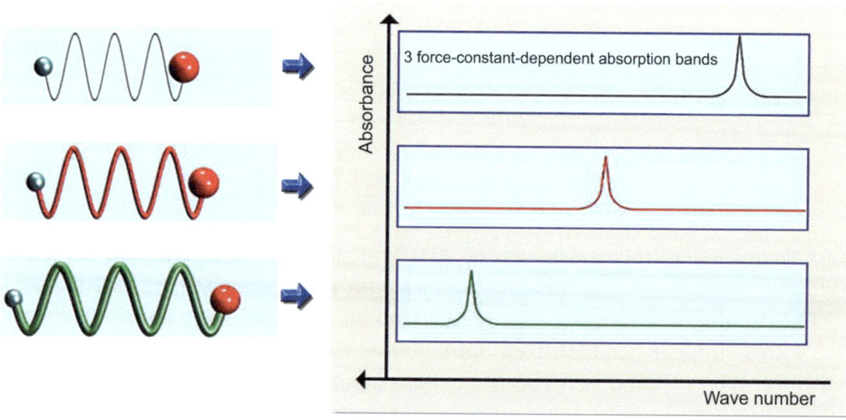

Fig. 2.23 Force constant-dependent position of the absorption bands in the spectrum. (Hart et al. 2002)

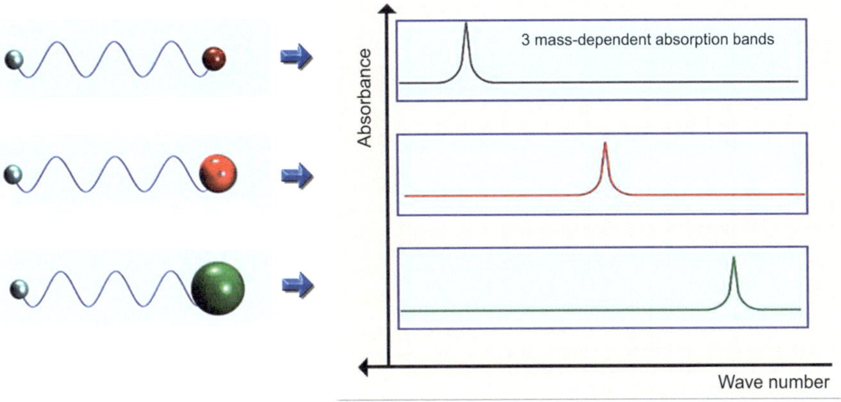

Fig. 2.24 Mass-dependent position of the absorption bands in the spectrum. (Hart et al. 2002)

However, energy cannot be absorbed in arbitrary amounts. If it is a molecule with a dipole moment, it can only absorb quantized energy amounts through absorption, which either correspond to $h \cdot \nu$ or multiples thereof (Günzler and Heise 2003).

The oscillation frequency Fig. 2.25 of the light must match the natural frequency of the molecule.

The energy of the atomic system must be calculated with the Schrödinger equation; it fully describes the behavior of electrons—and thus bonds—in molecules.

It is now possible to describe the state of the matter field by standing waves that are confined to this space, where the amplitude of the matter field or the wave function $\psi(x)$ changes from point to point within the space and is practically zero outside (Alonso and Finn 2000).

The Schrödinger equation for all three spatial directions (Alonso and Finn 2000; Atkins 2013):

$$-\frac{h^2}{8\Pi^2 \mu}\left(\frac{\partial^2 \psi}{\partial x^2} + \frac{\partial^2 \psi}{\partial y^2} + \frac{\partial^2 \psi}{\partial z^2}\right) + E_{\mathrm{pot}}(x)\,\psi = E\psi \qquad (2.14)$$

This can be transformed using Eq. 2.13 as follows:

$$\frac{\partial^2 \psi}{\partial x^2} + \frac{\partial^2 \psi}{\partial y^2} + \frac{\partial^2 \psi}{\partial z^2} + \frac{8\pi^2 \mu}{h^2}\left(E - 2\pi^2 \mu v_0^2 x^2\right)\psi = 0 \qquad (2.15)$$

By rewriting Eq. 2.15, the following results:

$$\frac{\partial^2 \psi}{\partial x^2} + \frac{\partial^2 \psi}{\partial y^2} + \frac{\partial^2 \psi}{\partial z^2} + \left[\frac{2E}{hv_0} - \frac{4\pi^2 \mu v_0 x^2}{h}\right]\psi = 0 \qquad (2.16)$$

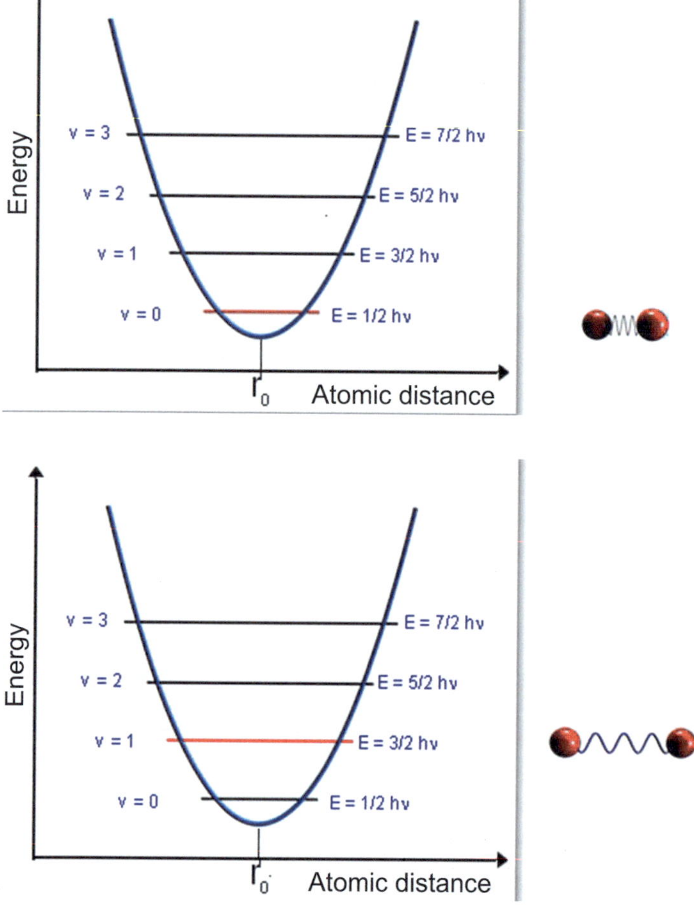

Fig. 2.25 Potential curve of the harmonic oscillator with discrete energy states or vibration levels. (Hart et al. 2002)

The solution to the differential equation Eq. 2.16 is a mathematically difficult problem that will not be discussed here. It can be shown that the allowed energy values correspond to Eq. 2.17 (Alonso and Finn 2000):

$$E(n) = h \cdot v_0 \left(n + \frac{1}{2} \right) = \cdot \frac{h}{2\pi} \cdot \sqrt{\frac{k}{\mu}} \left(n + \frac{1}{2} \right) \tag{2.17}$$

If we now set the vibration quantum number n to zero, the ground state corresponding to the quantum number has the energy $E/2$. This corresponds in Fig. 2.28 to the distance between the apex of the parabola and the energy level $n = 0$.

The distances of the energy levels are equidistant in the case of the harmonic oscillator, which can be shown by modifying Eq. 2.17 by dividing by $h \cdot c$:

$$\frac{E(n)}{h \cdot c} = \frac{v_0}{c}\left(n + \frac{1}{2}\right) = \tilde{v}\left(n + \frac{1}{2}\right) \tag{2.18}$$

In addition, only certain transitions with $\Delta n = \pm 1$ are allowed. These selection rules are based on the time-dependent Schrödinger equation, which is not listed here.

The energy difference between two adjacent vibrational states corresponds to the energy of a light quantum.

The harmonic oscillator is sufficient for the theoretical clarification of the interactions between the vibrating molecular system and electromagnetic radiation.

However, the equidistant spacing of the term scheme of the energy levels does not correspond to reality. When transitioning from the mechanical model to the biatomic linear molecule, some phenomena, such as dissociation, i.e., the breaking of the bond between the atoms with increasing amplitude, cannot be explained. In addition, the restoring force in a compression (nucleus-nucleus repulsion) is different from an extension of both masses by the same amplitude (nucleus-electron shell attraction), which also makes the slopes of the two parabolic branches different (anharmonic). As a result, the terms of the higher energy states in the so-called anharmonic oscillator move closer together (Naumer and Heller 1997).

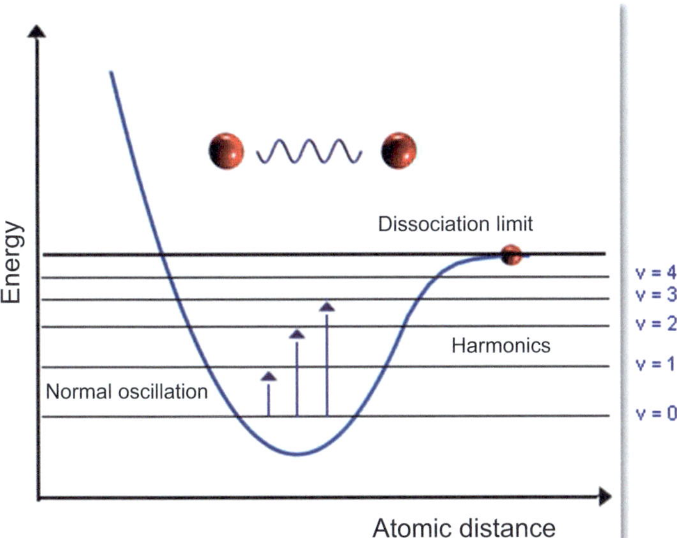

Fig. 2.26 Potential curve of the anharmonic oscillator with discrete energy states or vibration levels. (Hart et al. 2002)

The energy values of the anharmonic oscillator change taking into account the dissociation energy according to (Naumer and Heller 1997):

$$E(n) = h \cdot v_0 \left(n + \frac{1}{2}\right) - \frac{h^2 \cdot v_0^2}{4 \cdot D} \left(n + \frac{1}{2}\right)^2 \tag{2.19}$$

From Eq. 2.19 it can be derived that the term distances with increasing energy levels become smaller and smaller (Fig. 2.26) and finally converge to zero (Günzler and Heise 2003).

Another difference between the harmonic and anharmonic oscillator is that in the latter, transitions of the energy levels of $\Delta n = \pm 2, \pm 3$ etc. are also allowed (Günzler and Heise 2003).

The transition from $n = 0$ ($n = $ vibrational quantum number) to $n = 1$ is referred to as the fundamental vibration, from $n = 0$ to $n = 2$ as the first overtone, and from $n = 0$ to $n = 3$ as the second overtone (Günzler and Heise 2003).

Exciting an overtone requires more energy compared to exciting a fundamental vibration. The logical consequence is the absorption at higher wave numbers $\tilde{v}$.

However, the probability of such a quantum leap with $\Delta n > 1$, and thus the intensity of the absorption band, decreases significantly with increasing distance of the energy levels to be skipped.

Molecular Rotation

The energy absorption in a molecule with a dipole moment occurs not only through vibration but also through rotation. A permanent dipole moment in the ground state and its temporal change during a transition of the energy levels are prerequisites for the absorption of energy in a rotating molecule (Spangenberg n.d.).

In a molecule with a center of symmetry, all vibrations that occur symmetrically to the center are IR-inactive, as the dipole moment does not change. As a result, homonuclear diatomic molecules, such as H_2, O_2 or N_2, do not have an IR spectrum (Hesse and Meier 2016).

From classical mechanics, the rotational energy E_{Rot} of a molecule follows (Spangenberg n.d.; Alonso and Finn 2000):

$$E_{Rot} = \frac{1}{2} \cdot I \cdot \omega^2 \tag{2.20}$$

The moment of inertia I is calculated from the sum of all masses m_i with the distance r_i to the axis of rotation:

$$I = \sum_i m_i \cdot r_i^2 \tag{2.21}$$

The angular frequency ω is calculated as follows:

$$\omega = 2 \cdot \pi \cdot v_{Rot} \tag{2.22}$$

The moment of inertia I is calculated as follows, if r_1 and r_2 represent the respective distance of the atoms from the center of mass of the molecule (Günzler and Heise 2003):

$$I = m_1 \cdot r_1^2 + m_2 \cdot r_2^2 = \mu \cdot r^2 \tag{2.23}$$

It should be noted that this arrangement is a rigid rotator. That is, the dumbbell model demonstrated in Fig. 2.25 is abandoned and replaced by the rotation of the reduced mass μ around a fixed axis at a distance r.

As previously in the case of vibration, the energy of the molecular rotator is also quantum mechanically calculated via the Schrödinger equation (Spangenberg n.d.; Alonso and Finn 2000; Atkins 2013):

$$\frac{\partial^2 \psi}{\partial x^2} + \frac{\partial^2 \psi}{\partial y^2} + \frac{\partial^2 \psi}{\partial z^2} + \frac{8 \cdot \pi^2 \cdot \mu \cdot r^2}{h^2} E_{Rot}\, \psi = 0 \tag{2.24}$$

The detailed solution of the differential equation is not discussed here. It can be proven that the equation only provides solutions for the following discrete energy eigenvalues (Spangenberg n.d.; Alonso and Finn 2000; Atkins 2013):

$$E_{\text{Rot}} = \frac{h^2}{8\pi^2 \mu r^2} J(J+1) \tag{2.25}$$

The following conclusions can be drawn from Eq. 2.25:

- In the rotational ground state with $J = 0$ (J = rotational quantum number), the molecule—unlike in the vibrational state—contains no rotational energy.
- The larger the moment of inertia $I = \mu \cdot r^2$, the smaller the rotational energies become.
- The molecule can only rotate at certain molecular speeds.

By extending Eq. 2.25 with the relationship $\frac{E}{h \cdot c} = \tilde{v}$, the equation for the sequence of rotational energies is obtained:

$$F(J) = \frac{E_{\text{Rot}}}{h \cdot c} = \frac{h}{8\pi^2 \cdot c \cdot I} \cdot J(J+1) = B \cdot J(J+1) = \tilde{v} \tag{2.26}$$

$$B = \frac{h}{8\pi^2 cI} \rightarrow \text{Rotationskonstante}$$

The wave number of the absorbed radiation can be directly determined from the term difference for an energy transition.

The situation can be formulated in an equation as follows (Spangenberg n.d.; Alonso and Finn 2000; Atkins 2013):

$$\tilde{v} = F(J') + F(J) = BJ'(J'+1) - BJ(J+1) \tag{2.27}$$

$F(J')$ Rotational term of higher energy

$F(J)$ Rotational term of lower energy

For the change in the rotational quantum number J, the selection rule applies:

$$\Delta J = \pm 1$$

For an IR-active molecule, the absorption energy is composed of the rotational and vibrational energy (Naumer and Heller 1997):

$$E_{\text{Gesamt}} = h \cdot c \cdot \tilde{v}_0 \left(n + \frac{1}{2} \right) + h \cdot c \cdot B \cdot J(J+1) \tag{2.28}$$

Due to the significantly lower occupation energy of the rotational states compared to the vibrational transitions, a molecule in a vibrational level is in several rotational levels.

A vibrational transition thus takes place taking into account the selection rules for the quantum numbers $\Delta n = \pm 1$ or $\Delta J = \pm 1$, from the rotational states of a vibrational level to the rotational states of the higher vibrational level.

This fact is reflected in the term scheme of a molecule (Fig. 2.30). Starting from the central line, the forbidden state, due to the non-fulfillment of the selection rules, since $\Delta J = 0$, the transition $\Delta J = +1$ follows at a distance of $2B$ to its right on the so-called R-branch.

Again, $2B$ to the left of the central line, you find the transition $\Delta J = -1$ on the P-branch.

The distance of $2B$ between two absorption bands of a rotational-vibrational transition remains constant.

To prove this statement, the rotational-vibrational transition from

$$J = 0, n = 0 \rightarrow J = 1, n = 1 \tag{2.29}$$

is first calculated:

$$\tilde{v}_{1,\,n=0,\,J=0 \rightarrow n=1,\,J=1} = \left[\frac{3}{2} \cdot \tilde{v}_0 + B \cdot 1(1+1) \right] - \left[\frac{1}{2} \cdot \tilde{v}_0 + B \cdot 0(0+1) \right]$$

$$\tilde{v}_1 = \tilde{v}_0 + 2B$$

To obtain the distance between two rotational-vibrational bands, the wave number of the adjacent band must be calculated, namely the transition from

$$J = 1, n = 0 \rightarrow J = 2, n = 1 \tag{2.30}$$

$$\tilde{v}_{2,\,n=0,\,J=1 \rightarrow n=1,\,J=2} = \left[\frac{3}{2} \cdot \tilde{v}_0 + B \cdot 2(2+1) \right] - \left[\frac{1}{2} \cdot \tilde{v}_0 + B \cdot 1(1+1) \right] \tag{2.31}$$

$$\tilde{v}_2 = \tilde{v}_0 + 4B$$

The distance between the bands is obtained by forming the difference of Eqns. 2.31 and 2.29:

$$\Delta \tilde{v} = \tilde{v}_2 - \tilde{v}_1 = (\tilde{v}_0 + 4B) - (\tilde{v}_0 + 2B) = 2B \tag{2.32}$$

Fig. 2.27 Term scheme of an HCl molecule. (Alonso and Finn 2000)

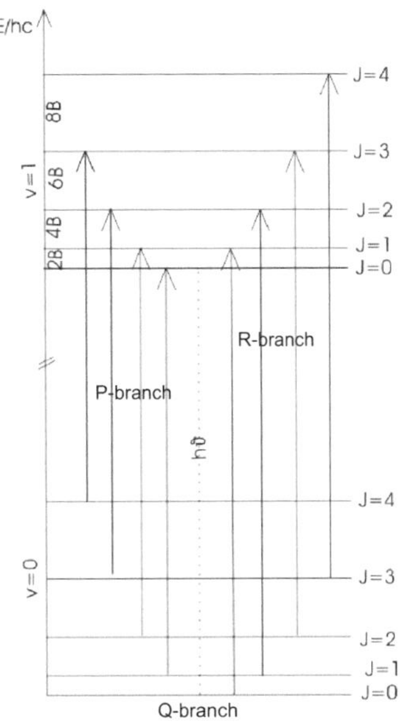

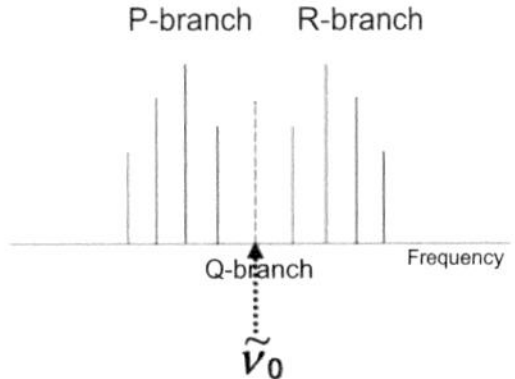

From $B = \frac{h}{8\pi^2 c I}$ and $I = \mu \cdot r^2$ it is now possible to directly calculate the atomic distance r in a diatomic molecule like HCl by reading the wave numbers of two adjacent bands in the spectrum. For linear or non-linear molecules like CO_2 and H_2O, the bond lengths can also be determined from the corresponding moments of inertia ($I = 2m_O r^2$ or $I = 2m_H r^2$) and the associated rotational constant (Atkins 2013). The rotational constant can be measured from a high-resolution IR spectrum ($\Delta\tilde{v} = cm^{-1}$). In the HCl molecule, whose term scheme and the resulting rotational vibration spectrum are shown in Fig. 2.27, one obtains a $\Delta\tilde{v}$ average value between the bands of 20 cm^{-1} and a $\tilde{v}_0 = 2889$ cm^{-1}. According to Eq. 2.32, one obtains a rotational constant $B = 10$ cm^{-1}. If one inserts the reduced mass of HCl: $\mu = (m(H)*m(Cl))/(m(H)+m(Cl)) = 1.6138E{-}27$ kg into the formulas given above, one obtains in addition to the spring constant $k = 478.6$ N/m (measure for the bond strength) the moment of inertia $I = 2.79E{-}47$ kg m^{2}and consequently also the bond length in the ground state with $r_0 = 1.3$ Å.

In the case of ATR spectroscopy of plastics, there is naturally also water vapor in the sample chamber, depending on humidity and carbon dioxide. Both are

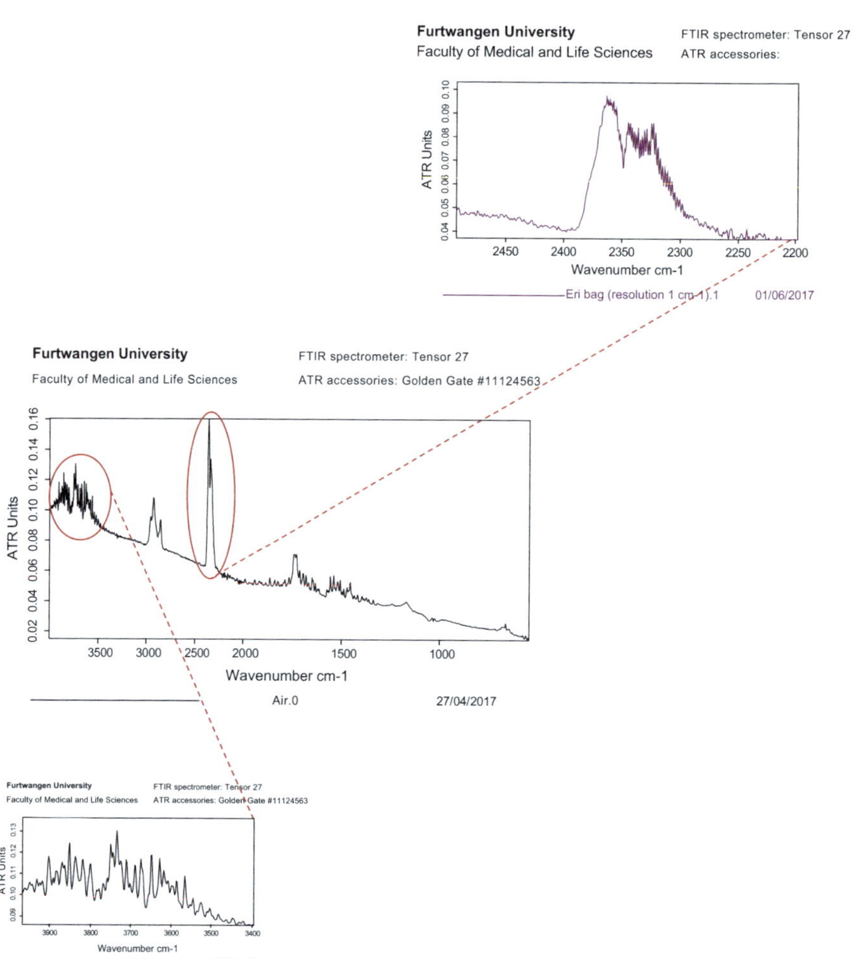

Fig. 2.28 Infrared spectrum of an uncleaned sample chamber

Table 2.6 Normal vibrations of water (g) and carbon dioxide. (Günzler and Heise 2003)

	H_2O (g)		CO_2	
Vibration	Wavenumber [cm^{-1}]	Type of vibration	Wavenumber [cm^{-1}]	Type of vibration
ν_1	1285	ν_s; IR-inactive	3657	ν_s; IR-inactive
ν_2	2349	ν_{as}; IR-active	1595	δ; IR-active
ν_3	667	δ; doubly degenerate; IR-active	3756	ν_{as}; IR-active
ν_4	667			

IR-active molecules. All other gases in the air are not IR-active. In the gas phase, water and CO_2 molecules can experience a vibrational-rotational excitation, and the corresponding R and P branches of both molecules appear in the IR spectrum. The spectrum shown in Fig. 2.28 shows the fine splitting of the rotational vibration spectra of water vapor and carbon dioxide.

Since the positions of the absorptions are known (Table 2.6), an automatic compensation can be carried out using the software, so that these bands are eliminated in the spectrum and they do not cover any other characteristic bands.

A molecule of any structure made up of N atoms has 3N degrees of freedom due to the independent spatial coordinates of each atom (Fig. 2.29). Of these, three degrees of freedom are due to translational motion along the x, y, and z directions, where the atoms do not move relative to each other, but all move in the same direction, changing the center of mass. Thus, no interaction with electromagnetic radiation is possible (Günzler and Heise 2003).

Furthermore, three more degrees of freedom are due to rotations around the principal axes of inertia.

The number of actual vibrational degrees of freedom n is thus reduced to:

$$f = 3 \cdot N - 6 \quad \left(\text{Freiheitsgrade nichtlinearer Moleküle}\right) \quad (2.33)$$

In linear molecules, only two rotational degrees of freedom exist, as rotation around the molecular axis is not associated with any movement of the atoms or the center of mass (Günzler and Heise 2003).

Therefore, this type of molecule has an additional vibrational degree of freedom n:

$$f = 3 \cdot N - 5 \quad \left(\text{Freiheitsgrade linearer Moleküle}\right) \quad (2.34)$$

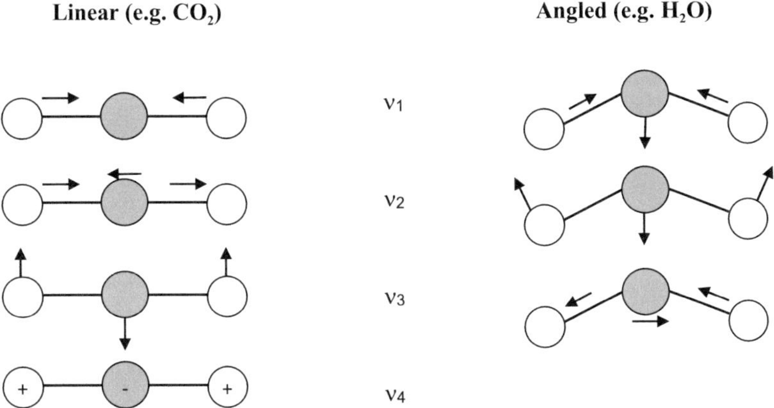

Fig. 2.29   Vibrations of triatomic molecules

The number of vibrations calculated in this way is called the normal or fundamental vibrations of a molecule, which can be excited independently of each other.

The atoms involved in the normal vibration oscillate with the same frequency and a fixed phase relative to each other. Each normal vibration is assigned a specific vibration frequency or wave number.

Complex molecules naturally have a large number of vibration possibilities. For the interpretation of spectra, vibrations that can be approximated to individual bonds or functional groups of a molecule, i.e., localized vibrations, are particularly useful. These occur in the spectrum in the range of wave numbers from 4000 cm^{-1} to 1000 cm^{-1}.

Localized vibrations are distinguished according to their form of vibration as follows:

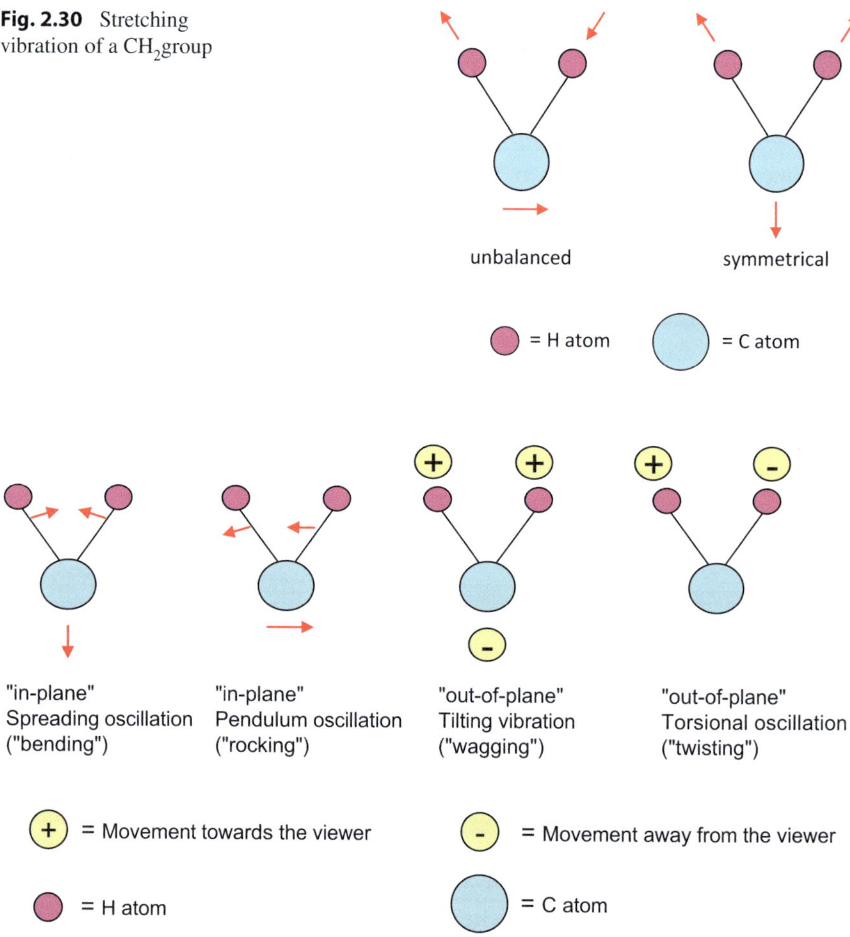

Fig. 2.30 Stretching vibration of a CH$_2$ group

Fig. 2.31 Deformation vibrations of a CH$_2$ group

- *Valence vibrations* ν (Fig. 2.30): The atoms are deflected in the direction of the bond axis. The bond distance changes periodically.
- *Deformation vibrations* δ, τ, γ (Fig. 2.31): The bond angle between the atoms changes periodically, while the bond distances remain approximately constant.

Furthermore, the different forms of a deformation vibration are distinguished. Here, the spreading *(bending)*or pendulum vibration *(rocking)* are bending vibrations in the plane with the symbol δ. The torsional vibration τ *(twisting)* or the tilting vibration γ *(wagging)* protrude out of the plane.

Quantitative Spectrum Evaluation

The basis for quantitative evaluations in IR spectroscopy is the Lambert-Beer's law. It represents the linear relationship between the absorbance of the sample and its concentration.

The light transmittance is referred to as transmission T and is defined by (Hesse and Meier 2016):

$$T = \frac{I}{I_0} \cdot 100 \ \% \tag{2.35}$$

The light intensity experiences an exponential decrease when passing through a sample solution (Bruker Optik 2008):

$$I = I_0 \cdot \exp\left(-\tilde{\varepsilon} \cdot c \cdot d\right) \tag{2.36}$$

By rearranging Eq. 2.36 we get:

$$\ln\left(\frac{I}{I_0}\right) = -\tilde{\varepsilon} \cdot c \cdot d \tag{2.37}$$

The molar extinction coefficient and the transmission are not defined over the natural logarithm. However, since there is a linear relationship between the natural and decimal logarithm, the transformation corresponds to a factor in the equation. This is incorporated into the extinction coefficient. Thus, $\tilde{\varepsilon}$ becomes ε:

$$\lg\left(\frac{I}{I_0}\right) = -\varepsilon \cdot c \cdot d \tag{2.38}$$

By rearranging Eq. 2.38 we obtain the Lambert-Beer's law, which describes the relationship between the absorbance A and the concentration c of the absorbing substance as a linear function:

$$A = \lg\left(\frac{I_0}{I}\right) = \varepsilon \cdot c \cdot d \tag{2.39}$$

2.3.3   Applications of IR Spectroscopy to Plastics

Compared to mass and NMR spectroscopy, infrared spectroscopy is a relatively inexpensive and especially fast analytical technique for identifying hydrocarbons. However, in terms of sensitivity and quantification, it is inferior to the other two analytical techniques, especially since structural elucidation with IR spectroscopy is not possible. Only functional groups can be detected through the excited vibrations of the atoms involved in a bond (Sect. 2.3.2). Since different atoms and bonds are present in plastics, the interaction of infrared radiation in the form of vibrational excitation is ideal for quickly and easily identifying technical or synthetic polymers as solids using ATR-IR spectroscopy (Fig. 2.32; description in Sect. 2.4).

The type of plastic of microplastic particles <500 μm, as they occur in waters or digestive tracts (Leser 2015), can be identified with an infrared microscope due to its peak pattern compared to a stored known spectrum in the plastic database.

Furthermore, infrared spectroscopy is a tool for separating pure types of plastics via their specific absorption bands or the peak pattern within a recycling plant. An automated separation of pure plastics, which is not labor-intensive and therefore cost-effective, also determines whether our plastic waste is recycled or subjected to thermal recovery. Separation processes that exploit the different specific weights of various types of plastic in float-sink processes or hydrocyclones have been replaced by infrared spectroscopy (Weber 2002). Why only about 40% of plastic packaging that ends up in the yellow bag or the yellow bin is recycled, although technically a much larger percentage would be possible, is a question of economic viability.

One reason for the relatively low rate is the competition of recyclers with waste incinerators, which need the plastic as an important fuel for generating heat and

Fig. 2.32 Hose under ATR unit

electricity. The recycler loses this price battle for plastic waste as long as he cannot separate it into pure types in a cost-effective way. For this reason, the recycling rate stagnates more or less at 40%. A legal regulation, as demanded by NABU, for example an energy tax for waste incineration plants (MVA), RDF power plants (Ersatzbrennstoffkraftwerke), could counteract this. The difficulty of separating into pure types in order to obtain a recyclate, which is often mixed with primary plastic, lies in the fact that packagingor products like chip bags or toothpaste tubes are composed of several different types of plastic or the plastics occur in composite form like in Tetra Paks. These consist of several layers of paper, plastic, and aluminum. To achieve purity, plastic mixtures and composites would have to undergo a costly mechanical separation, which is saved in the thermal recovery of these mixed products. Even the pure packaging, which mainly consists of PP, PE, PS or PET, differs due to its additive cocktail, which limits the use of the recyclate and does not qualify it for higher quality products.

Some of these additives, such as plasticizers, can be detected simultaneously with the identification of the plastic type within an infrared spectroscopic examination of a plastic granule or a separated plastic chip from an imported product. A quick decision as to whether a plasticizing aid is used in a product, which is no longer allowed in Europe without a declaration obligation to the customer due to the REACH regulation, can be quickly determined here.

The three described contemporary and very current application areas of infrared spectroscopy– microplastic analytics (particle size and plastic type, IR microscopy), plastic recycling, and pollutant detection – justify an intensive presentation of this methodology.

Infrared spectroscopy (IR) is used to identify microplastics. Some working groups have already examined sediment samples using the technique of micro-FTIR spectroscopy (Vianello et al. 2013; Harrison et al. 2012). At the Alfred Wegener Institute on Heligoland, two methods for water and sediment samples have successfully established:

- (ATR-)IR(Attenuated Total Reflection-IR)-spectrometry for plastic particles with a size >500 μm
- Micro-Fourier Transform Infrared (FTIR) spectroscopy for plastic particles with a size <500 μm (Löder et al. 2015a).

For micro-Fourier transform infrared (FTIR) spectroscopy, infrared microscopes are used that make particles with a resolution of 1–100 μm visible through ATR-Imaging. With this imaging method, surfaces can be scanned within a mapping, thus creating a microplastic particle distribution on a filter paper (reflection measurement mode, aluminum oxide filter; transmission measurement mode).

To investigate waters for microplastics, an elaborate and time-consuming sample preparation is necessary. The filter residues of the water samples, which remain in the stainless steel filter after 1000 liters of water have been filtered, are enzymatically cleaned with SDS (protein denaturation), protease, cellulase, chitinase and also with H_2O_2 and divided into two size fractions. For the quantification and identification of microplastics, two different IR spectrometers are used. For the analysis of particles >500 μm, the *attenuated-total-reflection* (ATR)-IR spectrometer is used and for the analysis of particles <500 μm, the *micro*-Fourier-*transformed-infrared* (FTIR)-spectrometer is used.

In the project "Rheines Wasser", the concentrations of microplastics of different polymer types were analyzed using both methods. Before the results of this extensive investigation are presented in the further course of this chapter, additions to the basics of infrared spectroscopy (Sect. 2.4.2) regarding the special plastic particle analysis technique are appropriate.

In principle, two methods are distinguished by which IR spectrometers work:

1. Dispersive method
2. Fourier transform method

The essential element of the dispersive IR spectrometer (Fig. 2.33) is the monochromator. This consists of an entrance slit, dispersive element, and exit slit. A prism or diffraction grating is used as the dispersive element. Dispersive IR spectrometers work almost exclusively according to the two-beam principle. The light from the source is divided into two beam paths: the sample beam and the reference beam. If the two beams have different intensities, the reference beam is attenuated by means of an aperture until the intensities are equal. The position of the aperture is a measure of the transmission.

The monochromator is located behind the sample room to eliminate the occurring scattered light.

FTIR

The Fourier Transform Infrared Spectroscopy (FTIR spectroscopy) is a special variant of IR spectroscopy, which has largely replaced dispersive IR spectroscopy in the mid and far infrared range.

In a dispersive measurement, the IR spectrum is recorded by gradually changing the energy of the IR beam incident on the sample. To split the energy of the polychromatic radiation generated by a thermal radiation source, a Monochromator is used in dispersive spectrometers. The spectral decomposition in the monochromator is carried out either by a prism or a diffraction grating as a dispersing element. The IR spectrum is recorded by a gradual change in the wavelength focused into the sample by rotating the grating or the deflecting mirrors directly over the Detector.

In FTIR spectrometers, the beam splitting is done via an interferometer and the spectrum is calculated by a Fourier transformation of a measured interferogram.

Fig. 2.33 Schematic structure of a two-beam IR spectrometer

The interferometer essentially consists of a beam splitter for the incident broadband Infrared radiation, a fixed and a mobile mirror. If the distance L from the beam splitter to both mirrors is the same, no beam splitting occurs and the broadband radiation is preserved. If the movable mirror is shifted by the path x, then both beam halves differ by 2x when recombining at the beam splitter. Constructive interference of radiation of a certain wavelength λ is only achieved if the path difference of 2x is an integral multiple of this wavelength λ. For all other wavelengths, destructive interference occurs, so that the maximum radiation intensity at the interferometer output and input into the sample to be measured is exactly assigned to this wavelength. For all other wavelengths, the detector signal is much lower due to the destructive interference. The resulting intensities of the detector signals depending on the Deflection x of the movable mirror result in the so-called interferogram. The mathematical transformation of the interferogram using the Fourier transformation initially results in the single-channel spectrum, which shows with which intensities the individual wave numbers occur. The quotient of this spectrum compared to a reference spectrum without a sample results in a representation analogous to the dispersive measured spectrum (Günzler and Heise 2003).

The FTIR measurement has clear advantages over the conventional dispersive measurement in terms of the signal/noise ratio and the shorter measurement time. While in the dispersive measurement the different wave numbers are recorded one after the other via the setting of the monochromator and less than 0.1% of the radiation entering the Monochromator hits the Detector, in the FT spectrometer the entire spectral range contributes with little attenuated intensity for the maximum interference to the signal. The noise is therefore also distributed over the entire spectral range. This improves the signal/noise ratio and thus the sensitivity of the measurement (Günzler and Heise 2003).

Nowadays, almost exclusively Fourier-Transform-IR-Spectrometersare used. These are practically always single-beam devices and do not have a Monochromator. In comparison, they have better resolution than dispersive devices and make better use of the intensity of the radiation source. Instead of the

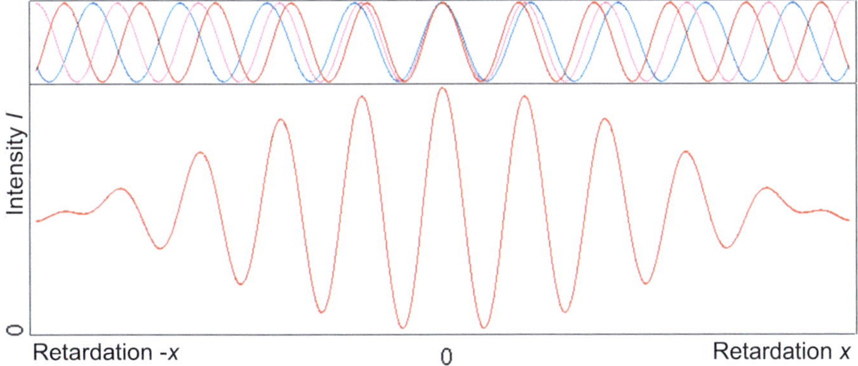

Fig. 2.34 Interference of two partial beams as a function of the runtime

monochromator, they work with a Michelson interferometer. Here, the polychromatic IR radiation is first split into two partial beams by a semi-transparent mirror. One beam is reflected by a stationary mirror and thus has a fixed path length.

The path length of the second beam, which is reflected by a movable mirror, can be continuously varied. When the partial beams are reunited, they interfere with each other and deliver a certain interference intensity as the sum over all wavelengths.

The interference intensity (Fig. 2.34) is registered as a function of the time in which the path length of one partial beam is continuously varied, and stored by a computer. This results in the so-called interferogram, which contains all the information of the spectrum in encrypted form.

Using the computer, the stored data can be transformed from the time domain to the frequency domain by Fourier transformation (Fig. 2.35). Fig. 2.36 illustrates the structure of the interferometer space.

With an empty sample space, an interferogram is recorded and Fourier-transformed. The result is the single-channel reference spectrum $R(v)$. A second

Fig. 2.35 Schematic structure of an FT-IR spectrometer

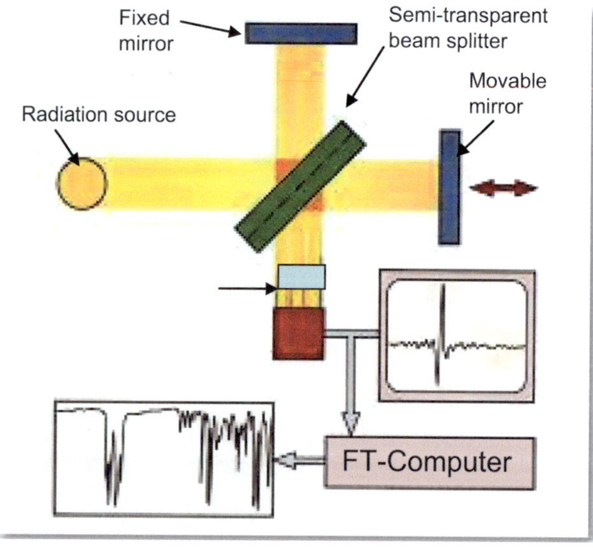

Fig. 2.36 Schematic representation of the interferometer space. (Bruker Optik 2008)

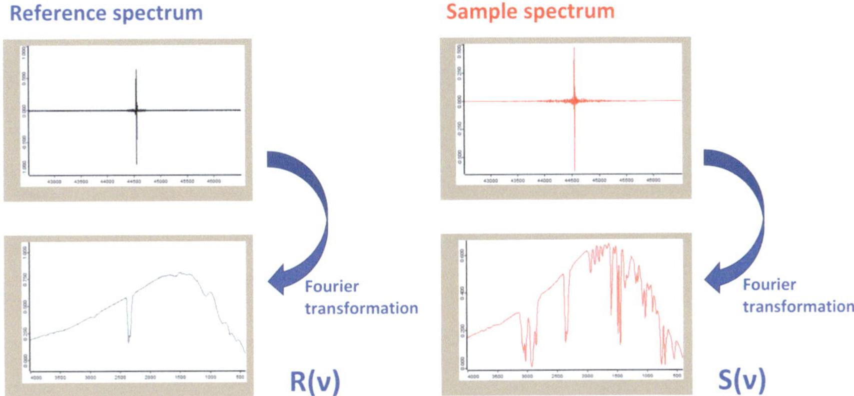

Fig. 2.37 Fourier transformation of the reference and sample spectra (Bruker Optik 2008). The transmission spectrum $T(v)$ is obtained by dividing the sample spectrum by the reference spectrum. (Eq. 2.40)

interferogram is recorded with the sample in the sample space and transformed. This results in the single-channel sample spectrum $S(v)$. The sample spectrum resembles the reference spectrum, but shows lower intensity in those wavenumber ranges where the sample absorbs radiation (Fig. 2.37) (Bruker Optik 2008). The transmission spectrum $T(v)$ is obtained by dividing the sample spectrum by the reference spectrum. (Eq. 2.40; Fig. 2.38)

$$T(v) = {S(v)}/{R(v)} \qquad (2.40)$$

Transmission FT-IR Spectroscopy

In transmission infrared spectroscopy, as illustrated in Fig. 2.39, the infrared beam completely penetrates, for example, a liquid sample within a cuvette. Within the Transmission spectroscopy, not only are liquids of an IR-active substance in a solvent such as Nujol irradiated, but also KBr pellets, hot-stamped plastic films or microtome sections of plastics provide information (concentration; substance class) about the material or the dissolved substance in a transmission measurement.

In this section, three different infrared spectroscopic transmission methods are presented and their accuracy compared using the example of quantifying the polybutadiene content in ABS (Section 2.4):

1. Quantification using ABS embossing foils
2. Quantification using KBr pellets
3. Quantification in solvent using transmission technique

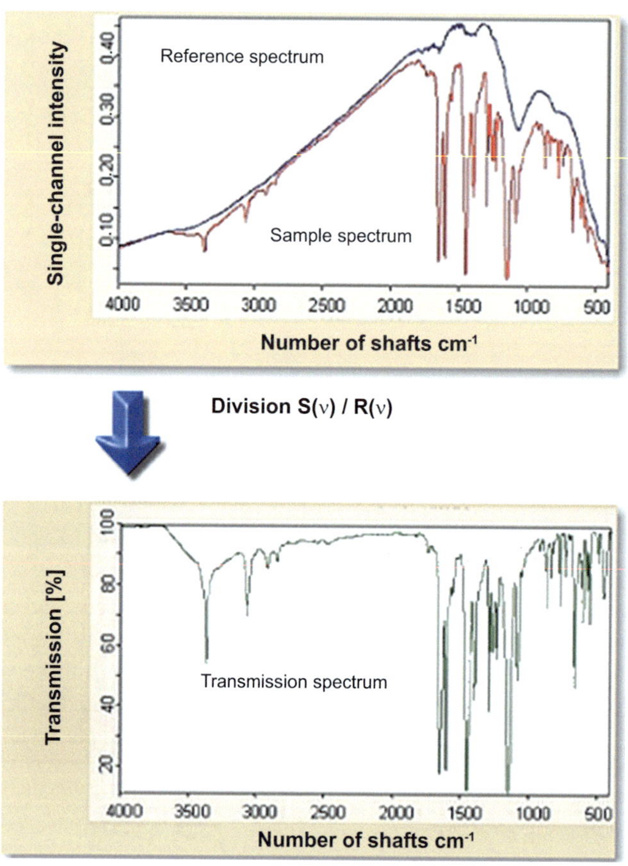

Fig. 2.38 The path to the transmission spectrum. (Bruker Optik 2008)

Fig. 2.39 Attenuation of the infrared radiation intensity in a Transmission of the radiation through a sample of concentration c with an absorption coefficient α

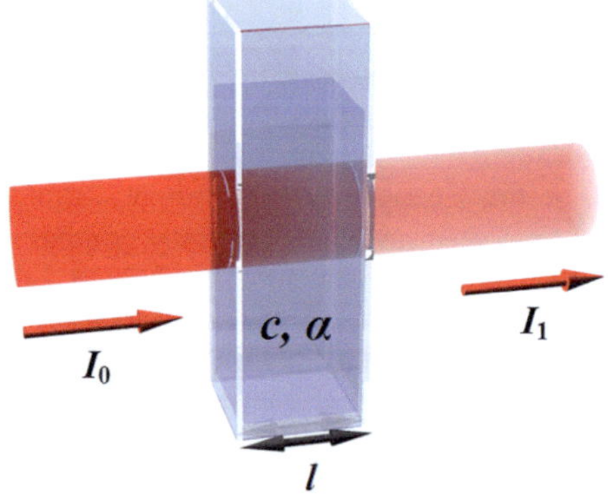

To enable quantification in the first place, the plastic supplier has provided six reference samples with different known polybutadiene contents. The calibration lines of the different measurement methods are created based on the theoretical polybutadiene contents.

An additional control instrument for checking the different measurement methods and results is the examination of the pure ABS plastics P2mc and Polylac 727. The polybutadiene contents of both plastics are approximately known, namely 20–24% for P2mc and 14–18% for Polylac 727. They serve to check the linear equation of the respective calibration line.

The classic method for quantifying the polybutadiene content in ABS is the Wijs method, which, together with the manufacturer's specification (theoretical value), serves to evaluate the results of the infrared transmission measurements accordingly.

The Wijs analysis is a titration method. The measurement principle consists of attaching iodine to polybutadiene and titrating back the excess iodine with 0.1 M sodium thiosulfate solution. This allows the polybutadiene content in the sample to be determined directly. The examination of the reference samples provided by the manufacturer for quantification leads to the result shown in Table 2.7.

The determination of the percentage proportions is determined using the following equation:

$$\%Pbu = \frac{(A - B) \cdot t \cdot 0.271}{Einwaage} \tag{2.41}$$

A Consumption of 0.1 molar sodium thiosulfate solution of the blank sample

Table 2.7 Polybutadiene contents of different ABS types according to the Wijs method

Sample	Weight in g	Thiosulfate in ml	%Pbu	Average value	%Pbu theoretical
Blank sample	0	16.1			
1.1	0.0735	11.6	16.6	16.6	15.7
1.2	0.0735	11.6	16.6		
2.1	0.071	10.75	20.4	19.4	19.8
2.2	0.071	11.3	18.3		
3.1	0.0735	11.2	18.1	18.5	17.8
3.2	0.073	11	18.9		
4.1	0.0707	10.4	21.8	21.7	22.1
4.2	0.0755	10.1	21.5		
5.1	0.0709	4.3	45.1	44.6	54.0
5.2	0.073	4.2	44.2		
6.1	0.0805	7.8	27.9	30.6	30.0
6.2	0.079	6.4	33.3		

B   Consumption of 0.1 molar sodium thiosulfate solution of the sample
T   Titer

Based on the product recipes, the manufacturer calculates the theoretical polybutadiene content.

As can be seen from Table 2.7, the determined values correspond to the theoretical values, with one exception. The cause of this may lie in the poor solubility of the polymer or there may be additional additives present that can be oxidized by iodine.

However, the problems of the analysis according to the Wijs method lie in the considerable time expenditure and the high solvent consumption of halogenated solvents.

The time required for analysis should be kept as low as possible so that the delivered batch can be examined directly from the transporter before filling the storage silo. This can keep the downtime for the transport company and thus the costs for the customer as low as possible.

Due to these stated reasons, the analysis using the Wijs method is no longer up-to-date and can be replaced by infrared spectroscopic methods.

Quantification With ABS Embossing Foils Using Transmission Technique

In the preparation for the transmission technique, approximately 20 mg of the plastic is pressed between an aluminum foil coated with magnesium stearate in a press heated to 220 °C (Fig. 2.40). With this production method, plastic foils with a few μm layer thickness are obtained.

Fig. 2.40 Heatable hydraulic press

Fig. 2.41   Plastic foil pressed onto aluminum foil (left) and fixed in holder

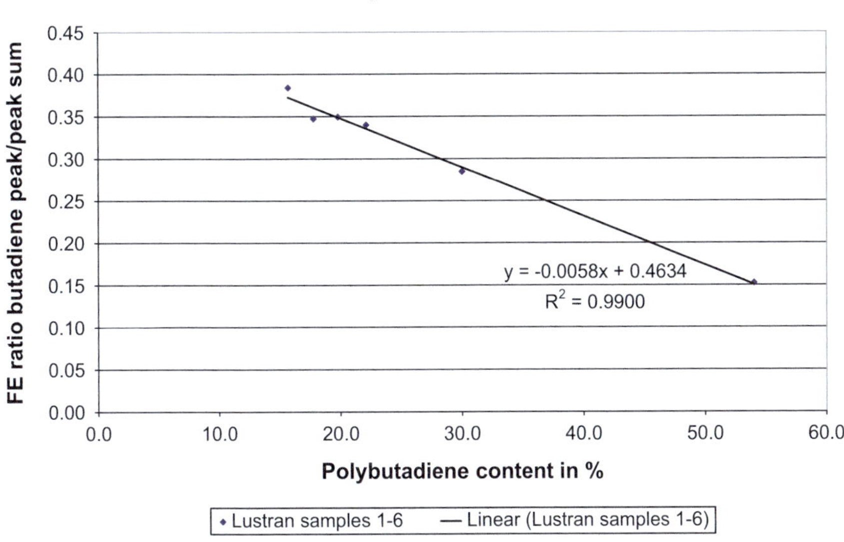

Fig. 2.42   Calibration line for polybutadiene quantification via the measurement of press foils

The plastic foil peeled off from the aluminum foil is fixed in the holder with magnets(Fig. 2.41).

Three foils are produced from each sample to check the reproducibility of the measurement.

For evaluation, the sum of the areas of the butadiene peak (1510.1–1478.27 cm^{-1}), the nitrile peak (2263–2217.7 cm^{-1}) and the styrene peak (987.89–946.69 cm^{-1}) is first formed. Subsequently, in Excel, the area of the butadiene peak is divided by the peak sum and plotted over the known theoretical butadiene contents of the reference samples provided by the manufacturer. The calibration line (Fig. 2.42) is created using the average of the three individual measurements. The mathematical evaluation of the peaks is carried out in Excel.

Using the line equation, it is now possible to determine the polybutadiene content of unknown samples. For this, the peaks used for evaluation must first be integrated with the software and the obtained area contents then related to each other in Excel.

The polybutadiene contents determined by the line equation for P2mc and Polylac 727 are in the ranges of 21.74% ± 1.43% (P2mc) and 15.7% ± 2.51%. Thus, the deviation is within the error range specified by the manufacturer and the method is therefore suitable for quantifying the polybutadiene fraction.

Quantification Using KBr Pellets

To determine the polybutadiene fraction in the reference samples, a KBr pellet must first be produced. To ensure a constant peak intensity in the spectrum, according to Lambert-Beer's law, the concentration of the substance to be measured and the layer thickness of the medium, here the KBr pellet, must be kept constant. Otherwise, no comparisons can be made between the different samples.

Therefore, for the investigation, 150 mg of KBr and 3 mg of the plastic sample are weighed in each case. The homogenization of the individual components takes place in an agate mortar (Fig. 2.43) with the addition of ten drops of dichloromethane, which causes the swelling of the plastic. Without the addition of dichloromethane, it would not be possible to distribute the sample homogeneously in KBr.

Three spatula tips of the sample, which has been homogenized and ground with KBr, are filled into the press tool (Fig. 2.44). In the press, the KBr pellet is produced under a pressure of ten tons.

Fig. 2.43 Plastic sample homogenized with KBr in agate mortar

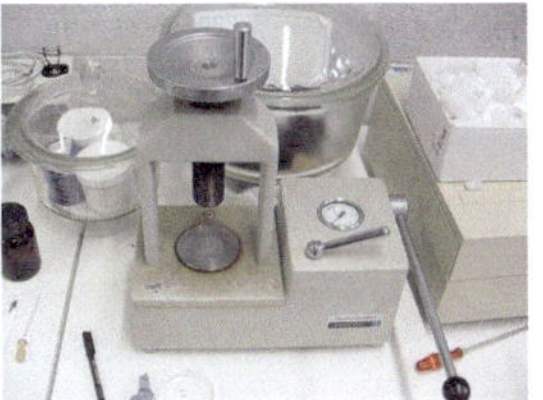

Fig. 2.44 Press tooland press for the production of KBr pellets

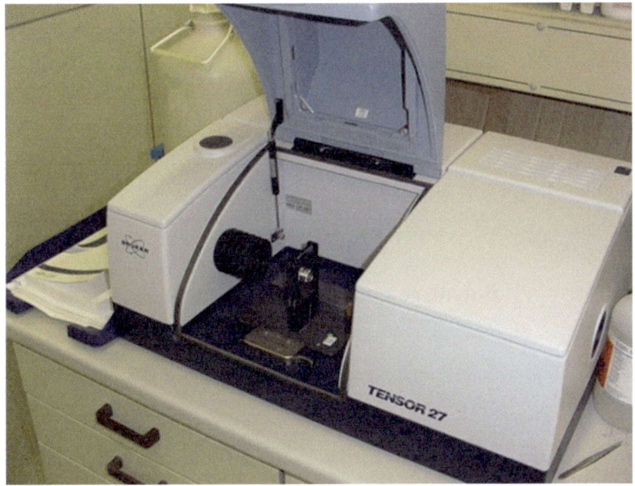

Fig. 2.45 IR device with a KBr pellet inserted in the beam path

After the pressing process, care must be taken not to touch the surface with the fingers, as this would lead to a falsification of the measurement result due to the applied finger grease.

It is also important to ensure that the pellet (Fig. 2.45) is stored in an airtight container, as KBr is extremely hygroscopic. The absorbed water would also interfere with the measurement.

The KBr pellet is clamped into the recording unit (Fig. 2.46) and measured. To obtain a representative result, five pellets of each type of granulate are measured, each of which is made from five different grains of the granulate. As with the methods previously examined, the average value formed from five individual measurements is used to create the calibration line (Fig. 2.47).

As with all three methods, the same peaks in the spectrum are analyzed, namely the butadiene peak (1510.1–1478.27 cm^{-1}), the nitrile peak (2263–2217.7 cm^{-1}) and the styrene peak (987.89–946.69 cm^{-1}). Thus, the results of all methods are comparable with each other.

Fig. 2.46 Clamping device with KBr pellet

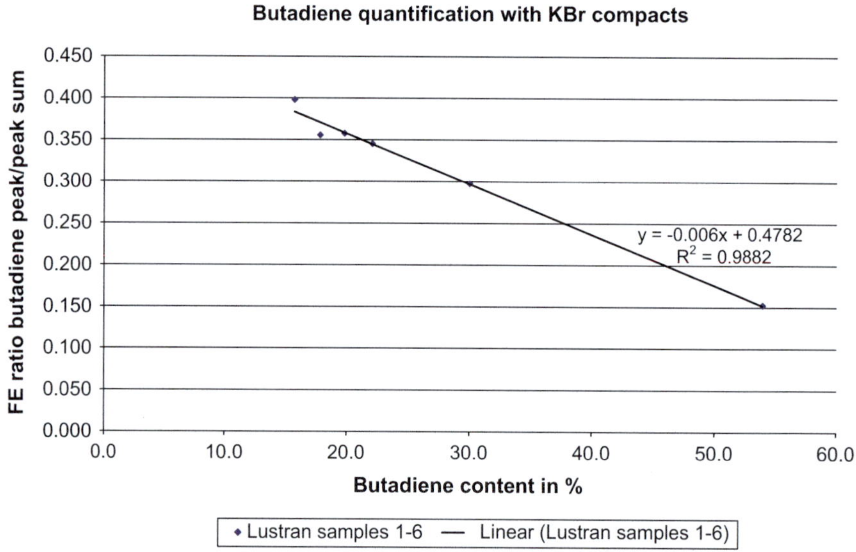

Fig. 2.47 Calibration line for polybutadiene quantification via the measurement of KBr pellets

The verification of the straight line equation of the calibration line with the plastics P2mc and Polylac 727 leads to a percentage polybutadiene content of 21.03% ± 3.78% in the case of P2mc and 16.53% ± 1.54% in the case of Polylac 727. The value of P2mc is at the lower end of the given window, considering the relatively large error, even below it. Therefore, this method cannot be used for the quantification of the polybutadiene content.

The calibration lines created in Excel for each measurement method can be directly compared with each other due to the identical peaks used for evaluation.

In the case of the measurement method with KBr pellets and press foils, the coefficient of determination of the calibration line is larger, but an additional investment in a hydraulic press and press tool would be necessary for implementation.

The standard deviations over the individual measurements are smallest in comparison in the case of the measurement of microtome sections, which speaks for the reproducibility of the measurement results with this measurement method.

Therefore, the method of choice is to revert to the measurement of microtome sections via ATR, which is sufficient with a coefficient of determination of the calibration line of 98.1%.

Quantification in Solvent by Transmission Technique

A further possibility for determining the polybutadiene content in ABS plastics is theoretically the analysis with a liquid cell. The liquid cell with corresponding KBr or CaF_2 windows can be assembled with different sample volumes by using so-called spacers (see the exploded view in Fig. 2.48).

The empty liquid cell is measured here as the background spectrum with 16 scans. To obtain the actual sample spectrum, the liquid cell filled with the solvent

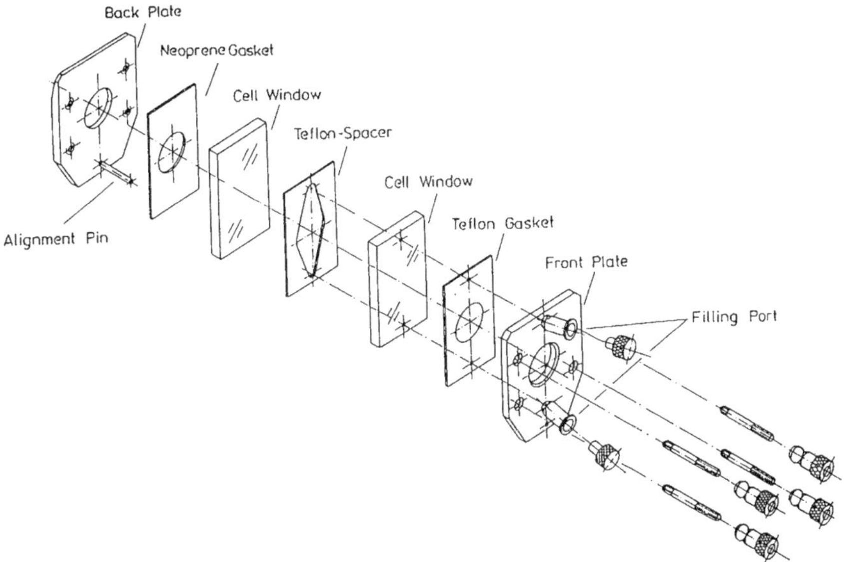

Fig. 2.48 Construction of the IR liquid cell (exploded view)

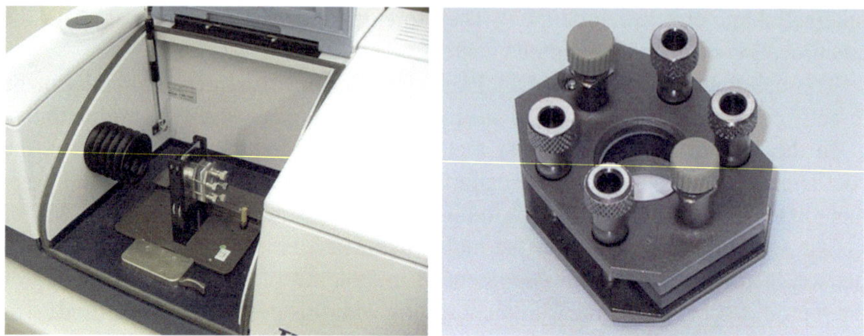

Fig. 2.49 Liquid cell inserted in the beam path Liquid cell

must first be measured as the sample spectrum. This is followed by the dissolved plastic sample in the solvent (Fig. 2.49).

The Opus 6.5 software can then perform a spectrum subtraction of the solvent spectrum from the spectrum of the plastic sample in the solvent, resulting in the pure IR spectrum of the plastic sample.

The literature recommends a concentration of 10–20% for measuring solutions, with a cell layer thickness of 100–200 μm.

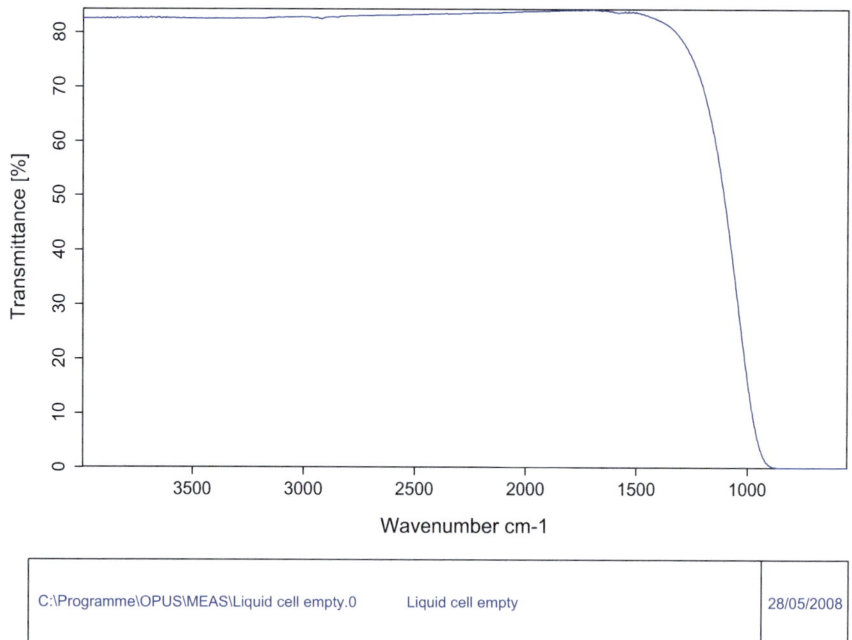

Fig. 2.50 IR spectrum of the calcium fluoride glasses

These specifications are met and a suspension with 10 mass percent is prepared by weighing 1.33 g of plastic granules into 10 ml of methylene chloride as a solvent. The layer thickness of the liquid cell during the measurement is 100 μm.

However, with this measurement method, quantification of the polybutadiene content is not possible, as the glasses used in the liquid cell are made of calcium fluoride (Fig. 2.50). At a wavenumber of approx. 1000 cm^{-1}, these go into total absorption, i.e., all the radiation is absorbed by the glasses. Therefore, an evaluation of the styrene peak needed for quantification is not possible. The peak sum of the three peaks used for the evaluation cannot therefore be formed.

ATR Spectroscopy and Microscopy

Unlike transmission infrared spectroscopy, where the infrared beam completely penetrates a sample, the infrared radiation in ATR spectroscopy only peripherally touches a solid sample. ATR *(attenuated total reflection)* means "reduced or attenuated total reflection", which can occur with reflection on absorbing materials. The method is therefore also referred to as Internal Reflection Spectroscopy (IRS) and is thus one of the reflection techniques.

The core of the ATR method is an IR-transparent crystal with a high refractive indexn. Zinc selenide (ZnSe), silicon (Si), germanium (Ger), or diamond (C) is used as a single crystal material with a high refractive index. The refractive index is a dimensionless number that describes the ratio of the propagation speed of the electromagnetic radiation in the vacuum to that in medium i. The refractive index is wavelength-dependent:

$$n_{i(\lambda)} = c_v / c_i \qquad (2.42)$$

When a light beam with the angle of incidence α passes from an optically denser to an optically thinner medium, it is refracted away from the perpendicular (β). This case is illustrated in Fig. 2.53 (Image 1). Since the angle of refraction β ($\beta >$ α, since $n_2 > n_1$) depends on the angle of incidence α, a limit case ($\beta = 90°$) arises with an ever-increasing angle of incidence, in which the refracted beam propagates along the interface. This behavior defines the critical angle α_g, which depends on the refractive indices of both media.

Using Snell's law of refraction, the light path can be described when it hits an interface of two media of different optical densities (Bruker Optik 2008).

$$n_1 \sin\alpha = n_2 \sin\beta \qquad (2.43)$$

for $\beta = 90°$, the critical angle $\sin\alpha_g = n_2/n_1$ or $\alpha_g = \sin^{-1}n_2/n_1$ results, as can be seen in Fig. 2.51 (Image 2).

Above a certain angle of incidence (Image 3 in Fig. 2.51), the critical angle α_g, total reflection occurs (Bruker Optik 2008). With an angle of incidence greater than α_g, the ATR crystal essentially acts as a light guide.

If an IR beam with an entrance angle $>\alpha_g$ is directed from a medium with a high refractive index (ATR crystal) to a medium with a low refractive index (plastic sample), the beam is reflected back into the original medium (ATR crystal) at the interface.

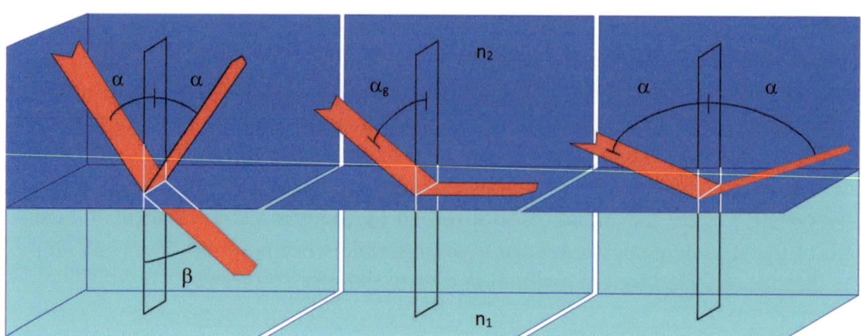

Fig. 2.51 The angle of incidence-dependent angle of refraction at the interface of two media with different optical densities $n_2 > n_1$. (Bruker Optik 2008)

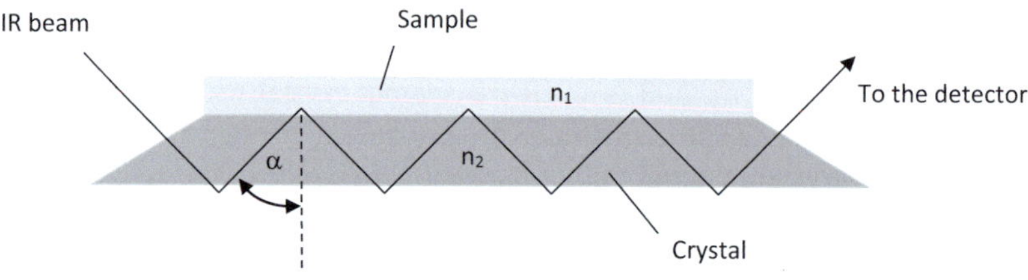

Fig. 2.52 Beam path of an infrared radiation through a medium (crystal) that is optically denser than the surrounding (plastic sample)

The infrared beam is also totally reflected on the opposite side to the sample application, so that the ATR crystal becomes a light waveguide, through which the radiation is guided in total reflection. In the ATR crystal, multiple reflections at both interfaces are possible in this design (Fig. 2.52).

However, a portion of the radiation penetrates into the optically thinner medium (sample) to be reflected back into the optically denser medium (ATR crystal) at a slight offset. This behavior, known as the Goos-Hänchen effect, is the basis for the ability to conduct spectroscopy using a light-conducting crystal (Bruker Optik 2008). The material sample pressed onto the internal reflection element absorbs a material-specific part of the incoming IR radiation (Fig. 2.55). Since a small portion of the infrared radiation is absorbed by the sample at all reflection points, the radiation that finally arrives at the detector is attenuated at some frequencies. This allows the measured reflected intensity to provide information about the absorbing medium.

According to Eq. 2.44, the penetration depth d_p of the radiation (Fig. 2.55) depends on the wavelength of the radiation guided in total reflection, the angle of incidence α, and the quotient of the refractive indices between the sample and the ATR crystal.

$$d_p = \frac{\lambda}{2 \cdot \pi \cdot n_1 \cdot \sqrt{\sin^2 \alpha - \left(\frac{n_1}{n_2}\right)^2}} \tag{2.44}$$

d_p Penetration depth in μm
λ Wavelength in μm
α Angle of incidence
n_1 Refractive index of the optically thinner medium
n_2 Refractive index of the optically denser medium

According to Eq. 2.44 in Fig. 2.53, the penetration depth d_p of the radiation depends on the wavelength of the radiation guided in total reflection, the angle of incidence α, and the quotient of the refractive indices between the sample and the ATR crystal.

In Table 2.8, refractive indices of some materials are compiled, and in Table 2.9, the calculated penetration depths are given depending on the wavenumber when using two different ATR reflection elements. The penetration depth is defined as the path of the radiation after the intensity has decreased by 1/e compared to the original value at the interface.

Due to the dependence of the penetration depth on the wavelength or the wavenumber, the ATR spectra differ from the transmission spectra in that the bands in the lower wavenumber range are more intense compared to high wavenumbers.

Fig. 2.53 Penetration depth d_p of an electromagnetic radiation of wavelength λ from an optically thinner to a denser medium

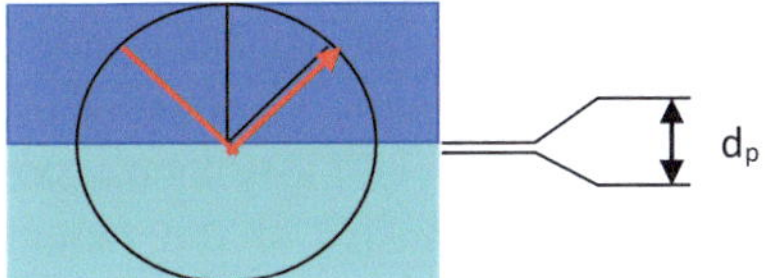

$$d_p = \frac{\lambda}{2 \cdot \pi \cdot n_1 \cdot \sqrt{\sin^2 \alpha - \left(\frac{n_1}{n_2}\right)^2}}$$

d_p: Penetration depth in μm

λ : Wavelength in μm

α : Angle of incidence

n_1: Refractive index of the optically thinner medium

n_2: Refractive index of the optically thicker medium

Table 2.8 Refractive indices of various materials

Refractive indices	
Germanium	4.0 (4.1[a])
Silicon	3.35[a]
Zinc selenide	2.59[a]
Diamond	2.42
PS	1.58
PMMA	1.49
PC	1.58
Quartz glass	1.46
Water	1.33
Air	1.0

[a]The refractive index at 633 nm; all other values are determined with the sodium D-line at 587 nm as the standard wavelength

Table 2.9 Dependence of the penetration depth on the wave number of the ATR crystal and the angle of incidence α

Material	Germanium			Diamond
α	24° (= α_g)	36.9°	48°	48°
Penetration depth[μm]				
3000 cm^{-1}	1.676	0.296	0.212	0.646
1800 cm^{-1}	2.793	0.491	0.353	1.077
600 cm^{-1}	8.380	1.474	1.055	3.230

The penetration depth also depends on the angle of incidence of the radiation. For the study of hydrocarbons, an angle of incidence α around 45° is mainly used, as an angle of 60° would yield less intense spectra.

The fact that the intensity of the radiation in the optically thinner sample decreases exponentially with distance from its surface makes ATR spectroscopy an adequate method for surface analysis. The material analysis is thus independent of the sample thickness or sample size. The limiting factor for non-destructive surface analysis is only the sample holder with its pressing device, whose task is to ensure the best possible contact between the sample and the reflection element. A torque screw ensures that the crystal is not destroyed.

If liquid or pasty samples are analyzed, it is not necessary to fix the torque screw.

In a more industrial context, ATR technology is used to examine all organic substances, such as polymers, fats, waxes, oils, pastes, plasticizers, additives, etc.

By converting the analysis unit, it is possible to examine liquid samples in a liquid cell as well as KBr pellets and pure plastic films using the transmission technique on the same spectrometer (Fig. 2.54).

In the transmission technique, due to the permeability, quartz or optical glass cuvettes, as commonly used in UV-VIS spectroscopy, cannot be used throughout the entire MIR spectrum (Günzler and Heise 2003). However, some salts, such as

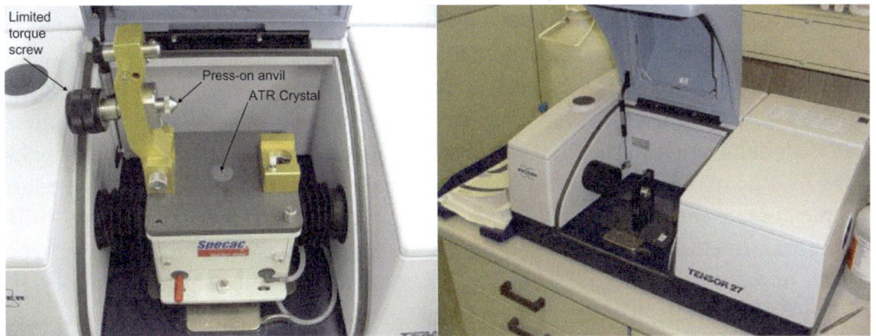

Fig. 2.54   Built-in measuring units in the IR spectrometer

KBr, NaCl, CaF_2, BaF, CsJ, etc., are characterized by good transparency in the IR range and also have the property of sintering into a quasi single crystal at relatively low pressures. Therefore, dry, powdery substance samples can usually be rubbed with KBr and pressed into a tablet (pellet). Since this preparation technique is very time-consuming and prone to errors, it has been replaced by more modern ATR reflection techniques.

Another advantage of this technique is that it also makes non-transmissible materials, such as plastics, accessible to IR measurement. A much less used technique is to dissolve mostly non-polar samples in an adequate, non-polar solvent, fill the solution into appropriate IR-transparent cuvettes (made of the aforementioned salts), and measure it. However, this technique is fraught with two major disadvantages: On the one hand, there is no IR-inactive solvent, which is why the spectra always contain the absorption bands of the solvent as an artifact, and on the other hand, the IR-transparent cuvettes are very sensitive to moisture, as they are made of water-soluble salts. Therefore, when comparing measured spectra with those from the literature, the sample preparation technique must always be taken into account and included in the comparison.

The ATR spectroscopy or infrared reflection absorption spectroscopy (IRRAS) can be used with little preparatory effort at moderate costs in surface analysis. With it, molecular groups can be detected with an information depth in the µ-range. This allows both monolayers and organic contaminants on surfaces to be examined, as well as information about the sample material itself to be obtained in deeper layers (multilayers). These properties qualify ATR spectroscopy for the structure elucidation of substances as well as for quality assurance in coating technology within error analysis. Whether a substrate has been properly cleaned for further coating process and the adsorption layer has been removed can be quickly determined using ATR spectroscopy. ATR spectroscopy can also be used in an imaging manner with a µATR microscope *(IR Imaging),* and with calibration and suitable sample preparation, quantification of molecular groups is also possible, as the example in the following section shows.

Quantification Using Thin Section Method via ATR Technology

The quantification of the polybutadiene content of the ABS granule grain cannot be carried out using ATR-IR spectroscopy. The granule grain does not have the required homogeneity for the surface-near examination method. Due to the spherical or globular granule shape, a concave or convex contact surface with air inclusions results, making a spectrum of the material not reproducible, as not the entire sample contact surface is equally covered by all granule grains. Consequently, different peak intensities are obtained for different granule samples of the same type, as can be seen in Fig. 2.55. The integration of the butadiene peaks would yield three different results for the same material.

However, this problem can be solved by suitable sample preparation, so that a quantitative analysis of plastic components and additives becomes possible with ATR spectroscopy. The suitability of this method is demonstrated by comparing the polybutadiene content in ABS with the results of transmission infrared spectroscopy.

The sample preparation is carried out with a microtome cutting device (Fig. 2.56), which separates sample foils with a thickness of 30 µm from the granule grain or base body.

The procedure ensures a homogeneous contact surface on the ATR crystal and leads to almost constant peak sizes when measuring different samples of the same material.

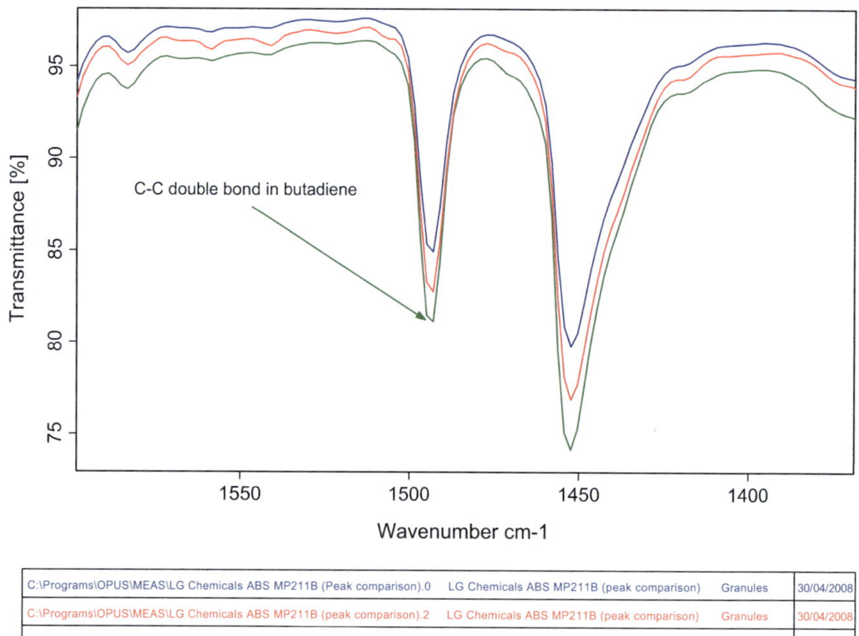

Fig. 2.55 Comparison of the peak size of ABS granules

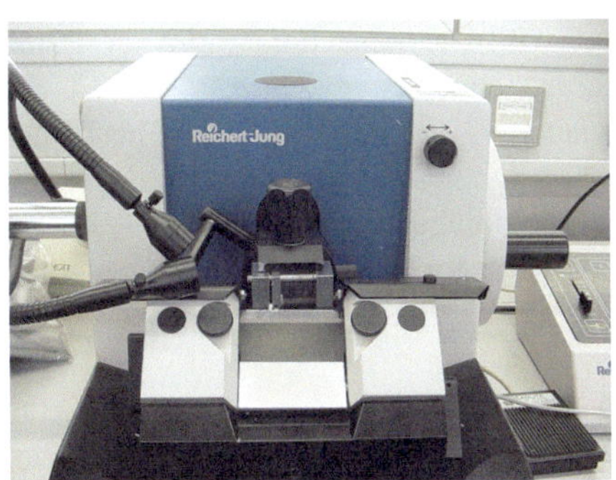

Fig. 2.56 Microtome cutting device with clamped granule sample

To illustrate the reproducibility of the measurement results, five grains of the pure ABS plastic P2mc are measured and again the butadiene peak (1510.1–1478.27 cm^{-1}), the nitrile peak (2263–2217.7 cm^{-1}) and the styrene peak (987.89–946.69 cm^{-1}) are used for evaluation.

Both the standard deviation of the ratio of the butadiene peak to the peak sum of all three peaks in the grain between the five microtome cuts and between the five grains is formed.

With a standard deviation of 1.2% over five grains, the peak ratio used for evaluation can be used for the creation of the calibration straight line (Fig. 2.57). For the creation of the calibration straight line, the mean value from five individual measurements per sample is also used in this measurement method.

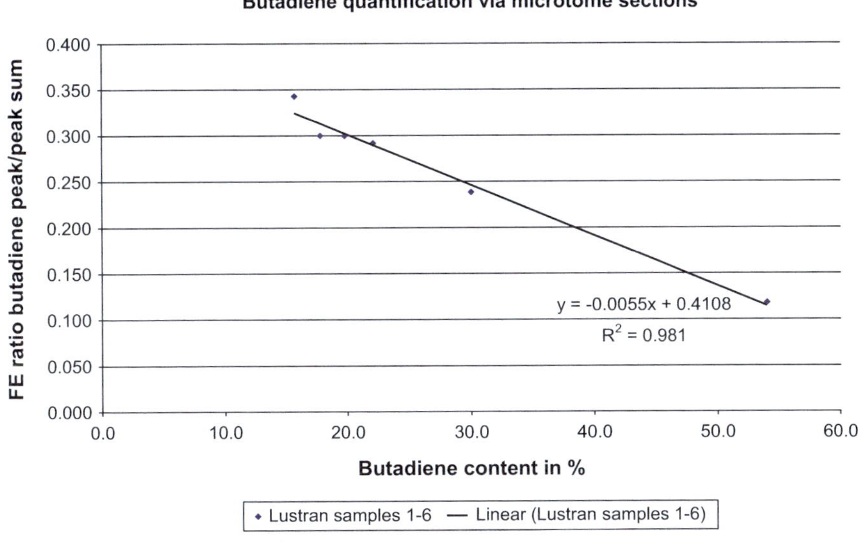

Fig. 2.57 Calibration straight line for polybutadiene quantification via the measurement of microtome cuts

By checking the straight line equation with the peak area ratios of P2mc or Polylac 727, the percentage polybutadiene contents of 22.69% ± 1.2% for P2mc-ABS or 17.24% ± 1.29% for Polylac-727-ABS are obtained. The values are within the range specified by the manufacturer. Therefore, the measurement method can be used for the quantification of the polybutadiene content.

The calibration straight lines created in Excel for each measurement method can be directly compared with each other due to the identical peaks used for evaluation.

In the case of the measurement method with KBr pellets and press foils, the coefficient of determination of the calibration straight line is somewhat larger, but an additional investment in a hydraulic press and press tool would be necessary to carry it out.

However, the standard deviations over the individual measurements are smallest in the case of the measurement of microtome cuts, which speaks for the reproducibility of the measurement results with this measurement method.

Therefore, the method of choice is to measure microtome sections via ATR, which is sufficiently accurate with a coefficient of determination of the calibration line of 98.1%.

2.3.4 FTIR-Imaging With ATR

One way to visualize and identify microscopically small plastic particles within a matrix, such as a filter cake, which also contains inorganic impurities like sand, is offered by IR imaging. IR imaging is an imaging method in which an infrared microscope is coupled with an IR spectrometer, as seen in Fig. 2.58.

Using a two-dimensional arrangement of radiation detectors of a so-called FPA (Focal Plane Array) detector and the IR spectrometer via the microscope, the spatial distribution of the sample composition can be visualized.

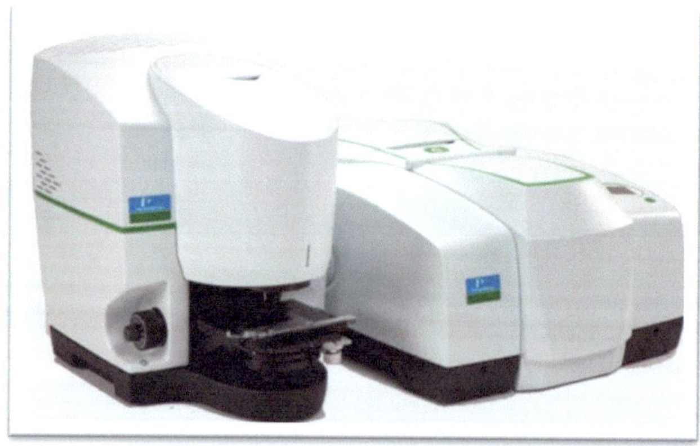

Fig. 2.58 Infrared microscope (left) with attached FTIR spectrometer (right)

The core piece in the IR microscope, which provides better resolution than ordinary light microscopes, is the ATR objective. In Fig. 2.59, the contact of the measuring head (ATR crystal) with a sample on a controllable sample table can be seen.

The working method and the beam path between the mirror lens objective (Cassegrain reflection objective) and the ATR crystal is graphically represented in Fig. 2.60. For a high-resolution examination (in the µm range) of a sample, a measuring point is first selected using an autofocus, which consists of a light microscope with a camera. Due to the mirror-symmetrical arrangement in the mirror lens objective, the infrared radiation also hits the exact same measuring point via the identical focusing optics. To obtain information about the selected

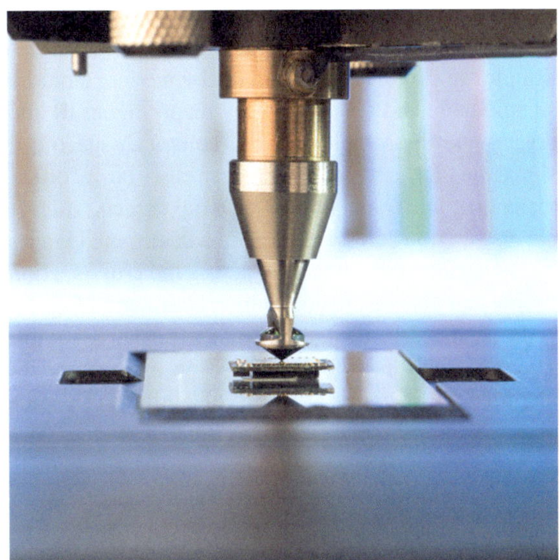

Fig. 2.59 Lowered germanium ATR crystal on a sample in ideal sample contact

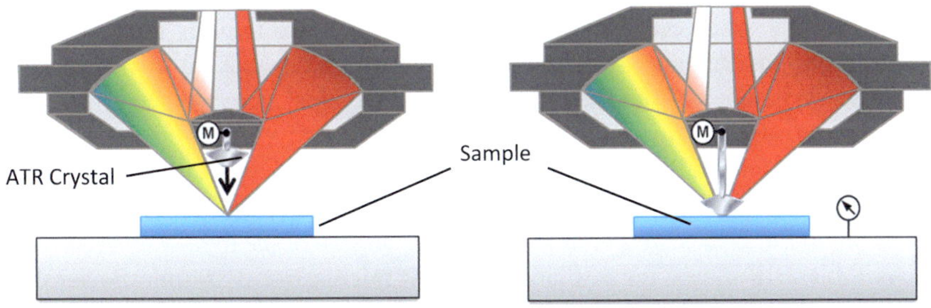

Fig. 2.60 Operation of the ATR objective. Left: Focusing of the measuring point. Right: Automated lowering of the ATR crystal to the measuring point

measuring point via the ATR reflection technique, a motor automatically lowers the ATR crystal to the measuring point. The sample is pressed for direct contact at the tip of the crystal with a diameter of 600 μm. A pressure sensor in the sample table regulates the optimal contact pressure to obtain a reproducible spectrum of the radiation reflected in the crystal. For maximum efficiency of the infrared light, the crystal surface facing away from the sample is anti-reflective.

The IR radiation enters the crystal through this anti-reflective curved surface and undergoes an internal total reflection at the sample surface before it leaves the crystal again. In the space immediately below the crystal tip, there can be energy absorption by the sample. If this is the case, the reflected beam is slightly weakened and thus contains spectral information about the surface of the absorbing sample. With the light reflected from the sample surface, an image is created by the ATR crystal and the Cassegrain objective. An array detector is located in the image plane of the system for this purpose. The crystal and sample are scanned laterally under the objective and the image is moved along the detector so that all points of the sample are captured. This is done by moving a platform in the X-Y direction on the sample table and simultaneously recording an IR spectrum for each position of the crystal and the sample, until a "complete, spectral map" is created. For improved signal quality, several spectra can also be recorded in each position. Using this technique, individual points (material defects, voids, etc.) can be examined or particle distributions on surfaces can be determined via a so-called mapping.

By performing a line scan across the cross-section of a sample, a pictorial depth profile can also be created, which is shown in Fig. 2.61.

Each measurement point of the scanned sample surface is assigned a color selectable via the software based on a characteristic peak or peak pattern of the recorded IR spectrum, thereby obtaining the depicted topography. Characteristic peaks of different plastics are presented in Sect. 2.4.

The resolution d of an IR microscope can be calculated just like that of a light microscope using the Rayleigh criterion according to the following equation.

$$d = 1.22\lambda/2\,NA = 0.61\lambda/n\sin\theta \tag{2.45}$$

The numerical aperture describes the ability of an optical element, such as a biconvex lens (Fig. 2.62), to focus light or radiation. In lenses, the numerical aperture determines the minimum size of the light spot P' that can be generated in its focus as an image of an object P and is thus an important size limiting the resolution. According to the Rayleigh relationship, the resolution depends on the wavelength or the wave number of the focused radiation. A typical numerical aperture of a lens is $NA = 0.60$ and corresponds to an opening angle of the light cone $\theta = 37°$.

In a conventional microscope, the refractive index $n = 1$, as the medium between the lens and focus is air. A typical numerical aperture of a lens is $NA = 0.60$ and according to Eq. 2.45 corresponds to an opening angle θ of the light cone of 37°.

As Table 2.10 shows, the resolution of an IR microscope is 10 times lower than the wavelength of the light:

By applying the ATR measurement principle, the classical spatial resolution (10 μm at 1000 cm^{-1}) can be undercut. With ATR imaging, it is at 3 μm, and

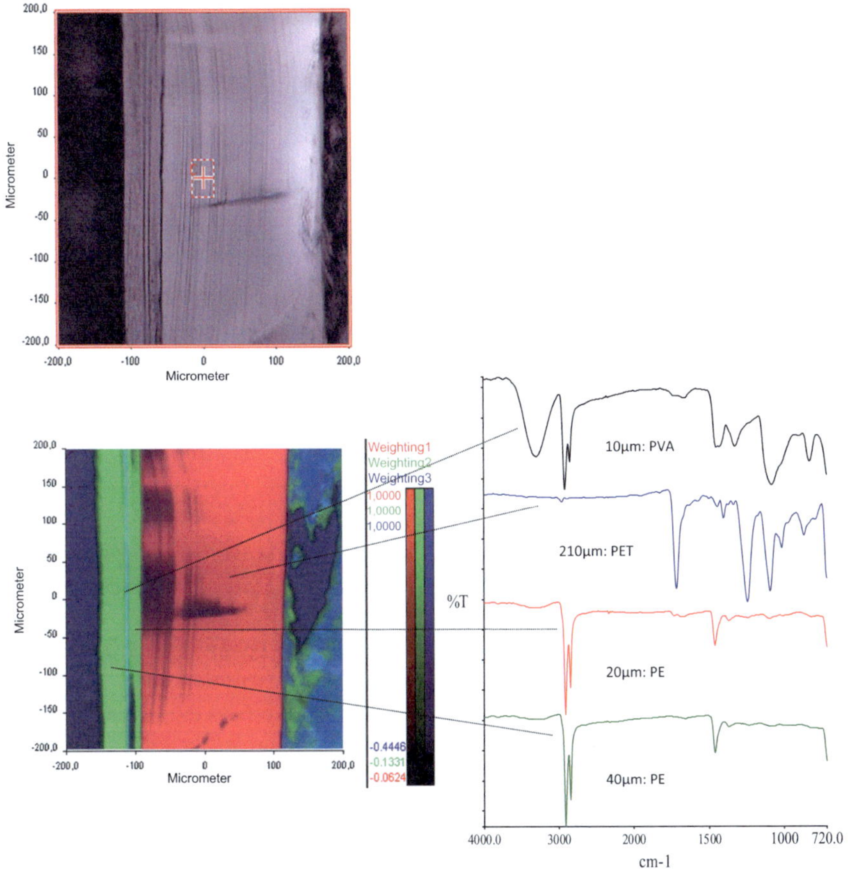

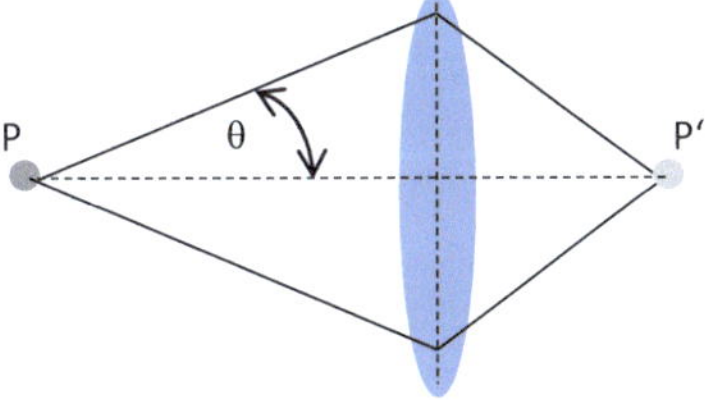

Fig. 2.61 Cross-section of a plastic film. Top: Light microscopic image. Bottom: ATR image. Right: Infrared spectra of individual measurement points and layer thicknesses of the color-differentiated plastic components in the film. Between the polyethylene layers is a thin layer of polyvinyl alcohol (PVA)

Fig. 2.62 Beam path through a biconvex lens

Table 2.10 Resolution of an IR microscope depending on the wave number

Wave number (cm^{-1})	d (μm)
4000	10
2000	5
1000	2.5

image pixels of 1.56 μm already provide a good IR spectrum, which is suitable for identification.

2.3.5 IR Spectra of Plastics for Identification

Compared to mass and NMR spectroscopy, infrared spectroscopy is a relatively inexpensive and especially fast analysis technique for identifying hydrocarbons. However, in terms of sensitivity and quantification, it is inferior to the other two analysis techniques, especially since structure elucidation with IR spectroscopy is not possible. Only functional groups can be detected via the excited vibrations of the atoms involved in a bond (Sect. 2.3.2). Since different atoms and bonds are present in plastics, the interaction of infrared radiation in the form of a vibrational excitation is predestined to quickly and easily identify technical or artificial polymers as solids using ATR-IR spectroscopy (Fig. 2.34; description in Sect. 2.4).

The type of plastic of the microplastic particles <500 μm, as they occur in waters or also in digestate (Leser 2015), can be identified with an infrared microscope due to its peak pattern in comparison to a stored known spectrum in the plastic database.

Furthermore, infrared spectroscopy is a tool to separate pure plastics via their specific absorption bands or the peak pattern within a recycling plant. An automated separation of pure plastics, which is not labor-intensive and therefore cost-effective, also decides whether our plastic waste is recycled or subjected to thermal recovery. Separation processes that exploit the different specific weight of various types of plastic in float-sink methods or hydrocyclones have been replaced by infrared spectroscopy (Weber 2002). Why only about 40% of plastic packaging, which ends up in the yellow bag or the yellow bin, is recycled, although technically a much larger percentage would be possible, is a question of economic viability (NABU n.d.)

The underlying methodology for the presented application areas of infrared spectroscopy—such as microplastic analytics, plastic recycling, and pollutant detection, is discussed in this chapter, based on the results of the "Rheines Wasser" project (Sect. 2.6). Material-specific IR absorptions of the microplastic particles filtered out of the Rhine are presented and interpreted.

The spectra are supplemented by other important technical plastics, which due to their high density were not detectable in the near-surface areas of the Rhine. This could be a reason why the universally applicable POM is likely to be found in deeper water layers or in the sediment. PU, although it has a similarly high density, is often used in its foamed variant. The enclosed carbon dioxide bubbles drive this plastic to the water surface. A depth profile and a sediment investigation would provide a more detailed picture of the microplastic distribution in waters.

Before taking a closer look at the polymers in the infrared spectrum, the importance of a fast, effective, and low-cost analysis should be emphasized again based on the question posed about the harmlessness of plastics.

Why are Plastics Not Harmless Despite Their Durability?
Durability There are several answers to this question. On the one hand, there are the various additives in plastics (Sect. 2.5), which are embedded in the inert polymeric matrix. These include organic and inorganic stabilizers based on heavy metals, polybrominated or polychlorinated flame retardants, inorganic pigments, organic UV absorbers such as benzophenones or benzotriazoles, and of course the multitude of different plasticizers, with which the ductility and impact strength of the artificial material can be adjusted. The additives are not chemically bound to the polymer chain, so they can be released into the aquatic environment or the atmosphere depending on environmental conditions. In addition to these low molecular weight additives, which can be extracted by the various waters, there are also so-called residual monomers, such as Bisphenol A, which are enclosed in the plastic matrix. Bisphenol A, PCBs and DEHP are among the *endocrine disrupting substances* (EDS), substances that are endocrine active (EAS). Endocrine disruptors are hormonally active substances with harmful effects. In addition to naturally occurring hormonally active substances (e.g., phytoestrogens), there are also the synthetically produced ones mentioned. They can impair the fertility of living beings. In the case of ethinylestradiol, a synthetic endocrine active substance is present, which is used in oral contraceptives ("birth control pill") to prevent pregnancy. In other cases, the hormone-like effects have been proven, but their impact is not fully researched (Manickum 2014).

Some representatives of the functional additives are demonstrably classified as SVHC substances and can be found in Table 2.11.

The SVHC list contains some chemicals that the compounder adds to the plastic and the granulate is supplied to the plastic processor. The supply chain from the monomers and additives to the finished plastic part is long and the information regarding the ingredients over global trade is sometimes incomplete. The safety data sheet of a product does not always contain the unique CAS number of all ingredients, as the formulation is either considered a trade secret and the know-how is not disclosed, or one has received sparse or no information from one's cheap sub-supplier. A material change on the part of a supplier can also remain undetected without analysis, as long as the quality requirements for the plastic end product are still met. To counter this uncertainty, which still exists despite different legal requirements (Reach; Corab, RoHs; Hauser etc.), the REACH radar network project was launched with the support of the DBU. The project module "Detect" is about the quick and cost-effective analysis of Reach banned substances in plastic

Table 2.11 SVHC (Substances of very high concern) substances and their area of application (from REACH candidate list)

CAS number	Name of the substance	Use
10043-35-3	Boric acid	Flame retardant
10099-74-8	Lead dinitrate	Unlikely in plastics
10108-64-2	Cadmium chloride	Electroplating
101-14-4	2,2′-dichloro-4,4′-methylenedianiline	Crosslinked in PU
10124-43-3	Cobalt(II) sulphate	Pigment (reddish)
10141-05-6	Cobalt(II) dinitrate	Unlikely (only as pigment in ceramics)
101-61-1	N,N,N′,N′-tetramethyl-4,4′-methylenedianiline (Michlerâ€™s base)	????
101-77-9	4,4′- Diaminodiphenylmethane (MDA)	As residue
101-80-4	4,4′-oxydianiline and its salts	Only in treated surfaces (corrosion protection)
10588-01-9	Sodium dichromate hydragenous	Unlikely
7790-79-6	Cadmium fluoride	Unlikely
10124-36-4	Cadmium sulphate (without crystal water)	Electrolyte Galvanic deposition
31119-53-6	Cadmium sulphate (including crystal water)	Electrolyte Galvanic deposition
15571-58-1	Dioctyltin bis (2-ethylhexyl), DOTE	Plastic processing
2687-91-4	N-ethyl-2-pyrrolidone, NEP	Solvent for plastics and paints
97-99-4	Tetrahydrofurfuryl alcohol, THFA	Polymerization initiator
3825-26-1	Ammonium pentadecafluorooctanoate, APFO	In production e.g. Teflon
68515-50-4	1,2-Benzenedicarboxylic acid, dihexyl ester, branched and linear, D	Plasticizer
10588-01-9	Sodium dichromate	Corrosion inhibitor for metal surfaces
3846-71-7	2-benzotriazol-2-yl-4,6-ditertbutylphenol	UV-Stabilization
25973-55-1	2-(2H-benzotriazol-2-yl)-4,6-ditertpentylphenol	Additive Aging Protection
15571-58-1	2-ethylhexyl-10-ethyl-4,4-dioctyl-7-oxo-8-oxa-3,5-dithia-4-stannatetra-dacanoate (DOTE)	Plastic Processing
106-94-5	1-bromopropane (n-propyl bromide)	Plastic Degreasing
107-06-2	1,2-dichloroethane	Vinyl Chloride (Solvent)
109-86-4	2-Methoxyethanol	Solvent for Paints and Varnishes
110-00-9	Furan	Solvent
110-71-4	1,2-dimethoxyethane; ethylene glycol dimethyl ether (EGDME)	Process Aid in Chemical Manufacturing

(continued)

Table 2.11 (continued)

CAS number	Name of the substance	Use
110-80-5	2-Ethoxyethanol	Intermediate also Solvent
11103-86-9	Potassium hydroxyoctaoxodizincatedichromate	Coatings (Aerospace)
11113-50-1	Boric acid	Flame Retardant and Antiseptic
111-15-9	2-Ethoxyethyl acetate	Adhesive Manufacturing, Solvent Coating
11120-22-2	Silicic acid, lead salt	Unlikely (Plaster, Glass, Cement)
111-96-6	Bis(2-methoxyethyl) ether	Reactive Solvent
112-49-2	1,2-bis(2-methoxyethoxy)ethane (TEGDME; triglyme)	Solvent and Auxiliary
115-96-8	Tris(2-chloroethyl)phosphate	Flame Retardant, Viscosity Adjustment
1163-19-5	Bis(pentabromophenyl) ether (decabromodiphenyl ether; DecaBDE)	Flame Retardant
117-81-7	Bis (2-ethylhexyl)phthalate (DEHP)	Plasticizer
117-82-8	Bis(2-methoxyethyl) phthalate	Plasticizer
120-12-7	Anthracene	Unlikely in plastic
12036-76-9	Lead oxide sulfate	Plastic compound
12060-00-3	Lead titanium trioxide	Unlikely in plastic
12065-90-6	Pentalead tetraoxide sulphate	Plastic additive lead reinforcement
120-71-8	6-methoxy-m-toluidine (p-cresidine)	Intermediate ???
121-14-2	2,4-Dinitrotoluene	Additive in PU
12141-20-7	Trilead dioxide phosphonate	PVC production, plastic additive
12179-04-3	Disodium tetraborate, anhydrous	Adhesive, flame retardant, ceramics
12202-17-4	Tetralead trioxide sulphate	PVC production, batteries
12267-73-1	Tetraboron disodium heptaoxide, hydrate	Washing and cleaning agent, flame retardant
123-77-3	Diazene-1,2-dicarboxamide (C,C'-azodi(formamide))	Bleaching agent, aging substance
12578-12-0	Dioxobis(stearato)trilead	Plastic compound
12626-81-2	Lead titanium zirconium oxide	Electronic and optical products
12656-85-8	Lead chromate molybdate sulphate red (C.I. Pigment Red 104)	Dye in various materials including plastics
127-19-5	N,N-dimethylacetamide	Solvent also plastic fibers
1303-28-2	Diarsenic pentaoxide	Metal hardening
1303-86-2	Diboron trioxide	Flame retardant, various other applications
1303-96-4	Disodium tetraborate, anhydrous	Flame retardant

(continued)

Table 2.11 (continued)

CAS number	Name of the substance	Use
1306-19-0	Cadmium oxide	Heat stabilizer
1306-23-6	Cadmium sulphide	Pigment
131-18-0	Dipentyl phthalate (DPP)	Plasticizer
1314-41-6	Orange lead (lead tetroxide)	Rubber protection, Pigment
13149-00-3	Cis-cyclohexane-1,2-dicarboxylic anhydride	Epoxy resin component, Plasticizer
1317-36-8	Lead monoxide (lead oxide)	Production of rubber protection, ceramics, batteries
1319-46-6	Trilead bis(carbonate)dihydroxide	PTC ceramics
1327-53-3	Diarsenic trioxide	Decolorization of glass and enamel, semiconductor
1330-43-4	Disodium tetraborate, anhydrous	Flame retardant
1333-82-0	Chromium trioxide	Electroplating
134237-50-6	Hexabromocyclododecane (HBCDD) and all major diastereoisomers identified	Flame retardant mainly for styrene-containing plastics
134237-50-6	Hexabromocyclododecane (HBCDD) and isomers	Flame retardant mainly for styrene-containing plastics
134237-51-7	Hexabromocyclododecane (HBCDD) and all major diastereoisomers identified	Flame retardant mainly for styrene-containing plastics
134237-51-7	Hexabromocyclododecane (HBCDD) and isomers	Flame retardant mainly for styrene-containing plastics
134237-52-8	Hexabromocyclododecane (HBCDD) and all major diastereoisomers identified	Flame retardant mainly for styrene-containing plastics
134237-52-8	Hexabromocyclododecane (HBCDD) and isomers	Flame retardants, especially for styrene-containing plastics
13424-46-9	Lead diazide, Lead azide	Explosive
1344-37-2	Lead sulfochromate yellow (C.I. Pigment Yellow 34)	Pigment (Rubber, Plastic, Paints)
13530-68-2	Acids from Chromium trioxide	Electroplating, rarely pigment
13814-96-5	Lead bis(tetrafluoroborate)	Electrolytic lead plating
140-66-9	4-(1,1,3,3-tetramethylbutyl)phenol	Production of polymeric mixtures, resins, rubber
14166-21-3	Trans-cyclohexane-1,2-dicarboxylic anhydride	Production of epoxy resins, plasticizers, insect repellents, alkyd resins
143860-04-2	3-ethyl-2-methyl-2-(3-methylbutyl)-1,3-oxazolidine	???

(continued)

Table 2.11   (continued)

CAS number	Name of the substance	Use
15245-44-0	Lead styphnate	Explosive
15606-95-8	Triethyl arsenate	Semiconductor manufacturing
17570-76-2	Lead(II) bis(methanesulfonate)	Production of circuit boards in electrolytic processes
1937-37-7	Disodium 4-amino-3-[[4'-[(2,4-diaminophenyl)azo][1,1'-biphenyl]-4-yl]azo] -5-hydroxy-6-(phenylazo) naphthalene-2,7-disulphonate (C.I. Direct Black 38)	Dye (Paper, Plastic, Leather, Wood)
1937-37-7	Disodium 4-amino-3-[[4'-[(2,4-diaminophenyl)azo][1,1'-biphenyl]-4-yl]azo] -5-hydroxy-6-(phenylazo) naphthalene-2,7-disulphonate	Dye (Paper, Plastic, Leather, Wood)
19438-60-9	Hexahydro-4-methylphthalic anhydride	Production of alkyd resins, Polyester
2058-94-8	Henicosafluoroundecanoic acid	Fluorinated Polymers
20837-86-9	Lead cyanamidate	???
2451-62-9	1,3,5-Tris(oxiran-2-ylmethyl)-1,3,5-triazinane-2,4,6-trione (TGIC)	Hardener for resins, adhesive, plastic stabilizer
24613-89-6	Dichromium tris(chromate)	Steel and aluminum coating
25155-23-1	Trixylyl phosphate	Use in polyurethane, PVC, TPE as well as in coatings and textiles, Flame retardant in plastic production, use as plasticizerPlasticizer
25214-70-4	Formaldehyde, oligomeric reaction products with aniline	Ion exchange resin, hardener for epoxy resin, adhesive
25550-51-0	Hexahydromethylphthalic anhydride	Flame retardant additive (polymer production)
25637-99-4	Hexabromocyclododecane (HBCDD) and all major diastereoisomers identified	Flame retardant additive (polymer production)
25637-99-4,	Hexabromocyclododecane (HBCDD) and isomers	Flame retardant additive (polymer production)
2580-56-5	[4-[[4-anilino-1-naphthyl][4-(dimethylamino)phenyl]methylene]cyclohexa-2,5-dien-1-ylidene] dimethylammonium chloride (C.I. Basic Blue 26)<em>[with â‰¥ 0.1% of Michler's ketone (EC No. 202-027-5) or Michler's base (EC No. 202-959-2)]	Dye (textiles, plastic, paper)
301-04-2	Lead di(acetate)	Intermediate, laboratory chemical, production of paints, thinners, putty

(continued)

Table 2.11 (continued)

CAS number	Name of the substance	Use
307-55-1	Tricosafluorododecanoic acid	Production of fluorinated polymers
3194-55-6	Hexabromocyclododecane (HBCDD) and all major diastereoisomers identified	Flame retardant additive (polymer production)
3194-55-6	Hexabromocyclododecane (HBCDD) and isomers	Flame retardant additive (polymer production)
335-67-1	Pentadecafluorooctanoic acid (PFOA)	Process aid, partially used in electroplating
3687-31-8	Trilead diarsenate	Application in metallurgical finishing processes (copper, lead; precious metals)
376-06-7	Heptacosafluorotetradecanoic acid	Production of fluorinated polymers
3825-26-1	Ammonium pentadecafluorooctanoate (APFO)	Partially used as a process aid in electroplating
48122-14-1	Hexahydro-1-methylphthalic anhydride	Production of Polyester and alkyd resins as well as plasticizers for thermoplastic polymers
49663-84-5	Pentazinc chromate octahydroxide	Coatings in the automotive industry
513-79-1	Cobalt(II) carbonate	Used in the production of catalysts
51404-69-4	Acetic acid, lead salt, basic	Production of fine chemicals
548-62-9	[4-[4,4′-bis(dimethylamino) benzhydrylidene]cyclohexa-2,5-dien-1-ylidene]dimethylammonium chloride (C.I. Basic Violet 3)<em>[with $\geq$ 0.1% of Michler's ketone (EC No. 202-027-5) or Michler's base (EC No. 202-959-2)]</em>	Paper dye
548-62-9	[4-[4,4′-bis(dimethylamino) benzhydrylidene]cyclohexa-2,5-dien-1-ylidene]dimethylammonium chloride (C.I. Basic Violet 3)<em>[with $\geq$ 0.1% of Michler's ketone (EC No. 202-027-5) or Michler's base (EC No. 202-959-2)]</em>	Paper dye
561-41-1	4,4′-bis(dimethylamino)-4″-(methylamino)trityl alcohol <em>[with $\geq$ 0.1% of Michler's ketone (EC No. 202-027-5) or Michler's base (EC No. 202-959-2)]</em>	Used for dyeing various materials
56-35-9	Bis(tributyltin)oxide (TBTO)	Among other things, used in polyurethane foams

(continued)

Table 2.11 (continued)

CAS number	Name of the substance	Use
57110-29-9	Hexahydro-3-methylphthalic anhydride	Production of Polyester and alkyd resins
573-58-0	Disodium 3,3'-[[1,1'-biphenyl]-4,4'-diylbis(azo)]bis(4-aminonaphthalene-1-sulphonate) (C.I. Direct Red 28)	Among other things, dye
573-58-0	Disodium 3,3'-[[1,1'-biphenyl]-4,4'-diylbis(azo)] bis(4-aminonaphthalene-1-sulphonate)	Among other things, dye
59653-74-6	1,3,5-tris[(2S and 2R)-2,3-epoxypropyl]-1,3,5-triazine-2,4,6-(1H,3H,5H)-trione (β-TGIC)	Used in solder resist lacquer, among other things, used as a lining material and stabilizer for plastics
60-09-3	4-Aminoazobenzene	Synthetic azo dye
605-50-5	Diisopentylphthalate	(DIPP) Plasticizer for plastics
62229-08-7	Sulfurous acid, lead salt, dibasic	Used in the production of PVC
625-45-6	Methoxyacetic acid	Used in the production of chemicals and chemical products
629-14-1	1,2-Diethoxyethane	Solvent
64-67-5	Diethyl sulphate	Among other things, reactant in polymer synthesis
6477-64-1	Lead dipicrate	Explosive
65996-93-2	Pitch, coal tar, high temp.	Pitch, coal tar, used in the production of electrodes
6786-83-0	α,α-Bis[4-(dimethylamino)phenyl]-4(phenylamino)naphthalene-1-methanol (C.I. Solvent Blue 4)<em>[with ≥ 0.1% of Michler's ketone (EC No. 202-027-5) or Michler's base (EC No. 202-959-2)]</em>	Formulation of printing inks
68-12-2	N,N-dimethylformamide	Catalytic functions, among other things, production of synthetic/artificial leather from polyurethane
683-18-1	Dibutyltin dichloride (DBTC)	Additive in many plastics (e.g. rubber, PVC, PU, …)
68515-42-4	1,2-Benzenedicarboxylic acid, di-C7-11-branched and linear alkyl esters	(DHNUP) Plasticizer in PVC
68515-50-4	1,2-Benzenedicarboxylic acid, dihexyl ester, branched and linear	Plasticizer in polymers
68515-51-5	1,2-benzenedicarboxylic acid esters with >0,3% of dihexyl phthalate	In plasticizers, in polymer oils and PVC components
68648-93-1	1,2-benzenedicarboxylic acid esters with >0,3% of dihexyl phthalate	In plasticizers, in polymer oils and PVC components

(continued)

Table 2.11 (continued)

CAS number	Name of the substance	Use
68784-75-8	Silicic acid (H $_2$Si $_2$O $_5$), barium salt (1:1), lead-doped <i> [with lead (Pb) content above the applicable generic concentration limit for 'toxicity for reproduction' Repr. 1A (CLP) or category 1 (DSD); the substance is a member of the group entry of lead compounds, with index number 082-001-00-6 in Regulation (EC) No 1272/2008]</i>	(Silicic acid) Coatings of pistons
69011-06-9	[Phthalato(2-)]dioxotrilead	Manufacture of plastic products
71-48-7	Cobalt(II) diacetate	Catalyst, among other uses in alloys
71888-89-6	1,2-Benzenedicarboxylic acid, di-C6-8-branched alkyl esters, C7-rich	(DIHP) Plasticizer in PVC
72629-94-8	Pentacosafluorotridecanoic acid	Plasticizer in PVC
7440-43-9	Cadmium	Element, among other uses in corrosion protection
75-12-7	Formamide	Among other uses as a plasticizer
75-56-9	Methyloxirane (Propylene oxide)	Uses in coatings; lubricants
7632-04-04	Sodium peroxometaborate	Bleaching and cleaning agents
7632-04-4	Sodium peroxometaborate	Bleaching and cleaning agents
7646-79-9	Cobalt dichloride	Use in surface treatment processes, intermediate product
77-09-8	Phenolphthalein	Indicator solutions
7738-94-5	Acids from Chromium trioxide	Use in metal finishing such as electroplating
7758-97-6	Lead chromate	Manufacture of dyes; embalming/restoration of art products
776297-69-9	N-pentyl-isopentylphthalate	Plasticizer in plastics
7775-11-3	Sodium chromate	Coating of metals
77-78-1	Dimethyl sulphate	Intermediate as a methylation agent
7778-39-4	Arsenic acid	Clarifying agent
7778-44-1	Calcium arsenate	By-products from metallurgical processes
7778-50-9	Potassium dichromate	Treatment and coating of metals, corrosion inhibitor
7784-40-9	Lead hydrogen arsenate	Primary biocide, among other uses in the production of plastic products
7789-00-6	Potassium chromate	Corrosion inhibitor

(continued)

Table 2.11 (continued)

CAS number	Name of the substance	Use
7789-06-2	Strontium chromate	Corrosion inhibitor (e.g., in coatings)
7789-09-5	Ammonium dichromate	Oxidizing agent
7789-12-0	Sodium dichromate	Areas of application: dye mordant, vitamin K production, glazes
78-00-2	Tetraethyllead	Fuel additives and fuel mixtures
7803-57-8	Hydrazine	Refinement of chemicals, fuel for spacecraft
79-01-6	Trichloroethylene	Cleaning and solvent
79-06-1	Acrylamide	Used in the synthesis of polyacrylamides
79-16-3	N-methylacetamide	Intermediate and laboratory reagent
8012-00-8	Pyrochlore, antimony lead yellow	Used in paints
81-15-2	5-tert-butyl-2,4,6-trinitro-m-xylene (musk xylene)	Fragrance or fragrance enhancer
838-88-0	4,4′-methylenedi-o-toluidine	Intermediate ???
84-69-5	Diisobutyl phthalate	Plasticizer for plastics
84-74-2	Dibutyl phthalate (DBP)	Used among other things in cellophane packaging
84-75-3	Dihexyl phthalate	Plasticizer for vinyl plastic
84777-06-0	1,2-Benzenedicarboxylic acid, dipenty-lester, branched and linear	Laboratory chemical for analytical purposes
85-42-7	Cyclohexane-1,2-dicarboxylic anhydride	Production of alkyd resins, plasticizers; intermediate product in chemical synthesis
85535-84-8	Alkanes, C10-13, chloro (Short Chain Chlorinated Paraffins)	Production of rubber products, textile coatings, sealants, adhesives and paints
85-68-7	Benzyl butyl phthalate (BBP)	Plasticizer for PVC and other Polymers
872-50-4	1-Methyl-2-pyrrolidone	Used as a solvent
88-85-7	Dinoseb (6-sec-butyl-2,4-dinitrophenol)	Production of plastic products
90-04-0	2-Methoxyaniline; o-Anisidine	Production of dyes, dyeing of polymers
90640-80-5	Anthracene oil	Oil for the production of industrial soot and anthracene
90640-81-6	Anthracene oil, anthracene paste	Oil for the production of industrial soot and anthracene
90640-82-7	Anthracene oil, anthracene-low	Oil for the production of industrial soot and anthracene
90-94-8	4,4′-bis(dimethylamino)benzophenone (Michlerâ€™s ketone)	Intermediate in dye production, possible addition to plastics

(continued)

Table 2.11 (continued)

CAS number	Name of the substance	Use
91031-62-8	Fatty acids, C16-18, lead salts	Production of plastic products, lead reinforcements
91995-15-2	Anthracene oil, anthracene paste, anthracene fraction	Production of dyes, reducing agent, impregnation
91995-17-4	Anthracene oil, anthracene paste,distn. lights	Production of dyes, reducing agents, impregnation
92-67-1	Biphenyl-4-ylamine	??
95-53-4	o-Toluidine	Intermediate and laboratory reagent
95-80-7	4-methyl-m-phenylenediamine (toluene-2,4-diamine)	Intermediate in the production of sulfur dyes
96-18-4	1,2,3-Trichloropropane	Paint and varnish remover, solvent or component of solvents
96-45-7	Imidazolidine-2-thione; (2-imidazoline-2-thiol)	For the production of rubber and rubber products
97-56-3	o-aminoazotoluene	???

granulates and plastic end products. The preparation and analysis technique is discussed in Sect. 2.4.13, "IR spectra of plastic additives".

Another potential hazard of plastic comes not only from its ingredients, but also from their combustion and bacterial decomposition products (biocorrosion).

In controlled thermal recovery of plastic waste, temperatures of over 1000 °C are reached in high-temperature furnaces in the waste incineration plants. At these temperatures, organic pollutants burn completely. What should not burn completely is eliminated in the flue gas cleaning plant via activated carbon filters from the exhaust air. In uncontrolled combustions such as in a house fire, neither such high temperatures are reached nor are filters in use. The harmful combustion products of plastics contained in the flue gas, such as dioxins and polycyclic aromatic hydrocarbons (PAH) (LfU 2016; Ortner and Hensler 1995). The fire gas pollutants can attach themselves to soot particles, spread through the air and settle or condense on cold surfaces. The soot particles and pollutants "rain" on buildings, furnishings and food. Since the mentioned substances as fire residues show a toxic, mutagenic or carcinogenic effect (Dekant 1994), they pose a danger to humans and the environment together with the contaminated firefighting water.

Microplastic particles and microfibers from waters are found not only in the gastrointestinal tract of fish (Fig. 2.63 and 2.64), but also under their scales and between their gills.

Table 2.12 Plastics and their use

Name	Short	Use
Polypropylene	PP	Cups, pipes, containers, machinery and vehicle construction, bicycle helmets
Polyethylene	PE	Plastic bags, Packaging, tubes, cosmetic containers
Polystyrene	PS	Insulation, Isolation, packaging (Styrofoam)
Polyamide	PA	Synthetic fibers (Nylon, Perlon), textiles (Fleece), toothbrushes
Polycarbonate	PC	CDs, DVDs, window glass replacement, eyeglass lenses, solar panels
Polyvinyl chloride	PVC	Packaging films, food packaging, hoses, flooring, cable insulation
Styrene acrylonitrile	SAN	Bowls, housings, kitchen appliances, reflectors, light guides (shower partitions)
Polyester epoxy	PEST	Paints, resins, coatings
Polyurethane	PU	Household sponges, paints, sealants and adhesives, mattresses
Acrylonitrile butadiene styrene	ABS	Metalized plastics, Lego blocks, automotive parts, snowboards, 3-D printers, housings
Polylactic acid	PLA	Based on Lactic acid; drinking straws, Packaging, office supplies
Polyoxymethylene	POM	Gears, ski bindings, hose couplings, lighter tanks, toys, snap-on toothbrushes
Polyethylene terephthalate	PET	Packaging, plastic bottles, films, textile fibers

The danger to humans when grilling a fish, if the plastic also burns over the embers and releases toxic gases, has not yet been investigated, as there is still too little knowledge about the amount of microplastics in limnic systems.

A conducted pyrolysis of ABS granulate (Fig. 2.67) at 200–300 °C shows which partially highly toxic substance composition can be released during thermal treatment.

Microplastic Pyrolysis

As an example, the results of the pyrolysis of an ABS granulate are presented here (Lewin-Kretzschmar n.d.; Fig. 2.65).

To investigate the release of pollutants, approximately 3 g of the granulate was heated in vials in a thermal block. To study the behavior upon heating, one sample was initially heated stepwise from 200 °C to 250 °C (sample A) and another from 250 °C to 300 °C (sample B) (total exposure time in each case 1 h). Subsequently, 2

Fig. 2.63 Asp

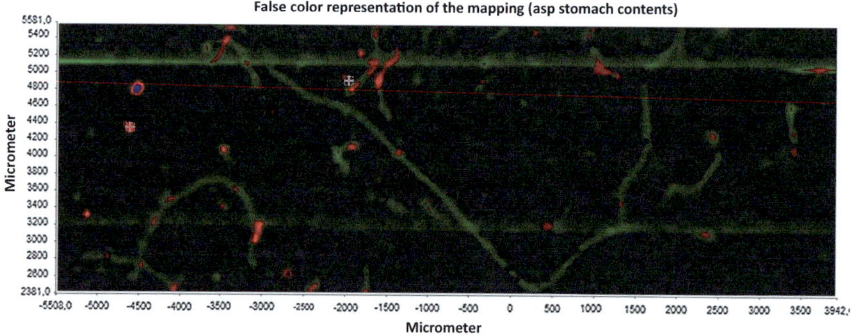

Fig. 2.64 Gastrointestinal tract of the asp

samples were tempered at 200 °C and 250 °C for 45 min (samples C and D). Then, 500 µl of gas was taken from the vapor space with a gas-tight syringe and subsequently chromatographed and detected by mass spectrometry. Morphologically, a

Fig. 2.65   ABS granulate (sample no. ALL: 2015_85053)

slight melting of the granulate was observed at 200 °C, which turned increasingly brown with rising temperature. Complete melting was not observed even at 300 °C.

Chromatographically, the main components detected in the vapor space in all experiments were a complex mixture of saturated and unsaturated aliphatic hydrocarbons, styrene, 4-vinylcyclohexene, and ethylbenzene. The concentration of these substances increased with increasing temperature and exposure time. In all experiments, traces of hydrogen sulfide and in experiments A, C, and D, cyanide were detected. However, ammonia was only detected in traces in experiment B.

Furthermore, depending on the thermal load, traces to medium concentrations of other aromatics (e.g., benzene, xylenes, toluene, cumene, propylbenzene, ethyltoluenes, trimethylbenzenes, 1,3-diphenylpropane), saturated and unsaturated aldehydes (e.g., acetaldehyde, acrolein (not in sample B), crotonaldehyde, propanal, benzaldehyde), ketones (e.g., acetophenone, 2-butanone, 2-pentanone, methylvinylketone), alcohols (e.g., methanol, ethanol, propanols), nitriles (e.g., acrylonitrile, propanenitrile), furan, and methylfurans were detected. In addition, there were indications of traces of other nitrogen compounds and thiophenes.

To summarize, there is a risk to humans and animals from microplastic contamination for the following four reasons:

1. due to damage to internal organs when consumed,
2. due to the ingredients of the plastic,
3. due to the release of toxic compounds during combustion, and

4. due to the accumulation of non-plastic-intrinsic pollutants from the aquatic environment, the soil, and the atmosphere.

Due to their small size, but in some cases very large surface area (Fig. 1.4) and low polarity, microstructured plastics are very capable of adsorbing organic substances. This property is also used in the form of passive sampling (Sect. 3.1) to determine organic pollutants by accumulating them on a plastic membrane. A fish that consumes microplastic particles may thus potentially ingest a higher toxic potential than if it were to drink a liter of water. Microplastic particles can be thought of as a magnet for organic substances. Particularly surface-active pollutants such as perfluorinated surfactants (PFT) "seek" large surfaces to attach to. The less polar the substances are, the higher their affinity to adsorb to microplastic particles, which are themselves nonpolar (Hüffer and Hofmann 2016).

Microplastic particles in bodies of water represent a new class of suspended matter, in addition to the known organic suspended solids such as phytoplankton and detritus, as well as inorganic suspended solids (e.g., sand). The mix of suspended solids is supplemented by tire wear, plant fibers, insect components, bacteria, and many more. Depending on their polarity and the polarity of the analyte, these suspended solids influence the analytical result of a water sample, which is analyzed directly after a microfiltration using HPLC/MS. If one examines PAH, the filter residue must be freed from the strongly adsorbed condensed aromatics using a suitable solvent by means of a solid phase extraction to make them accessible for analysis. This procedure is established for this class of compounds. For other substances such as perfluorinated surfactants, so far only the liquid phase has been examined. The addition of one gram of microplastic (POM granulate) to an aqueous solution of 120 ng/l PFOS shows a difference of 25% after the granulate has been filtered off. Since the standard deviation in the analysis of perfluorinated surfactants also lies in this range, a solid phase extraction does not make sense. The actual analytical influence of different microplastics on various substances is the subject of current investigations.

2.4 Manufacture, Use and ATR-IR Spectroscopic Identification of Plastics

2.4.1 Polyamide (PA)

Polyamides (Fig. 2.66) are produced by a polycondensation reaction between α,ω-diamines (e.g., hexamethylenediamine) and an α,ω-dicarboxylic acid (e.g., adipic acid). This results in a linear aliphatic polyamide (PA 6.6; Nylon). The two numbers indicate the number of carbon atoms in the two reactants. Polyamide fibers are mainly used in the textile industry or for cosmetic products such as toothbrushes. With special processes (e.g., Noviganth), polyamide can also be galvanized.

The ATR infrared spectrum in absorption mode in Fig. 2.67 shows characteristic vibrations at the wavenumbers listed in Table 2.13. The corresponding

Fig. 2.66 Structural feature of the polyamide

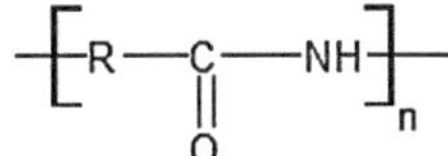

vibration within the molecule is assigned to them. It is noticeable that the spectrum shown here can be assigned to a polyamide without aromatic structural units (Sect. 2.3.2). The band at 3077 cm⁻1 is therefore not assigned to an aromatic C–H stretching vibration, but to a combination vibration between the two molecular vibrations indicated in Table 2.13.

Polyamides that contain aromatic phenyl groups are referred to as "polyarylamides" and abbreviated as PAA. In general, aromatic polyamides show higher strength properties and are used, for example, in razors as holders for the blades in the shaving head. For the designation of the individual types of vibration, refer to Sect. 2.3.2.

2.4.2   Polycarbonate (PC)

The thermoplastic polycarbonate (Fig. 2.68) is produced within a polycondensation from a diol, such as Bisphenol A (Fig. 2.69) and phosgene or the carboxylic acid diesters diethyl or diphenyl carbonate, and thus contains aromatic structural

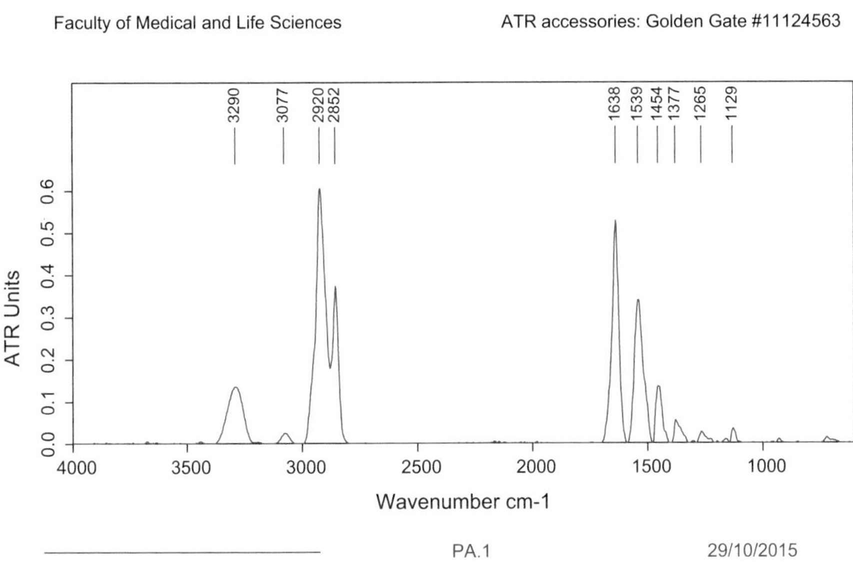

Fig. 2.67 ATR-IR spectrum of polyamide (granules)

Table 2.13 Characteristic IR vibration frequencies ($\tilde{v}$) of polyamide (PA)

$\tilde{v}$	Bond
3290	$v(N-H)$
3077	$v(C=O)+ \delta (N-H)$
2920	$v(C-H)$
2852	$v(C-H)$
1638	$v(C=O)$
1539	$\delta (N-H)$
1454	$\delta (C-H)$

units, which can also be recognized as weak peaks in the IR spectrum (Fig. 2.70). A typical characteristic of the most commonly used polycarbonate, based on the Bisphenol-A-educt in the IR spectrum, is the trident at 1200 wavenumbers. This absorption is attributed to the $C-O$ deformation and stretching vibrations. Below these bands, the $C-C$ stretch and deformation vibrations are found around 1000 cm^{-1}.

The aromatic vibration bands, which are introduced into the polymer by Bisphenol A (Table 2.14), are located at 3041, 1601, and 1504 as well as 829 cm^{-1}. They can be attributed to the $v(C-H)$-, $v(C=C)$-, $\delta(C-H)$- and $\gamma(C-H)$-vibrations in aromatic systems.

Characteristic for the functional group of the ester is a $C=O$ stretching vibration, which in saturated systems adsorbs between 1735–1750 wave numbers. In the case of polycarbonate, there is a diester, so that the carbonyl carbon is even more positively polarized by the double I-effect, and secondly, the $v(C=O)$ vibration shifts to higher energy ($v(\tilde{v}) = 1770$ cm^{-1}) due to the conjugation of the non-binding electron pairs of the two oxygen atoms with the adjacent aryl group.

Polycarbonate is characterized by its transparency and high impact resistance and is therefore used as a glass alternative due to its lower brittleness to avoid splintering. The lower abrasion resistance compared to glass can be compensated by sol-gel coatings, based on polysiloxanes with ceramic components, so-called Ormocer layers *(organic modified ceramics),*. Such "polycarbonate panes" are already used as rear windows in police vehicles.

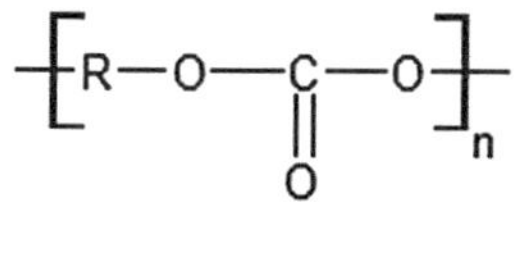

Fig. 2.68 Structural features of polycarbonate

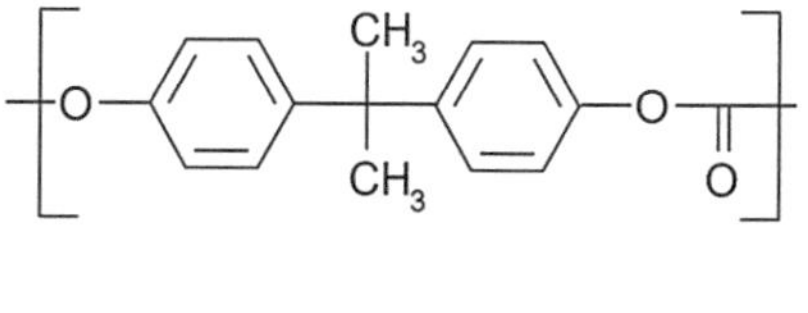

Fig. 2.69 Bisphenol A in the polyester chain

Furtwangen University

Faculty of Medical and Life Sciences

FTIR spectrometer: Tensor 27

ATR accessories: Golden Gate #11124563

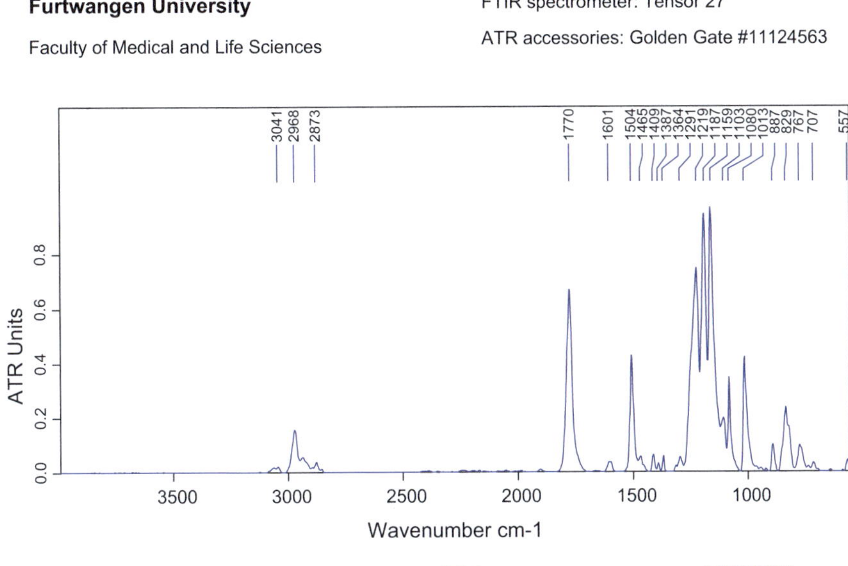

Fig. 2.70 ATR-IR spectrum of polycarbonate (granules)

Table 2.14 Characteristic IR vibration frequencies ($\tilde{v}$) of Polycarbonate (PC)

$\tilde{v}$	Bond
3041	$v(C-H)_{arom}$
2968	$v(C-H)$
2873	$v(C-H)$
1770	$v(C=O)$
1601	$v(C=C)_{arom}$
1504	$v(C=C)_{arom}$
1409	$\delta_{as}(CH_3)$
1387	$\delta_s(CH_3)$
1219	$v(C-O)$
1187	$\delta_{as}(C-C)$
1159	$\delta_s(C-C)$
1013	$v(C-C)$
829	$\gamma(C-H)_{arom}$

In plastic metallization (POP = *Plating On Plastics*), more than 80% of the thermoplastic base material used for injection molding is ABS (Acrylonitrile

Butadiene Styrene) and ABS blends. The blends mainly consist of additions of polycarbonate to the ABS.

The addition of PC does indeed worsen the plating ability, affecting the optical requirements for a high-gloss chrome surface, but on the other hand, the addition of PC improves the strength properties and the heat resistance of ABS from 95 °C (Vicat softening temperature) to 112 °C in Bayblend T 45 MN (Suchentrunk et al. 2007). More than 80% of the base material used for metallization is ABS (Acrylonitrile Butadiene Styrene) and ABS blends. The polycarbonate content in a blend can be quantified using IR spectroscopic methods based on characteristic peaks of the two components ABS and PC, based on a calibration with blends whose PC content is known (Neek et al. 2017).

2.4.3 Polyethylene (PE)

Polyethylene (Fig. 2.71) is produced by radical polymerization of ethylene. The radicals for the chain initiation can be generated thermally or photochemically. For the photochemically induced homolytic bond cleavage of common radical initiators, such as benzoyl peroxide ($\Delta_\mathrm{D}H°= 126$ kJ/mol) or azobisisobutyronitrile ($\Delta_\mathrm{D}H = 131$ kJ/mol), the energy of light is sufficient to start the chain reaction.

$$R\!\!-\!\!R \rightarrow 2\,R \cdot \text{Kettenstart mit Radikalbildung} \tag{2.46}$$

$$R \cdot + R = R \rightarrow R\!\!-\!\!R\!\!-\!\!R \cdot \text{Kettenfortpflanzung} \tag{2.47}$$

Chain propagation termination occurs through the recombination of two radicals or through a disproportionation. High molecular weight compounds with a chain length of n > 1000 or a molecular weight >10 kDa (1 Da = 1 u) are referred to as polymers.

Despite the same structural unit of ethylene, the properties of polyethylene can vary depending on the manufacturing process. In the low-pressure process, the harder and more heat-resistant HDPE *(High Density Polyethylene)* with higher density is produced, as the linear polymer molecule chains bind closely together via Van der Waals forces, while in the high-pressure process strong molecule branching occurs, which prevents the molecule chains from lying next to each other, thus producing a polymer with lower density, the so-called LDPE *(Low Density Polyethylene)*,. LDPE is softer and less heat-resistant.

In addition to the degree of branching, the chain lengths and the molecular weight distribution of the polymer chains are responsible for the properties of

Fig. 2.71 Structural feature
of polyethylene

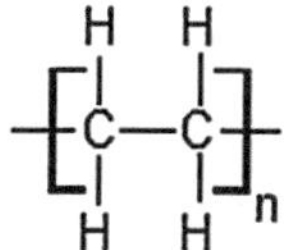

polyethylene. These can be adjusted or influenced by the reaction conditions such as pressure, temperature, and catalysts.

With the infrared spectrum in Fig. 2.72, PE can be easily identified due to the few vibrations. Since only two types of atoms, carbon and hydrogen, are present, the spectrum only shows their stretching and deformation vibrations. The characteristic signals are listed in Table 2.15.

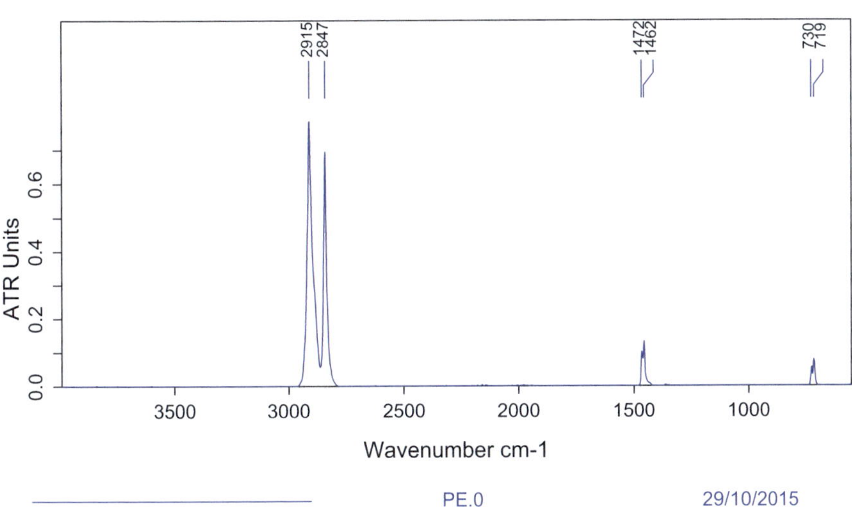

Fig. 2.72 ATR-IR spectrum of polyethylene (granules)

Table 2.15 Characteristic IR vibration frequencies ($\tilde{v}$) of polyethylene (PE)	$\tilde{v}$	Bond
	2915	$v(C-H)$
	2847	$v(C-H)$
	1472	$\delta_{as}(CH_2)$
	1462	$\delta_{s}(CH_2)$
	730	$\delta(CH_2)$rocking
	719	$\delta(CH_2)$rocking

2.4.4 Polypropylene (PP)

The polypropylene (abbrev. 2.73) is produced, like PE, via a radical polymerization. The additional methyl group in the monomer propene significantly changes the properties of the thermoplastic plastic compared to PE. The methyl group protruding from the side of the chain strand acts like a barb that can hook onto the neighboring chain, thus restricting the mobility of the linear polymer chains. The result is a higher hardness, strength and load capacity of the PP. Different reaction conditions can produce, in addition to the main product, the semi-crystalline isotactic PP (PP-I), in which all methyl groups point in one direction, also semi-crystalline syndiotactic PP (PP-S) and amorphous atactic PP. In the case of syndiotactic PP, the methyl groups are alternately opposite each other, and in the case of atactic, the position of the methyl groups is statistically distributed. This tacticity, together with the chain length and the molecular weight distribution, determines the properties of the PP polymer. The highest heat resistance ($T_m = 184$ °C) is exhibited by the isotactic isomer.

Like polyethylene, the PP polymer also shows few peaks in the infrared spectrum (Fig. 2.74). Here too, there are no heteroatoms present. The difference to the PE spectrum is due to the additional methyl group, which is noticeable in the additional $C-H$ stretching vibrations and the symmetric deformation vibration in the range of 1390–1370 cm^{-1}. In Table 2.16, the specific peaks are assigned to the individual vibrations.

2.4.5 Polyester

Polyester, for example Polyethylene terephthalate (PET) (Fig. 2.75), are synthesized through polycondensation reactions just like polyamides. Instead of a diamine, a diol reacts with the dicarboxylic acid. This means that on the product side in the equilibrium reaction, one molecule of water is produced per ester group. To further shift the equilibrium towards the product side, the resulting water must be removed from the system. This can be achieved by an azeotropic distillation, using toluene or xylene as both a solvent and a carrier. The resulting water can be distilled off during the reaction as an azeotrope with toluene (80:20) at a boiling temperature of 80 °C into a water separator. In the separator, the water separates from the carrier solvent in the cold, which flows back into the reaction flask, where it can continue to absorb water until the reactants are exhausted. If the terephthalic acid reacts with ethylene glycol, the polyester produced is PET, which

Fig. 2.73 Structural feature
of polypropylene

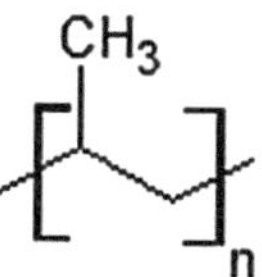

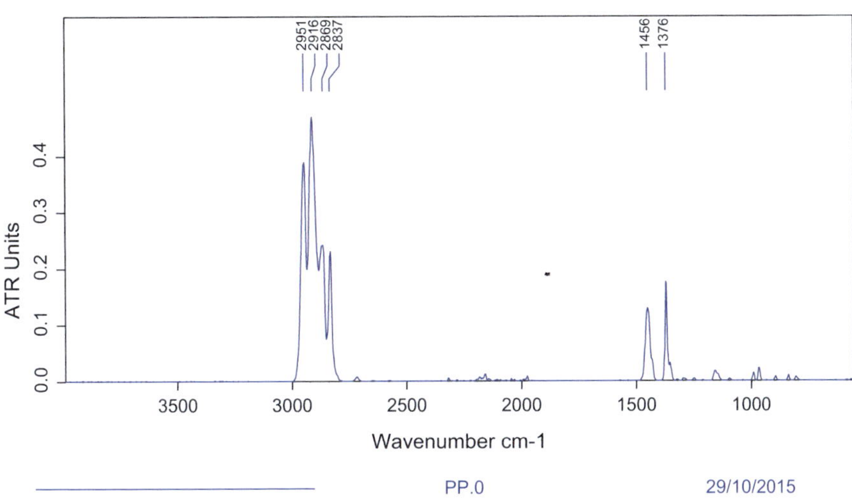

Fig. 2.74 ATR-IR spectrum of polypropylene (granules)

Table 2.16 Characteristic IR vibration frequencies ($\tilde{v}$) of polypropylene (PP)	$\tilde{v}$	Bond
	2961	$v(C-H)$
	2916	$v(C-H)$
	2869	$v(C-H)$
	2837	$v(C-H)$
	1456	$\delta(CH_2)$
	1376	$\delta_s(CH_3)$

Fig. 2.75 Structural feature of Polyethylene terephthalate

is often used for plastic surfaces. If 1,4-butanediol is used as the diol, the plastic PBT is obtained. Both aromatic thermoplastic polyesters are character-ized by high chemical resistance as well as high strength and rigidity. Thus, the polycondensates are superior to polyamide. In injection molding technology, PBT shows better cooling behavior and is therefore preferred over PET.

Aromatic and aliphatic C−H stretching vibrations are only very weakly recognizable in the spectrum (Fig. 2.76). The so-called benzene fingers and the aromatic C=C stretching vibrations around 1600 cm^{-1}are also faint (Table 2.17).

Characteristic for the functional group of an ester is the C=O stretching vibration at 1714 wavenumbers. In the infrared spectrum, PBT (see PBT spectrum) is difficult to distinguish from PET, as both polymers contain the same functional groups.

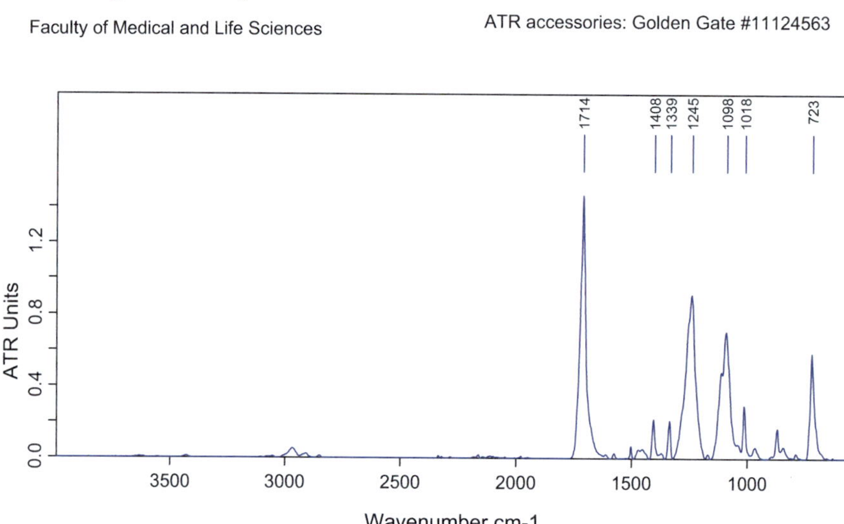

Fig. 2.76 ATR-IR spectrum of polyethylene terephthalate (granulate)

Table 2.17 Characteristic IR vibration frequencies ($\tilde{v}$) of polyethylene terephthalate (PET)

$\tilde{v}$	Bond
1714	$v(C=O)$
1408	$\delta_s(CH_2)$
1339	$\delta_{as}(CH_2)$
1245	$v(C-O)$
1098	$\delta(C-C)$
1018	$v(C-C)$
~800	$\gamma(C-H)$, 1,4 Subst.
723	$\delta(CH_2)$ *rocking*

2.4.6 Polymethyl Methacrylate (PMMA; Plexiglas)

Polymethyl methacrylate (PMMA) (Fig. 2.77), also known as acrylic glass or Plexiglas, is produced through radical polymerization, like all polymers whose monomers are unsaturated hydrocarbons. After the thermally initiated radical formation with dibenzoyl peroxide, the chain initiation with the unsaturated methacrylic acid methyl ester (MMA = Methyl methacrylate) occurs following the CO_2 elimination. Within the radical polymerization, an atactic, completely amorphous transparent polymer is formed from MMA. PMMA was one of the first thermoplastic polymers (1928). Due to its stability, toughness and UV resistance, but especially due to its high transparency or light transmittance, the polymer is mainly used for optical applications. The first contact lenses (1950) were made from PMMA due to its biological compatibility. However, due to the poor oxygen permeability, silicon acrylates are now used for contact lenses. The worldwide PMMA production increased to 1.8 million tons in 2012. The increase is due to the innovation engine of the electrical industry, which had an increasing demand for light guides, for example for the light guide plates in LED flat screens.

The IR spectrum of PMMA in Fig. 2.78 shows the typical aliphatic C−H stretching vibrations below 3000 wavenumbers and the C=O stretching vibration characteristic of an ester at 1722 cm^{-1}; also the deformation vibrations corresponding to the methyl groups. Below 1250 wavenumbers to 900 cm^{-1} are the fingerprint bands, where a diagnostic assignment to the corresponding molecular vibrations can only be suspected. Table 2.18 shows the most important wavenumbers.

2.4.7 Polystyrene (PS; Styrofoam)

To obtain polystyrene (Fig. 2.79), the monomer styrene, also called vinylbenzene, is polymerized radically. Depending on which catalysts are used, isomers with different properties are obtained. Without catalysts, the large phenyl groups along the polymer chain are statistically distributed on both sides. This atactic isomer has an amorphous structure and is transparent. With Ziegler-Natta catalysts, isotactic types are formed and with metallocenes as catalysts, syndiotactic polystyrene types with high heat resistance are obtained. Styrene polymers have high strength and are used as packaging materials for food (e.g., plastic lids of coffee-to-go

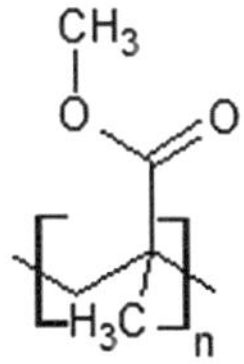

Fig. 2.77 Structural feature
of Polymethyl methacrylate

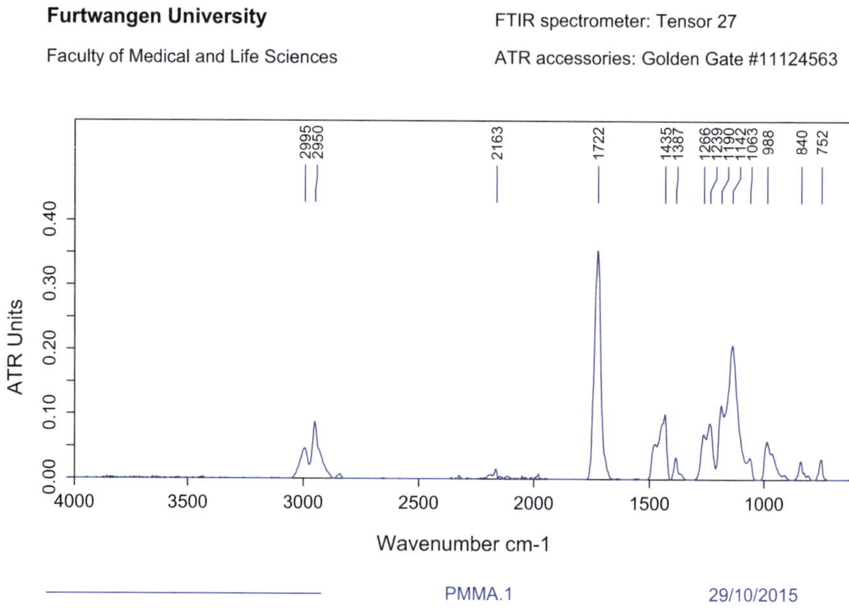

Fig. 2.78 ATR-IR spectrum of PMMA (granules)

Table 2.18 Characteristic IR vibration frequencies ($\tilde{v}$) of polymethyl methacrylate (PMMA)

$\tilde{v}$	Bond
2995	$v(C-H)$
2950	$v(C-H)$
1722	$v(C=O)$
1435	$\delta_{as}(CH_2)$
1387	$\delta_s(CH_3)$
1266	$\delta\ (C-O)$
1239	$v(C-O)$
1190	$\delta_{as}(C-C)$
1142	$\delta_s(C-C)$
1063	$v(C-C)$
840	$\delta(CH_2)$rocking
752	$\delta(CH_2)$rocking

cups) or for plastic cutlery. Foamed PS, which is used for sound and thermal insulation, also known as Styrofoam, is created by using a blowing agent such as carbon dioxide, which is added during the extrusion process.

Fig. 2.79 Structural feature of polystyrene

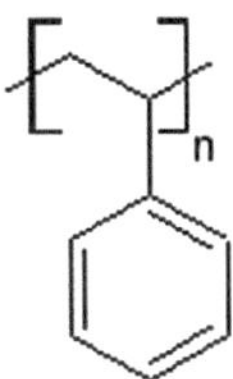

Styrene is an important component for the copolymerization with other monomers such as butadiene and acrylonitrile. This results in different important technical plastics within a block and graft polymerization such as butadiene rubber (SBR or SBS), styrene acrylonitrile (SAN) or with three different monomers the plastic ABS, which is used for example for Lego bricks.

Polystyrenes represent the fourth largest group behind PE (30%), PP (20%) and PVC (15%) with a share of 10% of the worldwide annual production (Abts 2014).

In the higher energetic range of the IR spectrum (Fig. 2.80), the aromatic C–H vibrations >3000 cm^{-1} are initially recognizable. These are followed by the aliphatic C–H vibrations. The energetic shift results from the lower bond strength of the longer bond between an sp^3-hybridized C atom and the 1s orbital of hydrogen compared to an sp^2-1s bond. The weak bands in the range of 1600–2000 cm^{-1}, the so-called benzene fingers, are due to overtones and combination vibrations of the aromatic bonds (Table 2.19). They are a signal, but not a guarantee for the presence of an aromatic compound. To be sure, further criteria such as the

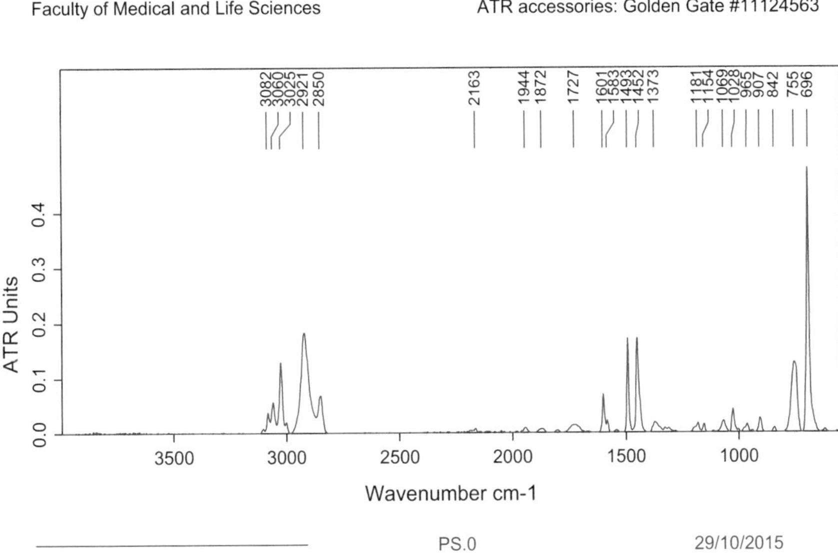

Fig. 2.80   ATR-IR spectrum of polystyrene (granules)

Table 2.19 Characteristic IR vibration frequencies ($\tilde{v}$) of polystyrene (PS)

$\tilde{v}$	Bond
3082	$v(C-H)_{arom}$
3060	$v(C-H)_{arom}$
3025	$v(C-H)_{arom}$
2921	$v(C-H)_{aliph}$
2850	$v(C-H)_{aliph}$
1727, 1872, 1944	"Benzene fingers"
1601	$v(C=C)_{arom}$
1493	$\delta_{s}(CH_2)$
1452	$\delta_{as}(CH_2)$
842	$\gamma(C-H)_{arom}$
907	$\delta(C-H)$
755	$\delta(CH_2)$rocking
696	$\delta(CH_2)$rocking

aforementioned aromatic C–H stretching vibration as well as the *out-of-plane* vibrations <900 cm^{-1}should be present.

2.4.8 Polyurethane (PU or PUR)

Polyurethane (Fig. 2.81) are produced by a polyaddition between diols or polyols and diisocyanates or polyisocyanates. This addition reaction forms urethane groups, and there is no elimination of by-products as in the polycondensation.

$$\mathrm{n\,HO-R^1-OH\,(Diol) + n\,O=C=N-R^2-N=C=O\,(Diisocyanat) \rightarrow Polyurethan} \tag{2.48}$$

For polyurethane foams, water is added as a chemical blowing agent together with the diol. The diisocyanate reacts with water to release CO_2 and form a diamine, which also acts as a reagent and contributes to cross-linking. Depending on the diol used, the degree of cross-linking and thus the hardness of the polyurethane can be adjusted. While glycol produces linear polyurethanes with diisocyanates, higher alcohols such as glycerin achieve spatially cross-linked structures. Long and short chain polyether polyols and polyester polyols (so-called soft segments) are often used. A common diol component is, for example, the polyester polyol, which is synthesized with adipic acid ($HOOC-(CH_2)_4-COOH$) and

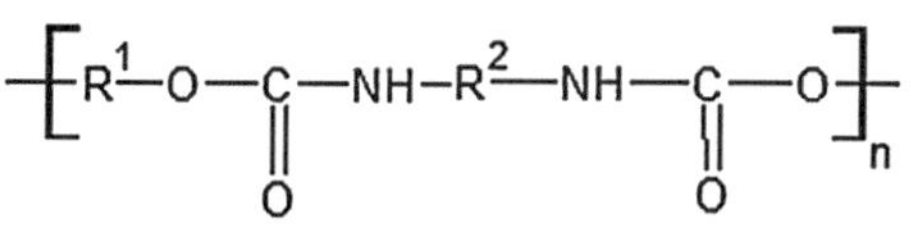

Fig. 2.81 Structural feature of polyurethane

1,4-butanediol ($HO-(CH_2)_4-OH$). The IR spectrum shown below belongs to this group of polyester polyurethanes. As diisocyanate components, mainly aromatic difunctional compounds are used, such as toluene-2,4-diisocyanate, toluene-2,6-diisocyanate (TDI) and diphenylmethane-4,4'-diisocyanate (MDI), or the aliphatic hexamethylene-1,6-diisocyanate (HDI).

Polyurethanes, also known as synthetic resins, are used as soft foams, mainly for upholstery in furniture and mattresses. The harder synthetic resin foams are used as insulation materials, for example to foam windows and doors tightly. Block foams are produced in the casting process (mattresses) and molded foams in closed tools in an injection molding process (electronic housings).

Household sponges are made from soft polyurethanes and dashboards from the harder ones. Another important area of application is OUR paints; for example, within a 2-component system, prepolymers (polyether or polyester polyols) are brought together with a hardener (diisocyanate), whereby the system polymerizes and cross-links very quickly at room temperature.

Characteristic for the polyurethanes is the N-H stretching vibration of the urethane group at 3323 wavenumbers (Fig. 2.82). This is followed by the aliphatic $C-H$ stretching vibrations and in the aromatic range >3000 cm^{-1} a weak peak, which suggests an aromatic polymer. This assumption is supported by the C=C-valence vibrations at 1508 cm^{-1}and 1597 cm^{-1}, which are typical in this range for aromatics. They cannot be assigned without doubt, as the amide-II band in urethanes, if at least one H is on the nitrogen, is also found between 1500 cm^{-1}and

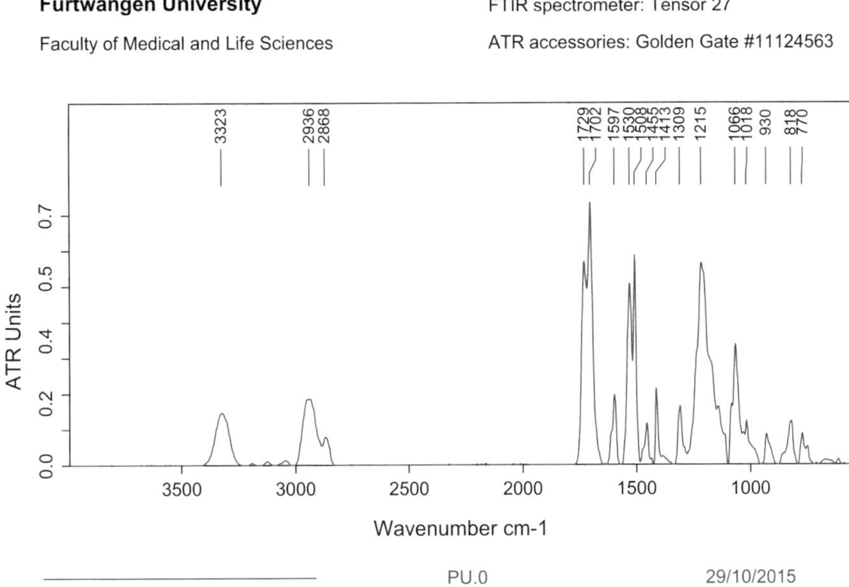

Fig. 2.82 ATR-IR spectrum of polyurethane (granulate)

1600 cm^{-1}. However, since only one band of the amide-II-(N$-$H-)deformation vibration can be assigned, this is an aromatic polyurethane. The *out-of-plane* vibrations in the IR range below 900 wavenumbers confirm this statement. Since most diisocyanate monomers, except HDI, are aromatic, the interpretation of the IR spectrum in Table 2.20 is plausible.

Notable in the present polyurethane spectrum is the double peak in the range of 1700 wavenumbers. Between 1660 cm^{-1} and 1850 cm^{-1}, depending on the functional group (acid chloride, ketone, ester, amide etc.), the stretching vibrations of the C$=$O bond absorb at different frequencies, depending on the bond order (Sect. 2.3.2). A +M effect of the adjacent amino group results in a mesomeric boundary structure, which overall reduces the bond order of the C$=$O bond. Compared to an isolated carbonyl group, this reduces the wavenumber to 1702 cm^{-1}. The band at 1729 indicates a carbonyl vibration, belonging to an ester. Accordingly, the present polyurethane is a polyester-polyurethane.

2.4.9 Polyvinyl Chloride (PVC)

Polyvinyl chloride (Fig. 2.83) is produced in a radical or ionic polymerization of the monomer chloroethene or vinyl chloride. Peroxides or AIBN (azobisisobutyronitrile cf. PE) serve as radical initiators. As with polystyrene, the atactic polymer is preferably formed due to the bulky side chains compared to the hydrogen in ethene. The isotactic polymer can be produced with Ziegler-Natta catalysts, to

Table 2.20 Characteristic IR vibration frequencies ($\tilde{v}$) of Polyurethane (PUR)

$\tilde{v}$	Bond
3323	v(N$-$H)
2936	v(C$-$H)
2868	v(C$-$H)
1729	v(C$=$O)$_{Ester}$
1702	v(C$=$O)$_{Amide\ I}$
1597	v(C$=$C)$_{arom}$
1530	δ(N$-$H)$_{Amide\ II}$
1508	v(C$=$C)$_{arom}$
1413	δ_{as}(CH$_3$); δ_s(CH$_2$);
1309	v(C$-$O)$_{Urethane}$
1215	v(C$-$O)$_{Ester}$
1187	v(C$-$O)
1066	δ_s(C$-$C)
1018	v(C$-$C)
818	γ(C$-$H)$_{arom}$
770	γ(C$-$H)$_{arom}$

Fig. 2.83 Structural feature of polyvinyl chloride

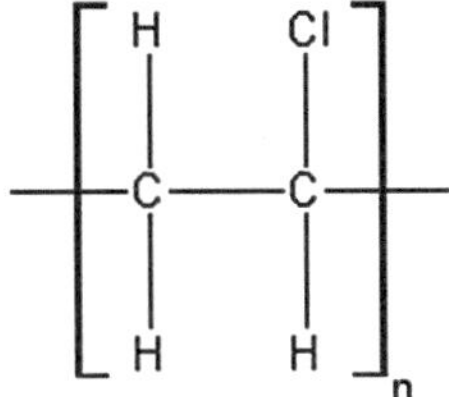

which the growing polymer temporarily attaches. This stereoregularity (tacticity = arrangement) has a strong influence on the properties of the plastic. It determines, together with other factors such as the degree of cross-linking or the chain length and chain length distribution, the hardness, brittleness, dimensional stability and the melting point or the glass transition temperature of the thermoplastic elastomer. The more uniform the arrangement of the side chains, the easier a crystalline structure can form. The degree of crystallinity, within which strong intermolecular interactions are present, thus also shapes the aforementioned macroscopic properties of the plastic.

PVC, known from vinyl records, is hard and brittle. It is used for floor coverings and pipes that transport liquids. However, in the chemical and metalworking industry, PVC is gradually being replaced by PP. In the event of a fire, for example caused by a gas explosion in a plastic galvanizing plant, the burning of PVC produces corrosive hydrochloric acid, which leads to severe corrosion damage to load-bearing steel elements.

To make the hard and brittle PVC more plastic, up to 50% of so-called plasticizers are added to the polymer. This expands the range of applications of the third most common plastic to all types of hoses and cable insulation.

These plasticizing additives lodge between the polymer chains and thus ensure that they get less entangled with each other, but can slide past each other. The plasticizers do not form a chemical bond with the plastic and can therefore, depending on the environment in which the soft PVC is used, be released into another medium, such as water. The water solubility and the mobility of the plasticizer in the polymer matrix play a crucial role in the quantity of release. Some of these plasticizers, such as certain phthalates, are harmful to health (Sect. 2.5.1). Although they are poorly soluble in water, they are capable of outgassing at high temperatures in the extruder or being absorbed by other aqueous solutions such as saliva. For this reason, for example, the use of DEHP (Diethylhexylphthalate) or DOP (Dioctylphthalate) has been banned in the EU since 1999 in toys for small children.

Meanwhile, in addition to DEHP, other phthalic acid esters such as DBP (Dibutylphthalate), DIBP (Diisobutylphthalate) and BBP (Benzylbutylphthalate) are listed on the REACH ban list (Annex XIV of the European pollutant regulation). All the substances mentioned are classified as reprotoxic.

Studies on the release of plasticizers from twist-off lids using PVC sealing compounds have shown that the migration of the plasticizer ESBO (epoxidized soybean oil) into fatty foods significantly exceeds the European limit of 60 mg

ESBO per kg of food and that phthalates are also used as plasticizers, which also pass into food in large quantities (BfR 2003b).

In addition to the health-endangering phthalates, less concerning alternatives such as the aliphatics DINCH or Pevalen are now also used (Fig. 2.84). Other plasticizers such as acetyl tributyl citrate, diethylhexyl adipate, and compounds of the substance class alkane sulfonic acid phenyl ester fulfill the function of plasticizing PVC (Federal Environment Agency 2011).

[3-pentanoyloxy-2,2-bis(pentanoyloxymethyl)propyl] Pentanoate CAS NO.15834-04-5

"Pevalen"

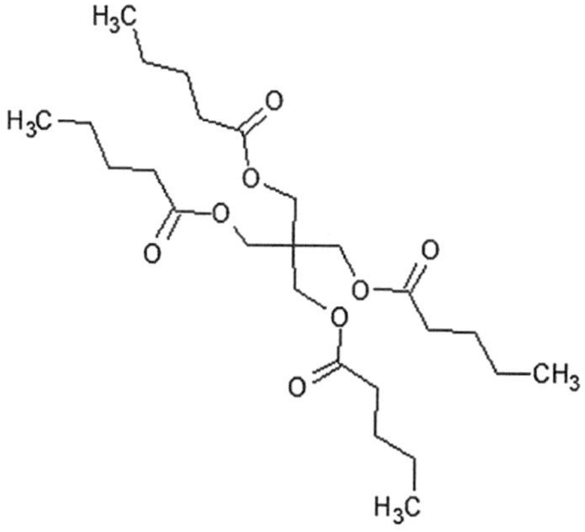

DINP (diisononyl phthlate) IUPAC CAS No.

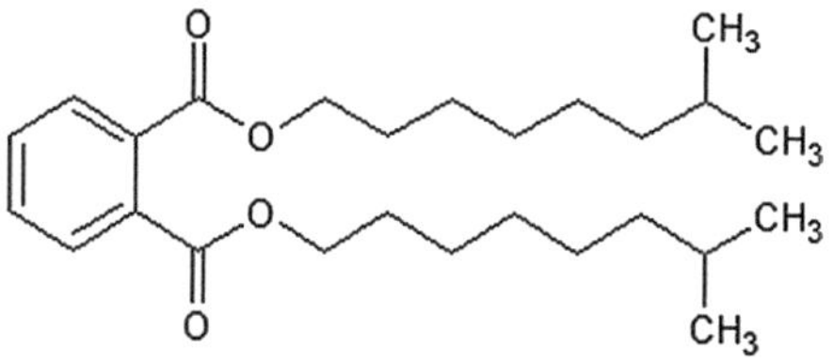

DINCH (IUPAC CAS)

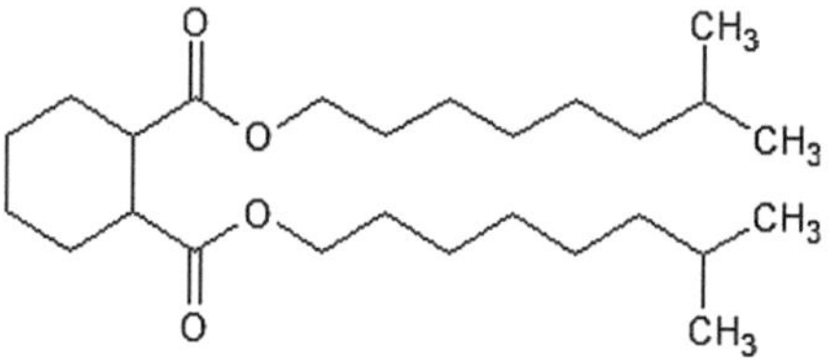

Fig. 2.84 Structure of different PVC plasticizers

Characteristic for rigid PVC (PVC-U; U for English *unplasticized*) are the symmetrical and asymmetrical C−H stretching vibrations. These are accompanied by the typical CH_2 deformation vibrations (Fig. 2.85; Table 2.21). In contrast to PE, PVC has an atom bound to a carbon that is 35 times heavier. This significant change is recognizable in the vibration spectrum at the CHCl deformation

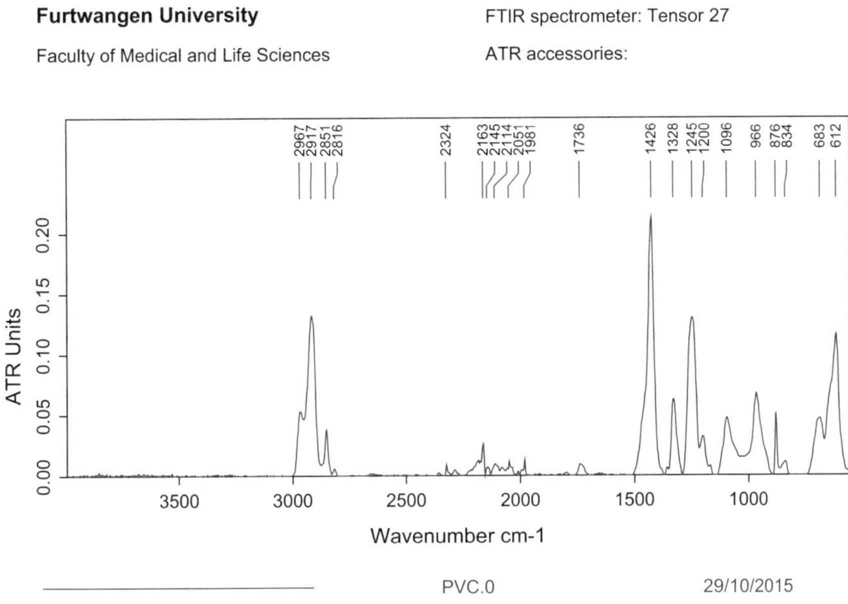

Fig. 2.85 ATR-IR spectrum of polyvinyl chloride (granules)

Table 2.21 Characteristic IR vibration frequencies ($\tilde{v}$) of rigid polyvinyl chloride (PVC-U)

$\tilde{v}$	Bond
2967	$v_{as}(C-H)$
2917	$v_{as}(C-H)$
2851	$v_s(C-H)$
2816	$v_s(C-H)$
1426	$\delta(CH_2)$
1328	$\delta_{as}(C-ClH)$
1245	$\delta_s(C-ClH)$
1066	$\delta(C-C)$
966	$v(C-C)$
683	$v_{as}(C-Cl); \delta(CH_2)_{rock}$
612	$v_s(C-Cl)$

vibration as well as at the C−Cl stretching vibrations in the low energy range of 600–700 wave numbers.

By comparing the peak of the spectrum of a soft PVC (PVC-P; P for English *plasticized*) with that of rigid PVC-U, which contains no plasticizer, it is at least unequivocally recognizable which type of plasticizer is contained in the PVC. That is, whether, for example, a phthalate-based plasticizing aid is included or not.

Using a calibration, the plasticizer content in the PVC matrix can be easily and quickly determined by IR spectroscopy (Harsch and Kirschner 2014). The depicted spectrum in (Fig. 2.86) shows the infrared vibrational absorptions of a soft PVC hose. The other two spectra show the respective IR bands for DINPand DINCH.

A comparison of the vibration absorption bands in rigid PVC versus those of soft PVC reveals some additional peaks (gray highlighted wave numbers in Table 2.22), which cannot be attributed to the polymer.

A comparison of the gray highlighted absorption bands with those of two commonly used plasticizers DINP and DINCH (Fig. 2.87) clearly indicates based on the aromatic vibrations and the matches in the fingerprint area (<1500 cm^{-1}), that in soft PVC the plasticizer DINP (Fig. 2.88) is a phthalate. The grey highlighted wave numbers are characteristic of DINP, which is confirmed by a comparison with the pure substance (Table 2.23).

The percentage plasticizer content can be determined by infrared spectroscopy after appropriate calibration (Harsch and Kirschner 2014).

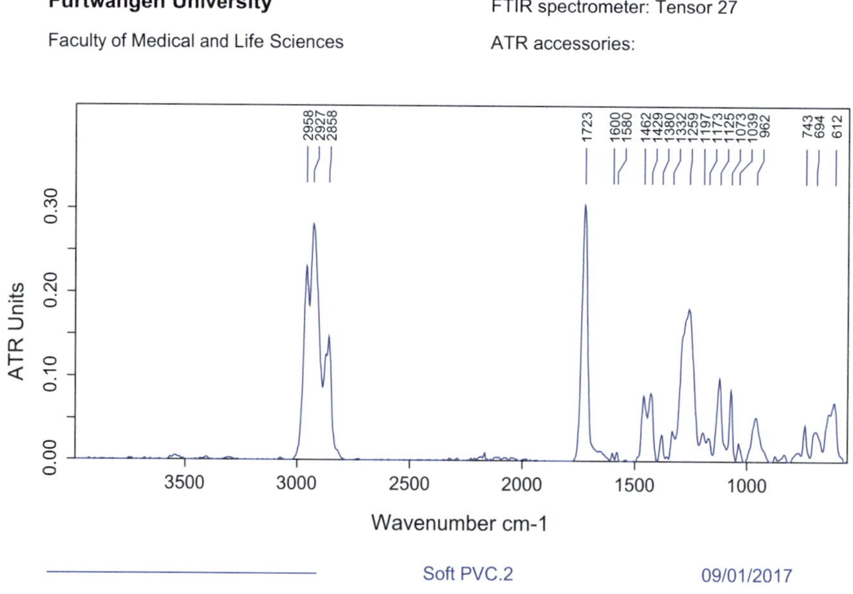

Fig. 2.86 ATR-IR spectrum of soft PVC (granules)

Table 2.22 Characteristic IR vibration frequencies (v) of soft polyvinyl chloride (PVC-P)

$\tilde{v}$	Bond
2958	$v_{as}(C-H)$
2927	$v_{as}(C-H)$
2858	$v_s(C-H)$
2816	$v_s(C-H)$
1723	$v(C=O)$
1600	$v(C=C)_{arom}$
1580	$v(C=C)_{arom}$
1462	$\delta(CH_2)$
1429	$\delta_{as}(C-ClH)$
1380	$\delta_s(CH_3)$
1332	$\delta_s(C-ClH)$
1259	$\delta_s(C-ClH); v(C-O)$
1125	$v(C-O)$; ober. + komb.
1073	$\delta(C-C)$; fingerprint
1039	$v(C-C)$; fingerprint
962	$v(C-C)$
743	$\gamma(C-H)_{arom}$
694	$v_{as}(C-Cl); \delta(CH_2)_{rock}$
612	$v_s(C-Cl)$

The comparison of the vibration absorption bands in Table 2.23 provides information about the structural similarities and also reveals the differences between the two plasticizer molecules.

2.4.10   (Poly-)Styrene Acrylonitrile (SAN)

Polystyrene acrylonitrile (Fig. 2.89) is a copolymer that is formed from two different monomer units, S (styrene) and A (acrylonitrile), within a radical chain polymerization. During the chain propagation, depending on the reaction conditions and concentration of the monomers, individual blocks of the same monomers are also formed, which is why we can also speak of block polymers. A polymer chain of the SAN could look like the following in sequence:

SSSSAASSSAAASSSAAASSSSSAA…

The ratio of styrene to acrylonitrile is about 70:30 parts. By copolymerizing styrene with acrylonitrile, the properties of the PS can be significantly improved. The result is a higher mechanical strength and better chemical resistance. Due to the higher rigidity, we find SAN in private households in the kitchen in the form of

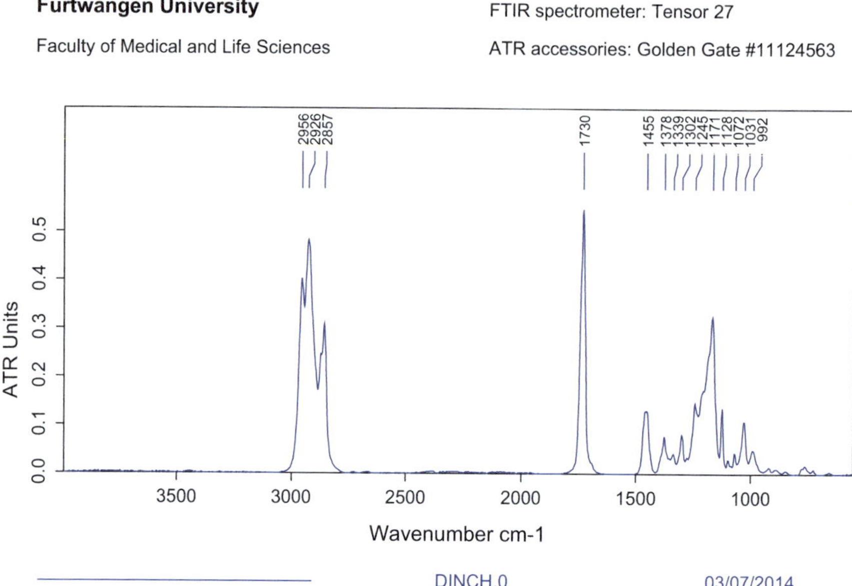

Fig. 2.87 ATR-IR spectrum of the plasticizer DINCH (oil)

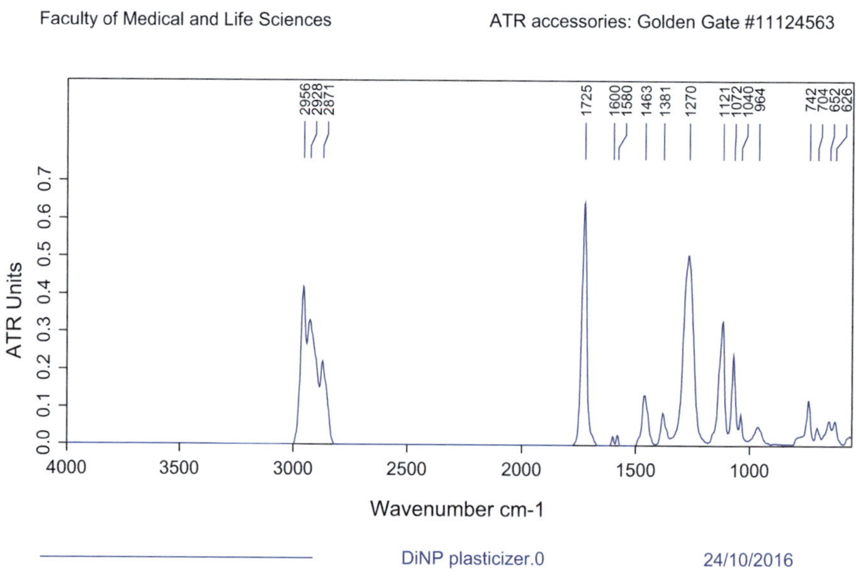

Fig. 2.88 ATR-IR spectrum of the plasticizer DINP (oil)

Table 2.23 Comparison of the vibration absorption bands of DINP and DINCH

DINP		DINCH	
$\tilde{v}$	Bond	$\tilde{v}$	Bond
–	$v(C-H)_{arom}$	–	–
2956	$v_{as}(C-H)$	2956	$v_{as}(C-H)$
2928	$v_{as}(C-H)$	2926	$v_{as}(C-H)$
2871	$v_s(C-H)$	2857	$v_s(C-H)$
1725	$v(C=O)$	1730	$v(C=O)$
1600	$v(C=C)_{arom}$	–	–
1580	$v(C=C)_{arom}$	–	–
1463	$\delta_s(CH_2); \delta_{as}(CH_3)$	1455	$\delta_s(CH_2); \delta_{as}(CH_3)$
1381	$\delta_s(CH_3)$	1378	$\delta_s(CH_3)$
–	–	1339	$\delta(CH_2)$, *twist, rock*
–	–	1302	$\delta(CH_2)$, *twist, rock*
1270	$v(C-O)$	1245	$v(C-O)$
–	–	1171	$\delta(C-C)$
1121	$\delta(C-C)$	1128	$\delta(C-C)$
1072	$\delta(C-C)$	1073	$\delta(C-C)$
1040	$\delta(C-C)$	1031	$\delta(C-C)$
964	$v(C-C)$	992	$v(C-C)$
742	$\gamma(C-H)_{arom}$	–	–
704	arom. Ringdef.	–	–

Fig. 2.89 Structural features of the two blocks in the SAN block polymer

cutlery, bowls, measuring cups and kitchen machines etc., and in the bathroom e.g. in the form of shower cubicle partitions.

The infrared spectrum of SAN (Fig. 2.90) is almost identical to that of polystyrene and therefore only shows slightly shifted adsorption vibrations (Table 2.24). The most striking distinction is due to the additional stretching vibration absorption of the triple bond in the nitrile group at 2237 cm^{-1}. To see this vibration band, it is necessary to perform an atmospheric compensation after the spectrum recording, as the IR-active deformation vibrations of the carbon dioxide in the measuring chamber can overlap the nitrile band.

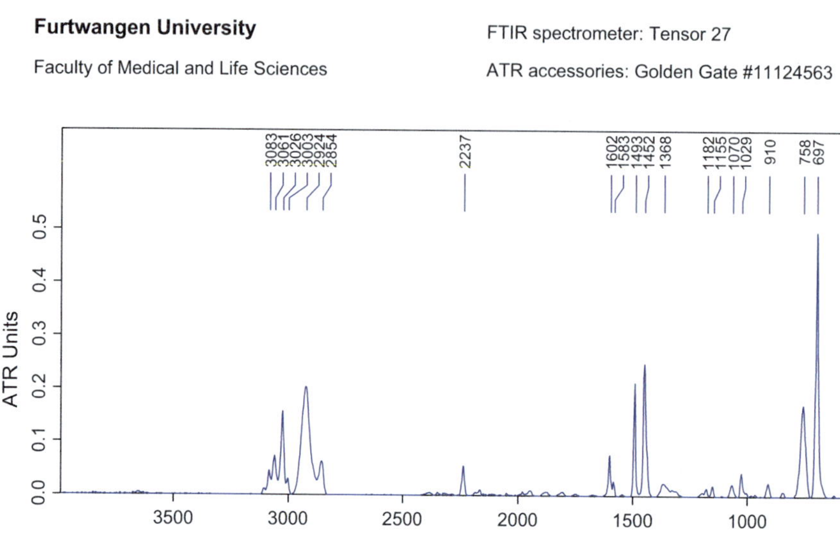

Fig. 2.90 ATR-IR spectrum of polystyrene acrylonitrile (granules)

Table 2.24 Characteristic IR vibration frequencies ($\tilde{v}$) of Polystyrene Acrylonitrile (SAN)

$\tilde{v}$	Bond
3083	$\nu(C-H)_{arom}$
3061	$\nu(C-H)_{arom}$
3026	$\nu(C-H)_{arom}$
3003	$\nu(C-H)_{arom}$
2924	$\nu(C-H)$
2854	$\nu(C-H)$
2237	$\nu(C{\equiv}N)$
1602	$\nu(C{=}C)_{arom}$
1583	$\nu(C{=}C)_{arom}$
1493	$\delta_s(CH_2)$
1452	$\delta_{as}(CH_2)$
1368	$\delta_s(CH_3)$
1029	$\nu(C-C)$
910	$\delta(C-H)$
(842)	$\gamma(C-H)_{arom}$
758	$\delta(CH_2)_{rocking}$
697	$\delta(CH_2)_{rocking}$

2.4.11 (Poly-)Acrylonitrile Butadiene Styrene (ABS)

The Styrene-Acrylonitrile Copolymer is very brittle, with a very low elongation at break. A SAN material is therefore poorly suited to deform under tensile and compressive forces without breaking. This property, measurable under the term "impact strength", limits the application range of the rigid SAN polymer. The introduction of a rubber phase in the form of a polybutadiene polymer into the styrene-acrylonitrile copolymer opens up a very wide range of variations for the copolymer now referred to as ABS (Acrylonitrile-Butadiene-Styrene). Important plastic material parameters such as, heat distortion resistance, impact strength, rigidity, modulus of elasticity, elongation at break, chemical resistance, stress crack sensitivity etc., can be adjusted by appropriate polymer proportions, polymer distribution and the manufacturing process.

By using 1,3-butadiene as a monomer, a reactive double bond remains for radical polymerization per monomer after its radical polymerization, regardless of whether a 1,2- or 1,4-polymerization takes place. In a 1,4-polymerization, the isolated double bonds are within the main chain (Fig. 2.91; *trans*-1,4-polybutadiene) and in a 1,2-polymerization in the side chain of the rubber (Fig. 2.92; syndiotactic 1,2-polybutadiene). Since the polymerization at these reactive centers generated by the polymerization sets in an additional direction (branching), this is also referred to as a graft polymerization.

In the ABS manufacturing process, a distinction is made between the joint and separate polymerization of the graft polymer (Polybutadiene) and the polystyrene acrylonitrile (SAN). In the joint process, all three monomers are mixed. This forms block polymers from the individual monomer units and grafts from copolymers on the butadiene blocks. The separate polymerization method, for example, is used to produce electroplatable ABS (Fig. 2.93).

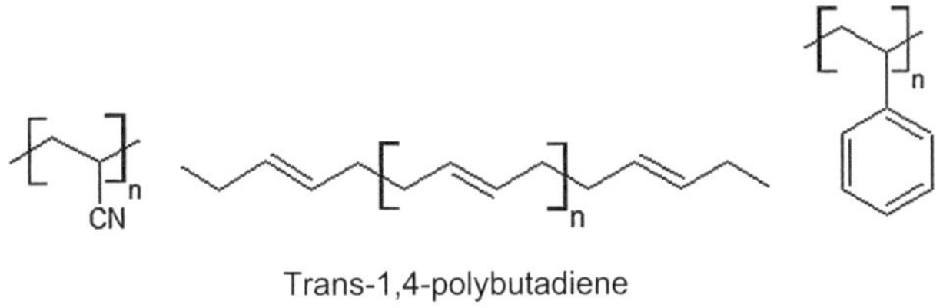

Fig. 2.91 *trans*-1,4-Polybutadiene

Fig. 2.92 Syndiotactic
1,2-Polybutadiene

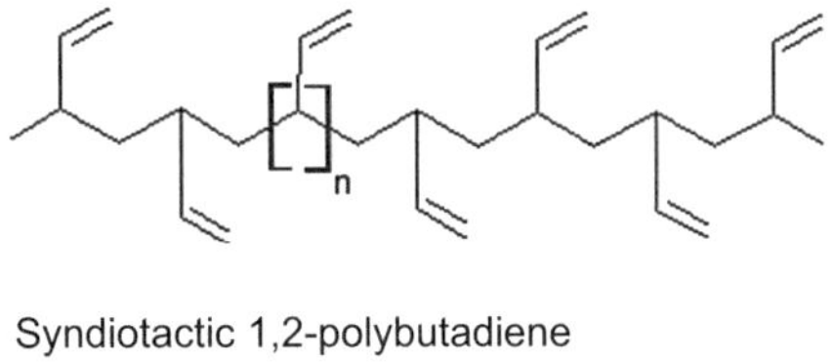

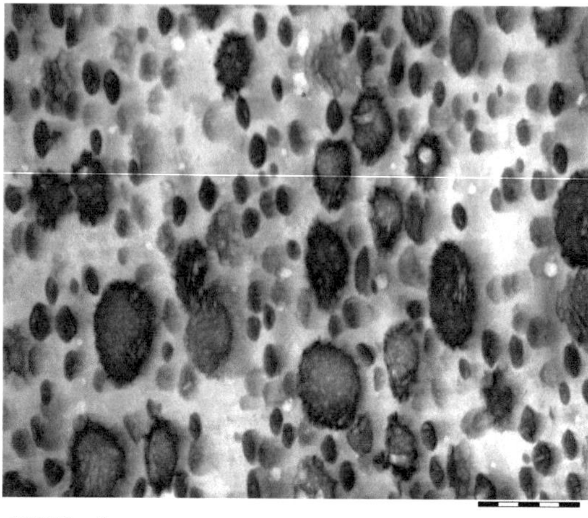

Fig. 2.93 SEM image of an electroplatable ABS plastic

50000 : 1

500nm

In this process, the butadiene is separately polymerized to the plastic rubber (Polybutadiene) (dark colored globular particles). In the next step, SAN (white areas) is grafted on, as can be seen from the white points in the dark areas. The grafting here serves the function that the rubber particles do not coagulate (clump or ball together). The strength-giving support matrix, the SAN (white spaces between the dark balls), is polymerized separately and then mixed with graft polymer depending on the requirements for the ABS properties. The diagram in Fig. 2.94 outlines this described second manufacturing process.

The softer and more deformable the ABS polymer needs to be, the higher the polybutadiene content. The function of the deformable rubber component becomes clearly visible in the following scanning electron microscopic image. Due to excessive pressure, the ABS injection molded part (Fig. 2.95) is severely deformed. The globular polybutadiene phases in the thermoplastic absorb the compressive stresses and deform into a crescent shape. However, from a certain load, as in this case, the compensation capacity of the soft components is exhausted

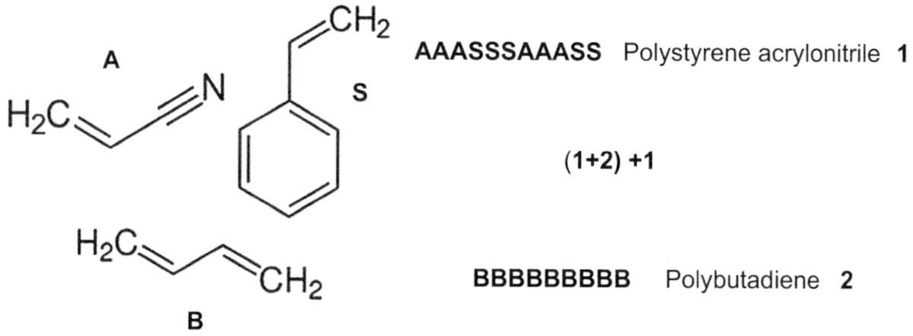

Fig. 2.94 Manufacturing process of polystyrene acrylonitrile

Fig. 2.95 SEM image of an
ABS injection molded part

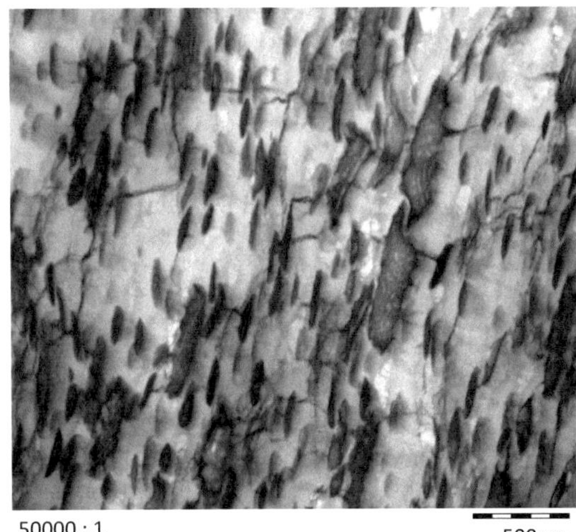

and the strength-giving and brittle SAN matrix breaks, as can be clearly seen in
the image from the cracks. Without the rubber, the cracking would begin at much
lower force.

For better heat distortion resistance, the rubber content must be reduced.
Without the polybutadiene content in ABS, ABS would not be electroplatable. The
polybutadiene content, which is important for the quality of the chrome-plated surface, is about 20% and can be quantified easily and without the use of chemicals
by means of IR spectroscopy (Sect. 2.3.3).

Of all the metallizable plastics, ABS and ABS blends (mainly PC) account for the
largest share of over 80%. Since the heat distortion resistance of the chrome-plated
plastic articles is often not sufficient for special applications, either polycarbonate
(ABS blend) is added or the bulkier α-methylstyrene is used as a styrene component.

To gain a better insight into the material properties of the most important base
plastic ABS for plastic metallization, we will go into detail here about its production process and molecular structure.

The plastic is synthesized from the monomers **A**crylonitrile, **B**utadiene, and
Styrene. Macromolecules that are made up of several different monomers are
summarized as copolymers (Briehl 2008). If the macromolecule consists of three
different monomers, as is the case with ABS, the copolymer is referred to as a ter-
polymer (Briehl 2008).

Polystyrene and polyacrylonitrile side chains are attached to a polybutadiene
main chain ("grafted"), therefore ABS is counted among the group of graft copolymers (Briehl 2008).

Due to the non-existent or low cross-linking of the individual chains in the ABS
plastic, ABS is classified as a thermoplastic. The low cross-linking explains the
large variation possibilities in the structure of the plastic and the associated different properties (Suchentrunk et al. 2007).

The higher the SAN content in the plastic, the more brittle it becomes (Fig. 2.96). In other words, the polybutadiene content is a measure of the toughness of the material. The proportion of polybutadiene in the matrix (Fig. 2.97) varies from

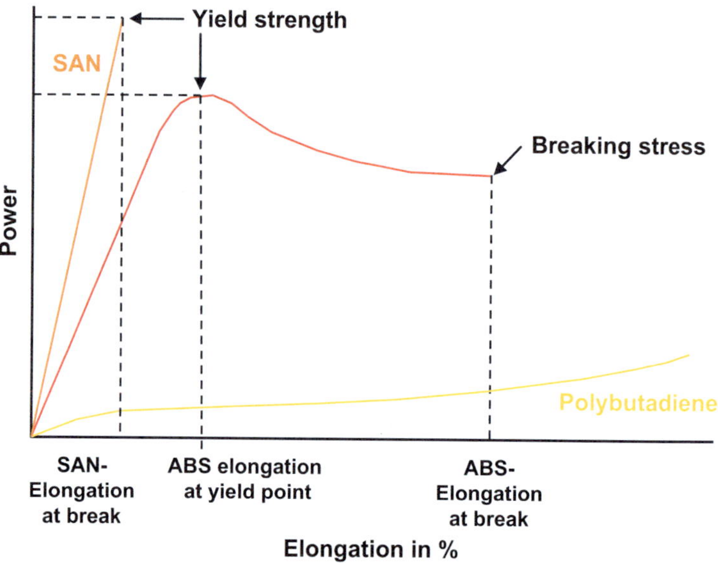

Fig. 2.96 Stress-strain diagram of ABS and its components. (Suchentrunk et al. 2007)

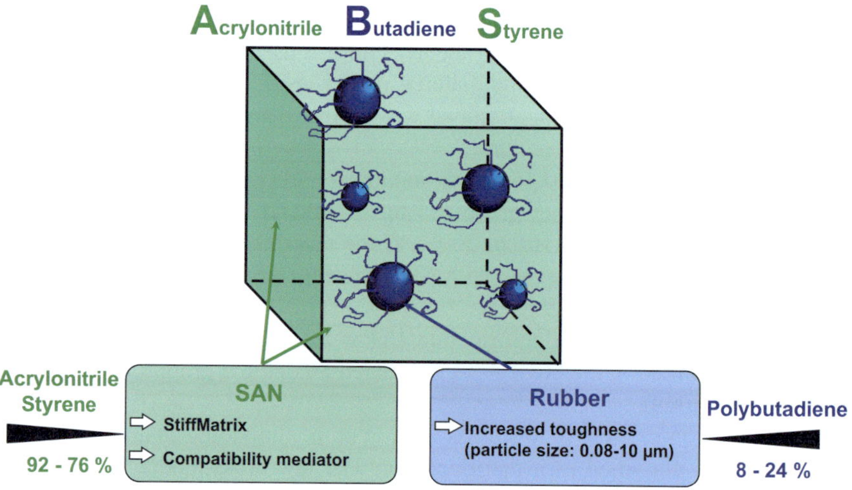

Fig. 2.97 Structure of the ABS matrix. (Suchentrunk et al. 2007)

8–24%. Accordingly, the SAN content is between 92% and 76%. The polybutadiene content is embedded in the support structure of the SAN.

The ABS polymer is produced via an emulsion process (Fig. 2.98), in which initially, in solution, SAN is grafted onto polybutadiene to ensure the solubility in the SAN matrix in a later step.

In terms of analysis with ATR-IR spectroscopy, the polybutadiene content in the plastic is of interest. This determines the time frame for the pretreatment in the process bath of the pickling in the electroplating. Here, the polybutadiene is dissolved out of the matrix by a chromosulfuric acid solution, leading to cavities with undercuts in the plastic. The higher roughness of the surface causes the adhesion of the metal layer in a later process step. The pickling process must not be too short or too long, as in both cases the adhesive properties of the plastic surface rapidly deteriorate. Too short dwell time means too few undercuts, the metal layer cannot adhere. Too long dwell time leads to the breaking out of the adhesion bridges that form between the dissolved polybutadiene (Fig. 2.99).

Of all the plastics capable of metallization, ABS and ABS blends (mainly PC) account for the largest share of over 80%. Since the heat distortion resistance of the chrome-plated plastic articles is often not sufficient for special applications, either polycarbonate (ABS blend) is added or the bulkier α-methylstyrene is used as a monomer for the styrene component.

The infrared spectrum of ABS (Fig. 2.100) is very similar to that of polystyrene and thus also to that of SAN and therefore only shows slightly shifted adsorption vibrations. The most striking distinction of ABS from the other two styrene

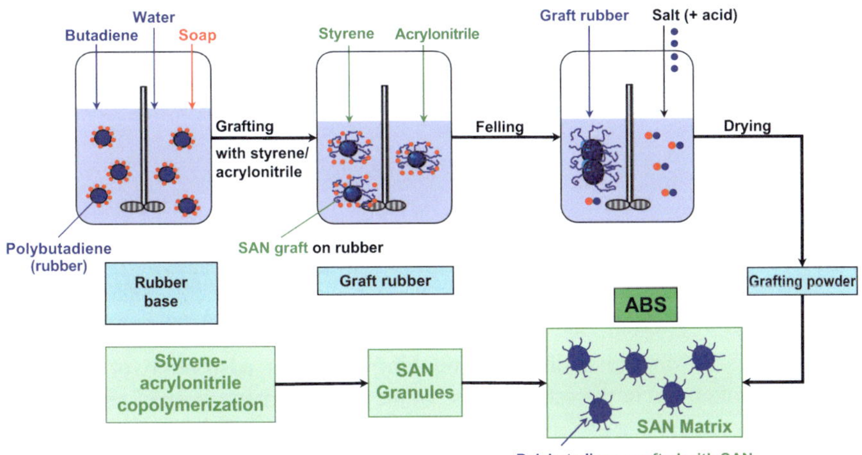

Fig. 2.98 Production process of ABS in emulsion

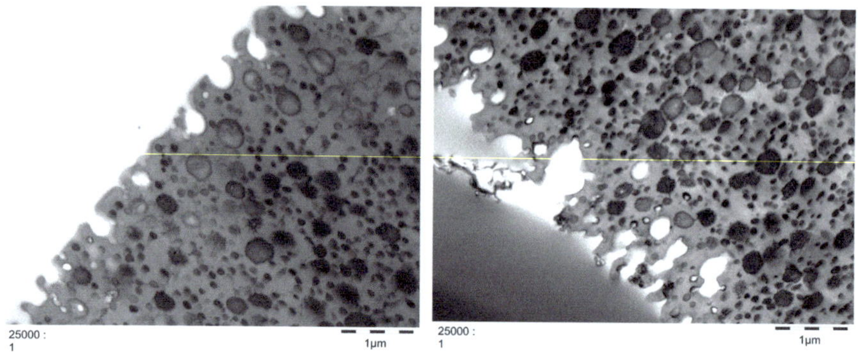

Fig. 2.99 Comparison of SEM images of correctly pickled (left) and over-pickled ABS surface (right)

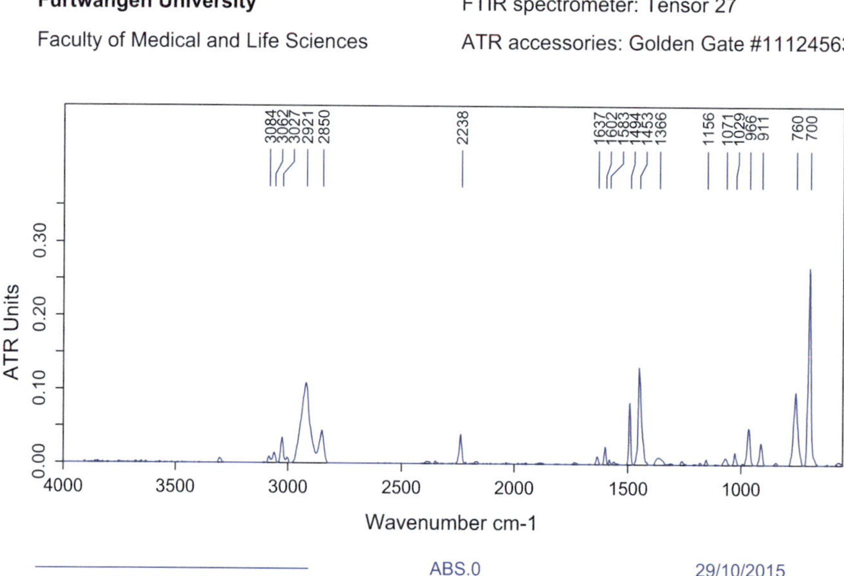

Fig. 2.100 ATR-IR spectrum of acrylonitrile butadiene styrene (ABS granules)

polymers is due to the additional stretching vibration adsorption of the only olefinic double bond in the rubber phase in the polybutadiene at 1637 cm^{-1} (Table 2.25).

Table 2.25 Characteristic IR vibration frequencies ($\tilde{v}$) of Polyacrylonitrile Styrene (ABS)

$\tilde{v}$	Bond
3084	$v(C-H)_{arom}$
3062	$v(C-H)_{arom}$
3027	$v(C-H)_{arom}$
2921	$v(C-H)$
2850	$v(C-H)$
2238	$v(C\equiv N)$
1637	$v(C=C)$
1602	$v(C=C)_{arom}$
1583	$v(C=C)_{arom}$
1494	$\delta_s(CH_2)$
1453	$\delta_{as}(CH_2)$
1366	$\delta_s(CH_3)$
1029	$v_s(C-C)$
966	$v_{as}(C-C)$
911	$\delta(C-H)$
(842)	$\gamma(C-H)_{arom}$
760	$\delta(CH_2)_{rocking}$
700	$\delta(CH_2)_{rocking}$

2.4.12 Polyoxymethylene (POM), Polyformaldehyde, Polyacetal

Due to its mechanical properties, Polyoxymethylene (POM) (Fig. 2.101) is very dimensionally stable. Moreover, this preferred construction material exhibits excellent sliding and wear behavior, making it a popular material for precision parts in precision engineering.

The Polyacetal is formed by radical or anionic Polymerization of the basic building block formaldehyde (H_2CO). The formed polymer chain is largely linear and, unlike Polyethylene, is not only made up of carbon atoms, but also of oxygen heteroatoms. Therefore, it is also referred to as a "heterochain". The Heteroatoms

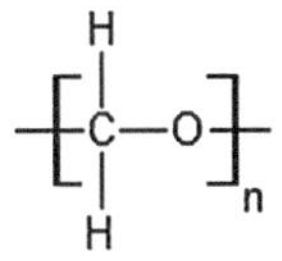

Fig. 2.101 Polyoxymethylene

have an influence on the Polarity of the chain. Due to the additional polar inter-actions between the linear chains, the Glass transition temperature or the melting point increases compared to the isochains (Briehl 2008). The predominantly linear chains in POM result in a high Degree of crystallinity, which is responsible for the chemical Resistance. Due to the enhanced Interaction and the associated small distances between the linear chains, POM achieves a high density of 1.41–1.43 g/cm^3.

To prevent a Depolymerization of the POM via the reactive hemiacetal end groups, these are either etherified or esterified.

The most characteristic feature in the infrared spectrum (Fig. 2.102), which distinguishes POM from other aliphatic plastics, results from the hetero chain. The C–H stretching vibrations of the methylene group between two oxygen atoms in the polyacetal are at 2791 cm^{-1} and are typical for this functional group (Hesse et al. 1987) in the polymer. Another characteristic of the acetal is the increase in the wave number for a C–O stretching vibration, which can be found in ethers (R–O–R') between the wave numbers 1070 and 1150 cm^{-1}. Due to an additional oxygen atom at the carbon atom O–C–O in the acetal, the energy of this vibration increases to 1235 cm^{-1} (Table 2.26).

Not only PVC, but also other plastics are added with plasticizers, as here with another POM material (Fig. 2.103). The typical phthalate peaks occur at 1732, 1597, 1532 and as a weak shoulder peak at 743 cm^{-1}.

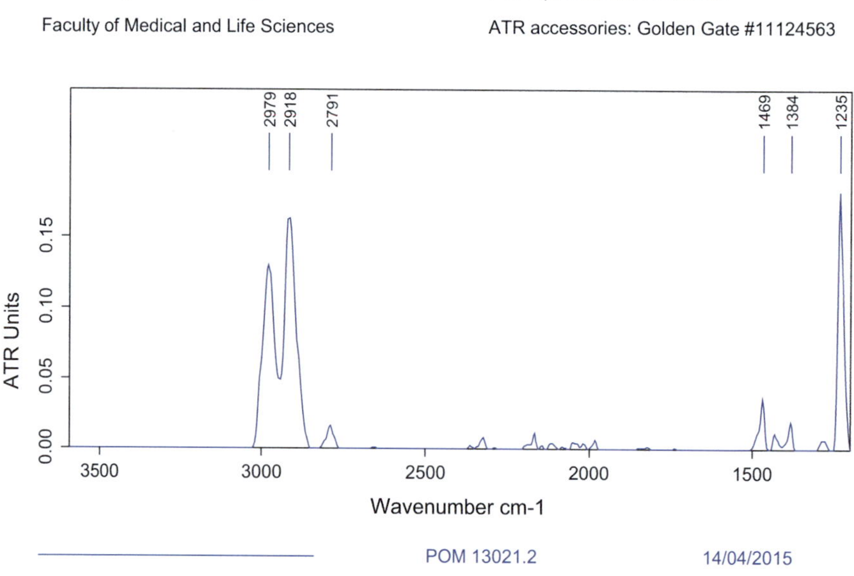

Fig. 2.102 ATR-IR spectrum of polyoxymethylene (granules)

Table 2.26 Characteristic IR vibration frequencies ($\tilde{\nu}$) of polyoxymethylene (POM)

$\tilde{\nu}$	Bond
2979	$\nu(C-H)$
2918	$\nu(C-H)$
2791	$\nu(C-H)_{acetal}$
1469	$\delta_s(CH_2)$
1384	$\delta_s(CH_3)$
1235	$\nu(C-O)$

Furtwangen University

Faculty of Medical and Life Sciences

FTIR spectrometer: Tensor 27

ATR accessories:

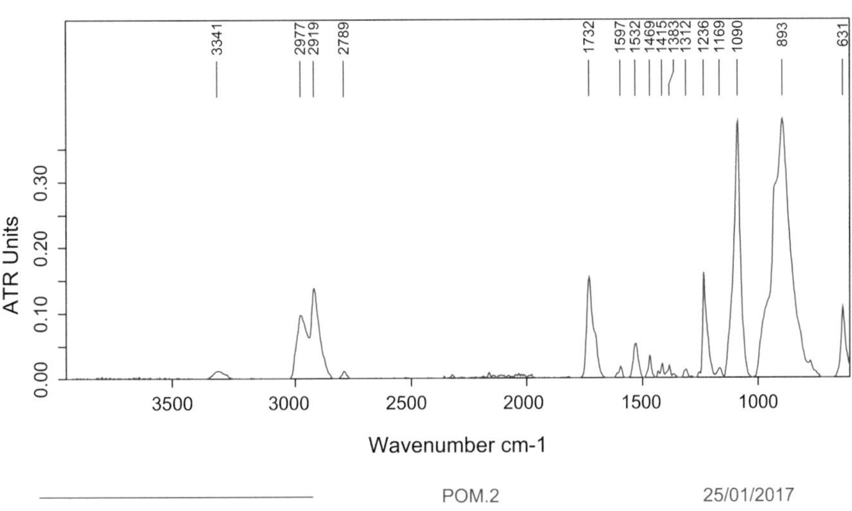

Fig. 2.103 ATR-IR spectrum of polyoxymethylene with plasticizer content (granules)

Fig. 2.104 Structural features of the stereoisomeric polylactic acids

2.4.13 Polylactic Acid (PLA; *poly lactic acid*)

Polylactic acid (PLA; Fig. 2.104) is a Biopolymer and, like the first bioplastic based on Cellulose, celluloid, belongs to the bioplastics. It is based on a renewable, i.e., regrowing Raw material, as can be inferred from the name, the Lactic acid. Everything that nature produces, it can also biologically degrade. The biological Degradability is, alongside the natural raw material source, a criterion that speaks for a bioplastic.

PLA is a Polyester, which is formed by a Polycondensation. To form an ester, a carboxylic acid group and an alcohol group are needed. Both are present in the lactic acid molecule. The resulting water is removed from the reaction equilibrium by azeotropic distillation.

Lactic acid or 2-hydroxypropionic acid is an intermediate product of energy metabolism and is produced during the microbial degradation of carbohydrates (sour milk). Lactic acid has a Chirality center and is therefore optically active. In an enzyme-catalyzed degradation, either the pure R- or S-enantiomers (Fig. 2.92) are formed (Metabolism) or the racemic mixture (microbial degradation).

The fact that Bacteria are capable of generating Lactic acid from sugar or starch is used for the biotechnological production of lactic acid. In this process, lactic acid bacteria ferment the substrate. The microbial Fermentation is the most important process for the production of over 250,000 tons of lactic acid annually (worldwide).

PLA is primarily used for food packaging or in agriculture as a cover film due to its Biocompatibility and in products that do not need to have a long Shelf life, such as implants (screws, threads), drinking straws or plastic cutlery. Besides ABS, PLA is the most commonly used plastic for 3-D printers.

The aliphatic Polyester is classified as a biodegradable plastic due to its molecular structure, but in reality, only industrial composting facilities achieve the necessary environmental conditions to initiate a slow decay with heat.

Whether the declared bioplastic PLA as microplastic particles and thus as a potential vehicle for Pollutants has no negative impact on living beings like the base material is still to be investigated.

Figure 2.105 shows the IR spectrum of polylactic acid. In the compilation of the main peaks in Table 2.27, the absorption bands listed in brackets stand out, as they cannot be assigned to the functional groups present in the polylactic acid molecule. Since Additives are often used in a polymer to adjust the necessary properties such as color, Ductility (cf. Plasticizer in PVC), Flame retardancy, chemical Resistance (antioxidants) or UV resistance, signals of these additives can also be seen, which can partially cover polymer signals. The values listed in brackets are aromatic signals, which do not occur in the pure polymer based on Lactic acid.

The $O-H$ stretching vibration at 3297 cm^{-1} is either due to free non-esterified hydroxyl groups or to bound water in the polymer. Characteristic for the Polyester is of course the $C=O$ stretching vibration (Table 2.27).

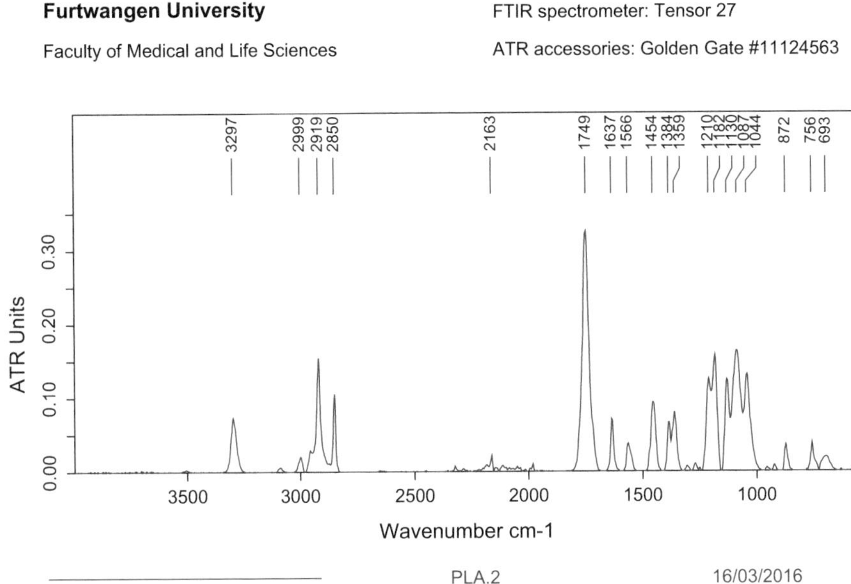

Fig. 2.105 ATR-IR spectrum of polylactic acid (PLA granulate)

Table 2.27 Characteristic IR vibration frequencies ($\tilde{v}$) of Polylactic Acid (PLA)

$\tilde{v}$	Bond
3297	$v(O{-}H)$
(2999)	$v(C{-}H)_{arom}$
2919	$v(C{-}H)$
2850	$v(C{-}H)$
1749	$v(C{=}O)$
(1637)	$v(C{=}C)$
(1566)	$v(C{=}C)$
1384	$\delta_s(CH_3)$
1359	$\delta_s(CH_3)$
1210	$v(C{-}OR)$
1130	$v(C{-}OH)$
1087	$\delta_s(C{-}C)$
1044	$v(C{-}C)$
(872)	$\gamma(C{-}H)_{Benzene}$
756	$\delta(CH_2)_{rocking}$
693	$\delta(CH_2)_{rocking}$

IR Spectra of Plastic Additives

Some examples of plastic additives, such as UV stabilizers, Flame retardants, Color pigments etc. were already introduced in Sect. 2.5. Since some of these additives are not harmless to the plastic processor, the end customer, and the water and the organisms living in it, it is necessary to carry out an identification and, if limit values are available, a quantification due to lack or insufficient data from the compounders about the ingredients of the plastic. Plastic granulate manufacturers, so-called compounders, consider the recipe of their product as a trade secret and sometimes do not disclose all data about its ingredients. A clear unmistakable identification of all ingredients via the CAS number would be desirable and is unfortunately not consistently practiced by all suppliers of the plastic processors.

In order to ensure sufficient protection of humans from harmful chemicals, the European pollutant regulation REACH was initiated in Stockholm. Within the project REACh-Radar-Verbund—Systematic identification and prioritization of substances of very high concern—in the corporate network in the electroplating of the Öko-Institut e. V. in cooperation with the Furtwangen University (HFU) and some small and medium-sized enterprises, the Faculty of Medical and Life Sciences (MLS) of the HFU at the Villingen-Schwenningen campus took over the project component DETECT, in which the use and operating materials of small and medium-sized enterprises (SMEs) are examined for substances of very high concern (engl. SVHC—*substances of very high concern*). These substances have so-called PBT properties due to their physical and chemical nature, i.e., they are persistent (hard to degrade), bioaccumulative (accumulate in organisms), and toxic. Due to these properties, the use of certain SVHCs in Europe is restricted or prohibited by the European chemical regulation REACh (engl. ***Registration, Evaluation, Authorisation and Restriction of Chemicals***); therefore, suitable substitute substances must be found that have less harmful properties. The corresponding SVHCs are listed in Annex XIV of the REACH Regulation (REACH Candidate List (Table 2.11).

The task of identifying and, if necessary, quantifying SVHC substances in plastic products using a suitable sample preparation and a fast and cost-effective method was initially limited to the so-called plasticizers based on phthalate. A look at the candidate list and also the SIN list *(Substitute It Now)* shows that some representatives of this substance class are already listed. A clear signal to obtain certainty about their use in products by means of a rapid test on the one hand, and to check an existing Reach conformity declaration from a supplier from abroad or also domestically on the other hand.

Plasticizers (Phthalates)

Plasticizers are substances that make plastics flexible; they are therefore added to them as additives. They lodge themselves between the polymer chains of the plastics and ensure their mobility. Because the plasticizers do not form a covalent electron pair bond with the mainly linear polymer chains, they can be dissolved out by a suitable solvent (e.g., mineral water in PET bottles) and thus enter the human body (Zhang et al. 2017). A common group of plasticizers are the so-called phthalates, which have reproductive toxic and endocrine effects on organisms (BfR 2010). Four of the frequently used phthalates are on the SVHC candidate list of Annex XIV of the REACH Regulation: These are benzyl butyl phthalate (BBP), bis(2-ethylhexyl)phthalate (DEHP), dibutyl phthalate (DBP), and diisobutyl phthalate (DIBP).

To identify these four plasticizers, which may be present in the submitted plastic granules, the IR spectrum of three of the four SVHC candidates was recorded as pure substance (Fig. 2.93).

Due to the multitude of different phthalic acid esters as plasticizers, it is sometimes very difficult to distinguish a dibutyl phthalate from a dihexyl phthalate, especially since the absorption bands of the matrix polymer can overlap those of the plasticizer. To be sure which plasticizer is actually present in the plastic, this can be determined by a GC/MS analysis, provided the reference substances are available.

The infrared spectra of the three different phthalic acid esters shown in Fig. 2.106 show both the typical signals for the substance class, such as the absorption peak for the carbonyl vibration of the ester group at 1725 cm^{-1}, but also characteristic differences. The additional phenyl group in benzyl butyl phthalate (blue spectrum) can be recognized by the aromatic C−H stretching vibrations in the range 3000–3100 cm[®1] and by the additional *out-of-plane* vibration in the range 700 cm^{-1}. DEHP can be distinguished from DBP only by the number of C−H stretching vibrations. Due to the branching and longer chain of the alkyl group, several vibration absorption signals can be seen in the range 2800–3000 cm^{-1}.

A comparison of the plasticizer signals (Fig. 2.107) with those of a plastic granulate quickly shows that none of the polyethylene examined contains any of the plasticizers stored in the established spectral database (Fig. 2.108).

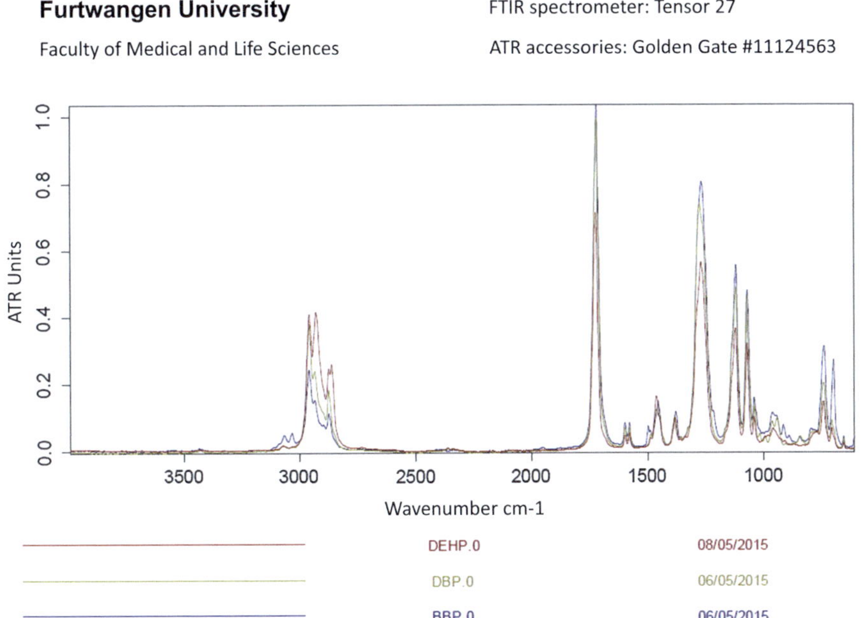

Fig. 2.106   Infrared spectrum of the plasticizers DEHP, DBP, and BBP

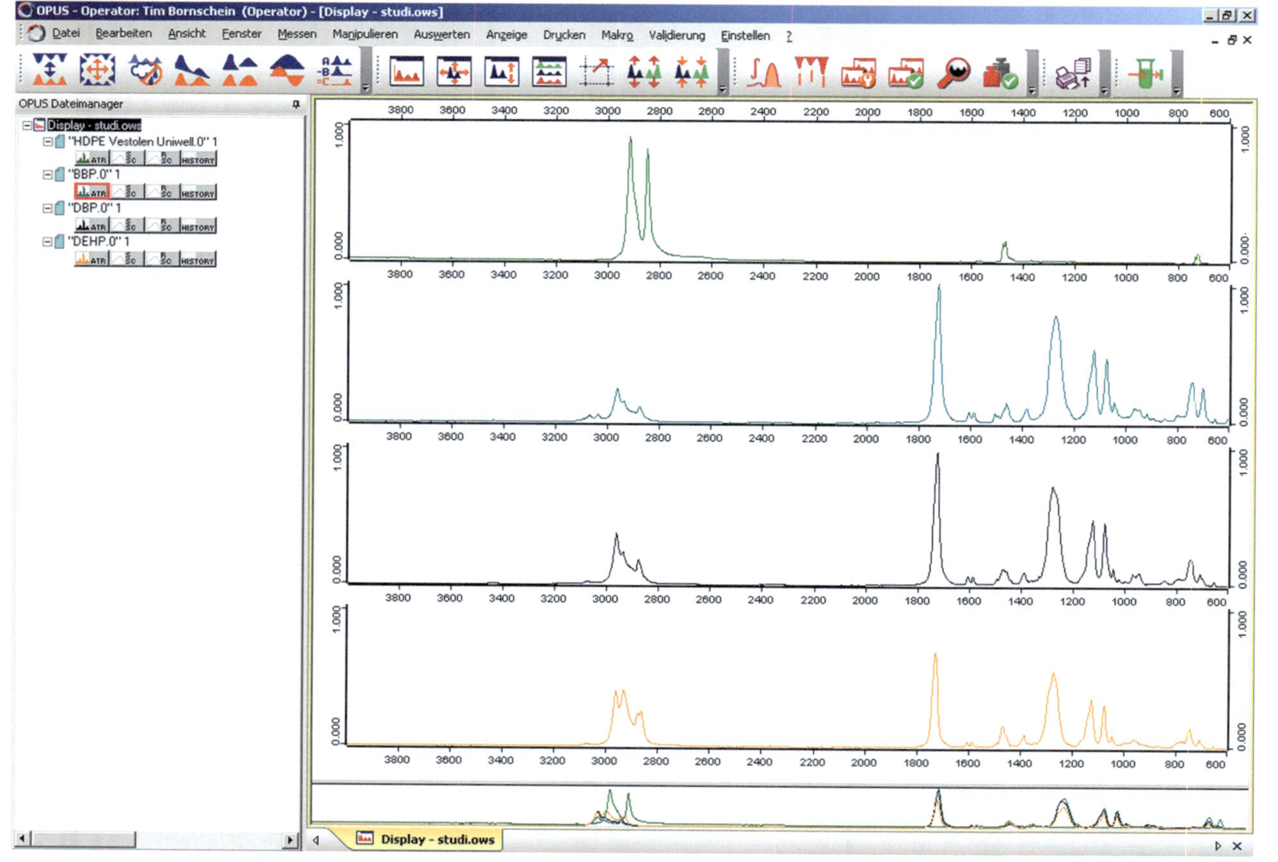

Fig. 2.107 Screenshot; check for plasticizer content in an HDPE material

Fig. 2.108 PVC granule grain Golden Gate

DINP Quantification in PVC Using FTIR Analytics

A quick method for identifying plastic additives is very good and can be practiced without a lot of time using ATR-IR spectroscopy, provided the comparison spectra of the additives are available. But also a quantification of the additives is possible with the infrared spectroscopy, provided a calibration can be created.

A quick and easy method to test the material requirements and, for example, to analyze the DINPplasticizer is provided by ATR-FTIR spectroscopy. For a granulate test in goods receipt, only a few granules from the batch need to be taken as samples. The actual measurement using ATR-FTIR is easy to perform. The individual granule only needs to be pressed onto the measurement crystal with a certain pressure (Fig. 2.95). Elaborate sample preparation is not necessary.

Quantification of diisononyl phthalate via the peak intensity or the peak area of the characteristic carbonyl stretching vibration in the phthalate ester is not reproducible and very inaccurate, as both peak height and peak area depend on the penetration depth of the infrared radiation into the sample. The sample shape, the surface structure of the sample, and the pressing pressure influence the scanned sample volume. As a result, the result can vary with each measurement. For a more accurate determination of the plasticizer content, the area ratios between selected characteristic peaks of the PVC polymer matrix are used quasi as an internal standard. The area ratios are independent of the peak height, which allows the above-mentioned influences to be eliminated and good reproducible results to be obtained.

For the determination of the DINP content in soft PVC, a typical absorption peak of the PVC, which can be clearly distinguished from the signals of the plasticizer, is integrated and used for the calculation of the plasticizer content.

The areas A_x (x = specific vibrational excitation of a functional group in the polymer molecule) of the two selected characteristic peaks are then set in relation using the mathematical relationship (Eq. 2.49) to obtain a peak area ratio (phthalate index). The integral of the carbonyl peak at 1722 cm^{-1} (ester-specific; plasticizer) for the C=O stretching vibration provides the area A_{CO} and the integral of the absorption peak for the C–Cl stretching vibration in the PVC polymer at 614 cm^{-1}provides the area A_{CCl}.

$$Peak-Fl\ddot{a}chenverh\ddot{a}ltnis = \frac{A_{CO}}{A_{CO} + A_{CCl}} \tag{2.49}$$

For calibration in peak comparison, four different soft PVC granulates with different and known DINP content are measured and evaluated (Fig. 2.109). PVC granulates with a DINP content of 21.9%, 32.5%, 39.1% and 44.6% are available. With the calibration line thus obtained with a correlation factor of more than 99%, the DINP content in PVC can be determined very accurately within the corresponding concentration limits in a few minutes.

The same "dry" ATR-IR spectroscopic analysis principle, which does not require a high consumption of solvents, has already been successfully used for the determination of the polycarbonate content in ABS blends (Neek et al. 2017) and for the determination of the polybutadiene content in ABS (Fedhahn 2008).

If there is no concentration information from the manufacturer or the plastic processor regarding their ingredients, which is necessary for the creation of a calibration curve, the ingredients must first be made accessible via other analytical methods.

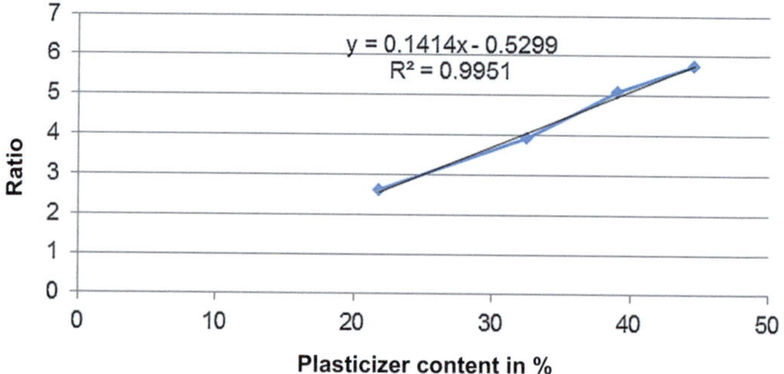

Fig. 2.109 Graphical representation of the relationship between the two peak heights at 1722 cm^{-1} and at 614 cm^{-1} and the actual plasticizer content

A suitable method for determining the plasticizer content in PVC is a gravimetric determination, provided the plasticizer, in this case DINP, is the only PVC additive in the percentage range.

First, the plastic granulate is completely dissolved in a nonpolar solvent. THF (tetrahydrofuran) is suitable for PVC. Subsequently, by successively adding a second, more polar solvent, the polymer is precipitated in such a way that the plasticizer remains in solution. The precipitation of the polymer is possible by adding water. However, the solubility of the phthalates in water is also very low. To avoid co-precipitation of the plasticizers, cyclohexane can be used as a suitable precipitant for PVC and a good solvent for DINP. In Fig. 2.110, the conventional soft PVC granulate can be seen on the left (blue) and the PVC precipitated with cyclohexane on the right (white).

An IR examination of the PVC precipitate shows by the absence of the carbonyl absorption band that no DINP is contained in the filtered PVC and that the plasticizer has completely remained in the cyclohexane solution (Fig. 2.111) (Harsch and Kirschner 2014).

After drying to constant weight of the precipitated PVC powder, a weight loss of 45.24% is obtained by differential weighing to the initial weight, which corresponds very well to the manufacturer's specification of 44.6% with a deviation of 2.5%. This deviation can be explained by systematic errors (residues in the precipitation vessel, etc.) or other additives such as pigments, which have remained in the cyclohexane solution (see color difference in Fig. 2.111).

With the help of the PVC granulate with a known DINP concentration, the quality of the developed method for quantifying the plasticizer content can be tested. The result of the method presented here has shown that the method is very well suited in this particular case.

For concentration determinations <1% share in plastic, more sensitive analysis methods must be used. To comply with REACH, a mass percentage share in the product of less than 0.1% is necessary. To analyze in this concentration range, the liquid phase can be separated by gas chromatography after the polymer has

Fig. 2.110 PVC with (left) and without (right) DINP

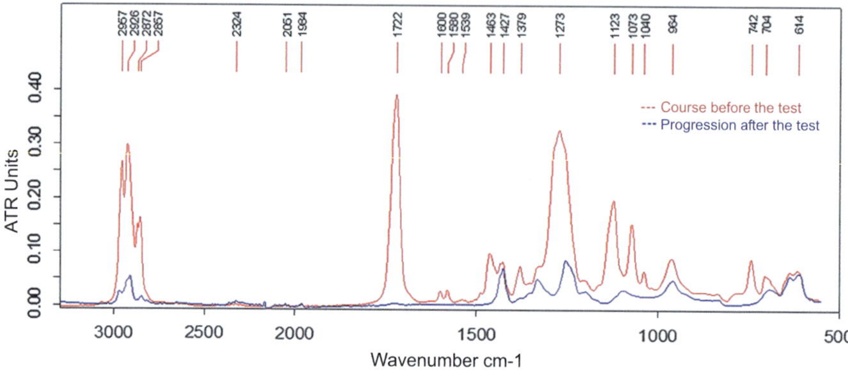

Fig. 2.111   Infrared spectrum of the PVC granulate after precipitation with cyclohexane from a THF solution of soft PVC (blue). The red spectrum shows the PVC sample with a DINP content of 44.6%

precipitated and analyzed by mass spectrometry. With infrared spectroscopy, lower concentrations of the plasticizer DINPin cyclohexaneor carbon tetrachloride can also be detected in the liquid cell (Fig. 2.48). In the transmission measurement through the liquid cell, the integral of a characteristic peak can be used alone. Since the path length within which absorption occurs is considered constant with the cuvette length, the concentration of the analyte is proportional to the absorption according to the Lambert-Beer's law. For the determination of the plasticizer content via the carbonyl band of the ester, a quartz cuvette cannot be used, as quartz already shows strong self-absorption from 2500 cm^{-1}. Instead, cell windows made of CaF_2, NaCl, KBr, CsBr or CsI can be used in the modular liquid cell, which do not show any absorption at 1700 cm$^{®1}$ (Wilhelm 2008). Neither the solvent cyclohexane nor carbon tetrachloride show overlapping bands in the analytically relevant wavenumber range (see IR spectrum of cyclohexane in Fig. 2.112). If no solvent is found that completely dissolves the plastic, possibly still with a glass fiber content, plasticizers or other ingredients such as polybrominated diphenyl ethers or amide wax can be extracted using a Soxhlet extraction. Before the plastic granulate is filled into the extraction sleeve, it is recommended to crush the granulate into powder in a cryogenic mill to increase the surface area. This can speed up the extraction process and the additives can be completely extracted from the polymer matrix. For organic plastic additives that need to be quantified via their C−H valence vibrations, CCl_4 (carbon tetrachloride) is suitable as a solvent. The disadvantage is that they are classified as toxic and carcinogenic.

Table 2.28 provides an overview of a selection of other types of plastic and their brand names by which they can be found.

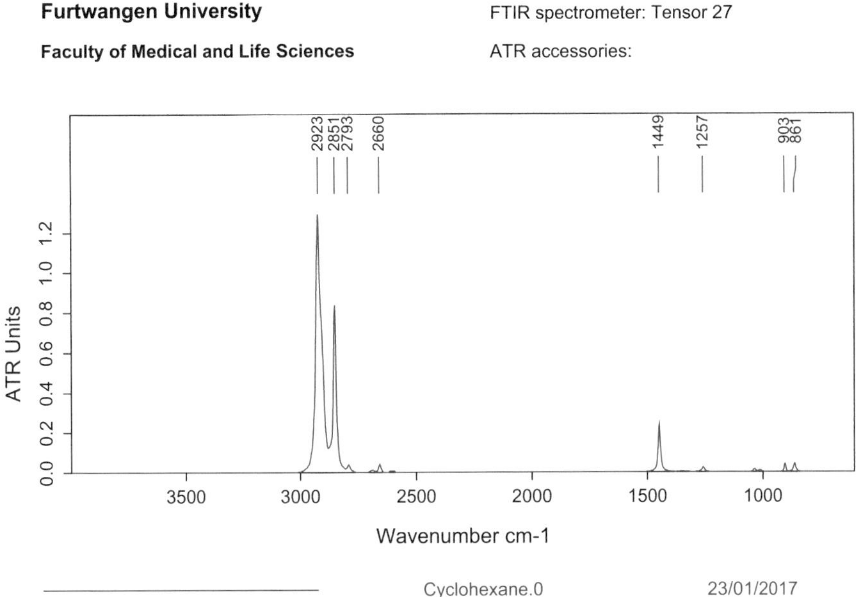

Fig. 2.112 Infrared spectrum of cyclohexane with C−H stretching and deformation vibrations

2.5   Plastic Ingredients (Additives): Properties and Uses

Plastics are typically not pure polymers, but they contain additives, which are added to alter or expand the mechanical, optical, tribological, and other properties of the polymers. This makes the range of applications of the artificial, petroleum-based material so large that the annual production of plastics has steadily increased to 300 million tons per year in recent years. The additives used often pose a potential hazard to humans and the environment. Together with microplastics, these substances enter the environment and also the gastrointestinal tract of various aquatic organisms. Since primary microplastics, which mainly come from cosmetic products, contain hardly any additives, the analysis of the additives is a method by which primary can be distinguished from secondary microplastics. This provides the opportunity to investigate the causes of microplastic contamination in waters. This chapter presents a selection of intrinsic pollutants that are added to the plastic during the manufacturing process. In contrast to those extrinsic pollutants that accumulate on the plastic surfaces during the exposure of microplastics in the water (adsorption) or are incorporated into it (absorption).

Table 2.28 Well-known brand names of some types of plastic

Abbreviation	IUPAC Name	Common/Brand Name
PA	Polyamide	Nylon, Perlon, Zytel®, Durethan®, Dinalon®
PE-HD	Polyethylene, high density	Hostalen
ABS	Acrylonitrile Butadiene Styrene	
PBT	Polybutylene Terephthalate	Crastin®
PC	Polycarbonate	Lexan, Makrolon
PE	Polyethylene	Hostalen, Lupolen, Vestolen
PE-LD	Polyethylene, low density	Hostalen
POM	Polyoxymethylene (Polyacetal)	Delrin®, Hostaform
PP	Polypropylene	Vestolen, Hostalen PP
PS	Polystyrene	Polystyrene
PVC	Polyvinyl chloride	Vestolit, Vinnolit
EP	Epoxy resin	
ETFE	Ethylene tetrafluoroethylene	
PAN	Polyacrylonitrile	
PE-MD	Polyethylene, medium density	
PMMA	Polymethyl methacrylate	Plexiglas, Acrylic glass, Paraglas
PPE	Polyphenylene ether	
PTFE	Polytetrafluoroethylene	Teflon®, Turcon, Gore-Tex®
PUR	Polyurethane	
PVDF	Polyvinylidene fluoride	Solef
SI	Silicones	
TPE	Thermoplastic elastomers	Riteflex, Hytrel®
VMQ	Silicone-Rubber	
CA	Cellulose acetate	
EPDM	Ethylene Propylene Diene Rubber	Buna
FKM, FPM	Fluoropolymer Rubber	Viton
LCP	Liquid Crystal Polymer	Vectra
NBR	Acrylonitrile Butadiene Rubber	
PAI	Polyamide Imide	Torlon, Tecator
PEEK	Polyether Ether Ketone	
PEI	Polyetherimide	Ultem
PE-LLD	Polyethylene, Linear, Low Density	Hostalen
PESU	Polyethersulfone	Ultrason
PET	Polyethylene Terephthalate	Impet, Rynite®

(continued)

Table 2.28 (continued)

Abbreviation	IUPAC Name	Common/Brand Name
PFA	Perfluoroalkoxy Alkane	
PFPE	Perfluoropolyether	Krytox, Fomblin, Galden, Solvera
PI	Polyimide	Vespel®, Kapton®
PP-C	Polypropylene, Copolymer	
PPS	Polyphenylene Sulfide	Fortron, Ryton
PS-E	Polystyrene, Expanded	Styrofoam
PSU	Polysulfone	
PVF	Polyvinyl Fluoride	
SBR	Styrene Butadiene Rubber	
UP	Unsaturated Polyester Resins	
PPO		Noryl
APE	Aromatic Polyester	
ASA	Acrylate Styrene Acrylonitrile	
BR	Butadiene Rubber	
CN	Cellulose Nitrate	Celluloid
COC	Cyclo-Olefin Copolymers	
CR	Chloroprene Rubber	Neoprene
CSM	Chlorosulfonated Polyethylene	
ECB	Ethylene Copolymer-Bitumen	Lucobit
EPM	Ethylene-Propylene Copolymer (Rubber)	
EVA, EVM	Ethylene Vinyl Acetate	
FEP	Perfluoro(Ethylene-Propylene-) Plastic	
FFKM, FFPM	Perfluorinated Rubber	Kalrez, ISOLAST®, Chemraz
FVMQ	Fluorosilicone-Rubber	
IIR	Butyl Rubber	
IR	Isoprene Rubber	
MF	Melamine-Formaldehyde Resin	
MPF	Melamine-Phenol-Formaldehyde Resin	
NR	Natural Rubber	
PAEK	Polyaryletherketone	
PCT	Polycyclohexylene Dimethylene Terephthalate	Eastar, Thermx®
PCTFE	Polychlorotrifluoroethylene	
PEC	Chlorinated Polyethylene	

(continued)

Table 2.28   (continued)

Abbreviation	IUPAC Name	Common/Brand Name
PE-HMW	Polyethylene, high molecular weight	
PEK	Polyetherketone	
PEN	Polyethylene naphthalate	
PE-UHMW	Polyethylene, ultra high molecular weight	
PF	Phenol-Formaldehyde Resin	Bakelite
PIB	Polyisobutene	
PMI	Polymethacrylimide	Rohacell
PMP	Polymethylpentene	TPX
PP-H	Polypropylene; Homopolymer	
PPSU	Polyphenylsulfone	
PPY	Polypyrrole	
PVAC	Polyvinyl acetate	Vinnapas
PVAL	Polyvinyl alcohol	
PVDC	Polyvinylidene chloride	
SAN	Styrene-Acrylonitrile	
UF	Urea-Formaldehyde Resin	

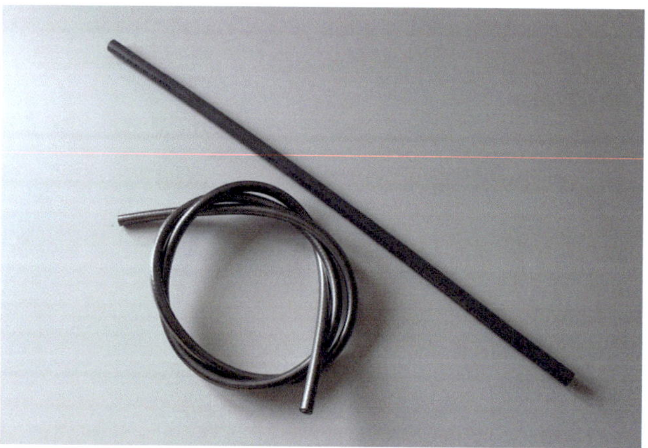

Fig. 2.113   Hard PVC pipe and soft PVC hose

2.5.1   Plasticizers

Plasticizers are substances that make the naturally hard and brittle plastic elastic. Flexible shower hoses or cable insulation, for example, are mainly made of PVC. Pure PVC, without plasticizers, can be used as a rigid and stable inflexible pipe

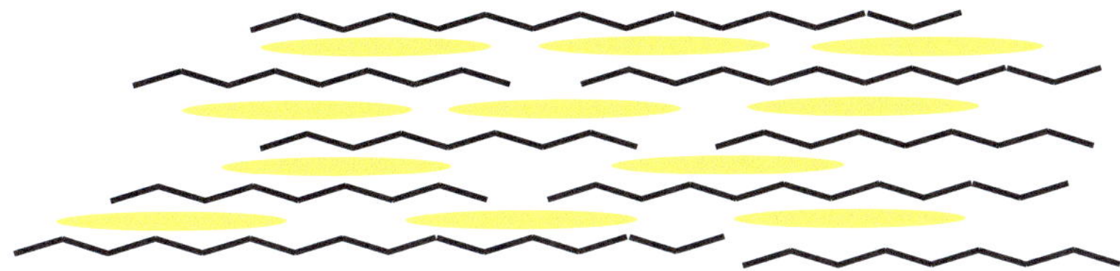

Fig. 2.114 Polymer chains (black) with interposed plasticizer molecules (yellow)

in piping systems, while the addition of plasticizers, also called plasticizing aids, turns the pipe into a flexible bendable hose (Fig. 2.113).

As plasticizers, hydrophobic flat organic molecules are preferably used. These substances are inserted between the bulky polymer chains (Fig. 2.114) and thus ensure, for example, the flexibility of a film or a hose. Along the plasticizer molecules, the polymer chains can slide past each other much more easily, without getting entangled. Since the plasticizers are not chemically bound to the plastic chains, they can naturally leave the polymer matrix and be released into the environment. The release depends on several factors such as steric hindrance, solubility, vapor pressure, temperature, K_{ow} value, and the Henry constants. The release of these plasticizers into the environment, into water, into nutrient solutions, or into the digestive organs in humans and animals represents a further hazard posed by the substances used as plasticizers due to their non-anchoring, which is also posed by plastics and thus by microplastics.

The group of *ortho*-phthalic acid diesters, briefly referred to as "phthalates", still represent the majority of all plasticizers for the various applications listed in Table 2.29. As an example, the structural formula of DINP (Diisononyl phthalate), which is frequently used in flexible hoses, is shown in Fig. 2.115. It is offered by plastic manufacturers as an alternative to DEHP.

Phthalate-containing plasticizers are used in 90% of cases primarily for the production of soft PVC. The plasticizer content can account for up to 50% of the PVC. In PVC, the plasticizer is distributed across the application areas cables (25%), films (22%), floor coverings (14%), hoses (11%), coated fabrics (10%), pastes (9%), others (9%) (Federal Environment Agency 2007).

DEHP as the most well-known representative of the phthalates is still among the most frequently used plasticizers. In total, over one million tons of phthalates are produced in Western Europe (Federal Environment Agency 2007).

Fig. 2.115 Structural formula of DINP

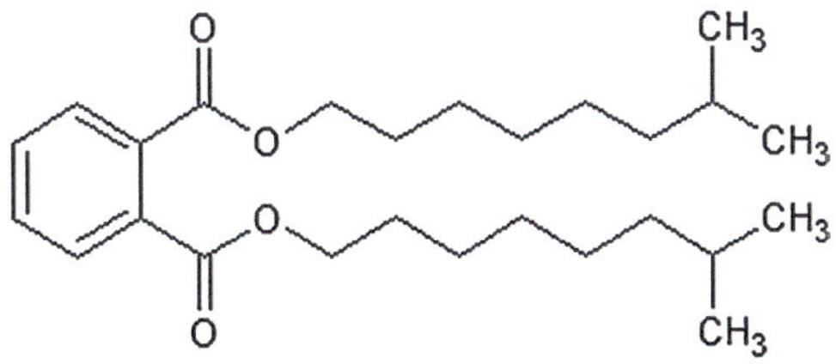

Table 2.29 Use of phthalates as plasticizers. (IPASUM n.d.)

Phthalate	Applications
DMP	Personal care products, perfumes, deodorants, pharmaceutical products
DEP	Personal care products, perfumes, deodorants, pharmaceutical products
BBzP	PVC e.g. transformers, floor coverings, pipes and cables, carpets, wall coverings, sealants, food packaging, artificial leather, food conveyor belts
DBP	PVC, cellulose plastics, dispersions, paints/varnishes (including nail polishes), adhesives (mainly polyvinyl acetates), foam inhibitors and wetting agents in the textile industry, personal care products, perfumes, deodorants, pharmaceutical products (*time-release* drugs), (food-)packaging
DEHP	PVC e.g. floor coverings, pipes and cables, carpets, wall coverings, shoe soles, vinyl gloves, automotive components, dispersions, paints/varnishes, emulsifiers, (food-)packaging
DnOP	PVC products (like DEHP)
DiNP	PVC e.g. floor coverings, pipes and cables, carpets, wall coverings, shoe soles, automotive components, dispersions, paints/varnishes, emulsifiers, (food-) packaging
DiDP	PVC e.g. floor coverings, pipes and cables, carpets, wall coverings, dispersions, paints/varnishes, emulsifiers, (food-)packaging

DMP = Dimethyl phthalate, DEP = Diethyl phthalate, BBzP = Butyl benzyl phthalate, DBP = Dibutyl phthalate (Di-*n*-butyl phthalate and Di-*iso*-butyl phthalate), DEHP = Di(2-ethylhexyl) phthalate, DnOP: Di-*n*-octyl phthalate, DiNP = Di-*iso*-nonyl phthalate, DiDP = Di-*iso*-decyl phthalate

The phthalic acid esters used are mostly compounds with identical alkyl groups (exception: BBP or BBzP. In di-(*n*-butyl)phthalate (DBP) the alkyl chains (*n*-butyl residues) are unbranched, while the octyl residues of di-(2-ethylhexyl)phthalate (DEHP) are branched at the C2 carbon. Since all four substituents at the C2 carbon are different, this results in the production of several stereoisomers, which do not necessarily have to be expensively purified by enantiomer separation for their use. However, different physiological effects of the individual stereoisomers (Fig. 2.116) are conceivable, but not yet investigated. Only the mode of action of the isomer mixture is known (NDR 2010; Koch 2006; van der Meer and Devine 2017; SWR2 2016; Fath 2010; Ivashechkin 2005; Federal Environment Agency 2007 2013; Braun et al. 2001; Thalheim 2016). When comparing the most important physico-chemical properties of DBP and DEHP, the following is found: Both substances have a low melting temperature (–35 °C for DBP and –50 °C for DEHP) and a high boiling point (340 and 385 °C respectively). Both compounds are liquid over a wide temperature range. Their consistency is oily and both compounds are colorless and odorless. Both substances have a low vapor pressure (DBP with about 10^{-2}–10^{-3} Pa; DEHP of about 10^{-5} Pa).

DEHP is therefore less volatile than DBP. The water solubility of the two phthalates is relatively low and is 9.9 mg/L for DBP and 2.49 µg/L for DEHP (Skrzypek 2003). The water solubility of the phthalates is very important for their

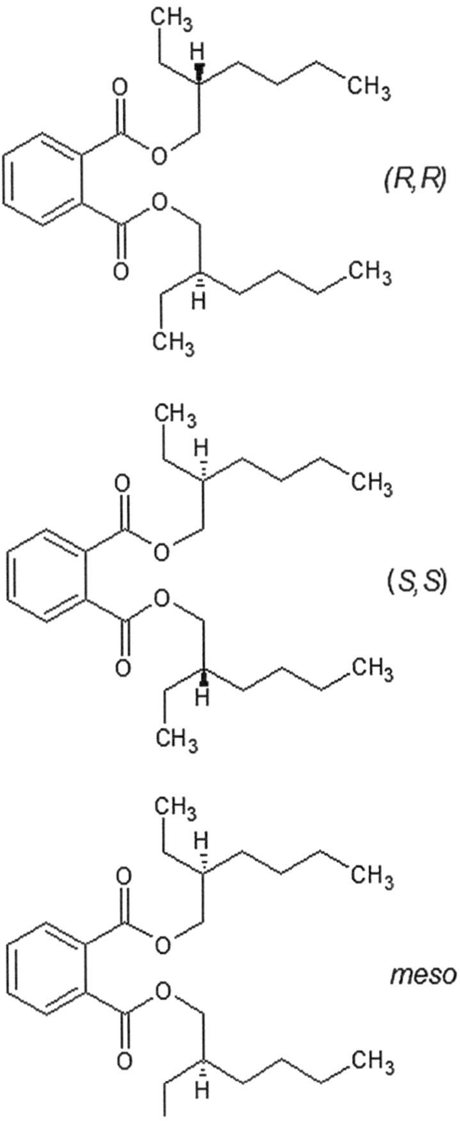

Fig. 2.116 Stereoisomeric structural formulas of DEHP

biological degradation, as it takes place in the dissolved state. The octanol/water distribution coefficient is significantly higher at 7.7 than that of DBP (4.3), due to the longer hydrophobic alkyl chains. From K_{OC} values, for example, the affinity to the organic material of the sediment and thus the sorption behavior can be explained. DEHP has stronger interactions with the organic components of the sediment than DBP due to its longer, branched alkyl chains. The adsorption behavior of phthalates is thus inversely related to their water solubility. Since the carbon content in plastics is generally higher than that of sediments with high mineral components such as lime or silicate, it is therefore to be expected that DEHP adsorbs even more strongly to microplastics.

It has been shown that in a heterogeneous mixture of microplastics and water, the concentration of DEHP on the polyethylene microplastic particles is 100,000 times greater than in the water 22.

Despite their low vapor pressure, significant amounts of phthalates can be released into the environment by outgassing from plastics, as they are not covalently bound to the plastic matrix.

Exposure and Toxicology

A variety of substances are used as plasticizers in plastics, but primarily substances from the class of phthalates. The lack of chemical bonding of the plasticizers to the polymer matrix causes them to spread throughout our environment. As long as phthalates are present in plastics and gradually "bleed out" from there, humans are constantly exposed to them. They can be found in our food, drinking water, air, and everyday objects. They already escape in the extruder during processing at temperatures of 200 °C and condense in the extraction lines, escape from floors and wallpapers, enter our groundwater via landfill leachate, packaging material preferably releases them to the fatty food packed in it, or they are ingested by toddlers through saliva when they put their toys in their mouths (Selke and Culter 2016). Phthalates enter our bodies through these routes and, as we know today, this can have serious health consequences.

The poor eco-efficiency and especially the toxicological properties of some phthalates (González-Castro et al. 2011; Dutescu 2011; BfR 2005) have led to some representatives of this substance class being found on the REACH-SVHC*(Substances of very high concern)*-candidate list (Table 2.11). For DEHP(Diethylhexylphthalate), DIBP (Diisobutylphthalate), BBP (Benzylbutylphthalate) and DBP (Dibutylphthalate), there has been a ban on use since 2015. These substances, which are subject to authorization and can be found in Annex XIV of the REACH Regulation, can only continue to be used with the appropriate authorization. A further signal from ECHA (European Chemicals Agency in Helsinki) to the industry to generally consider replacing phthalates and to replace them in the long term is provided by the SIN list *(Substitute it now),* on which further phthalates are listed. Sooner or later, once the toxicological studies on these substances are completed and they are most likely also to be classified as substances of very high concern (SVHC), these substances will also be entered into the candidate list. Due to the structural similarity of the phthalates, a similar structure-effect mechanism can be assumed.

Endocrine and teratogenic properties have been proven for some representatives of the phthalates, especially for the most intensively studied DEHP. As EDS (Endocrine Disrupting Substances), they can impair the function of hormones or have hormone-like effects themselves, so that certain plasticizers based on phthalates can, for example, cause infertility in men (NDR 2010; Thalheim 2016). Endocrine disruptors (ED) are substances that can disrupt the biochemical mode of action of hormones and consequently lead to growth and developmental disorders. An influence on reproduction and a susceptibility to specific diseases are possible due to this endocrine property. Not only DEHP and some other phthalates are among them, but also other plastic ingredients such as uncrosslinked Bisphenol A, which is used as a monomer for the production of polycarbonate.

In animal experiments on rodents, it was also found that reproductive ability is impaired. Since the concentrations of phthalates in waters are very low due to poor water solubility and because they also adsorb onto sewage sludge in sewage treatment plants, the impact on aquatic organisms such as fish is not conclusively proven (Cheng et al. 2013). Microplastic contamination (Thalheim 2016) now provides a new source of phthalates in waters, as microplastics, with the additives such as plasticizers they contain, are ingested by fish, mussels, and other aquatic organisms (Stryer 1990; Torre et al. 2016; EFSA 2016; Lart 2018; Collard et al. 2017; Van Cauwenberghe and Janssen 2014; Catarino et al. 2018; Lusher et al. 2017). This significantly increases exposure to phthalates, as studies have shown that microplastics adsorb pollutants in high concentrations (Hüffer and Hofmann 2016; Bakir et al. 2014; Hummel 2017) and then desorb much faster in a simulated gastrointestinal tract environment than in water (Bakir et al. 2014).

Phthalates are referred to as teratogens because they can cause malformations in unborn life (embryo) due to external influences. Both negative health effects resulted in most phthalates being banned in certain toys and baby items as early as 1999 (1999/815/EC). In 2004, the ban was extended to all toy and baby items (2004/781/EC). This was followed by a ban in cosmetic items and cosmetic additives (2004/93/EC). In 2007, the use of DEHP as a plasticizer in packaging for fatty foods was banned. From 2015, DEHP may no longer be used without authorization for the production of consumer products under the EU Chemicals Regulation REACH in the EU, which does not mean that we no longer come into contact with DEHP, as finished products, even food, can still be imported from outside Europe that contain DEHP, as the example of blood bags shows.

The member states of the European Union (EU) have not only classified DEHP, but also the phthalatesDIBP, DBP, and BBP as reproductive hazards. According to 67/548/EEC, the following hazard warnings are therefore used for handling DEHP, DBP, BBP, and DIBP:

- R 60: May impair reproduction
- R 61: May harm the unborn child

Preparations containing more than 0.5% of the mentioned phthalates must be labeled with the letter T (Toxic) and the poison symbol throughout the EU. Attempts are being made to substitute phthalate-containing plasticizers, but so far this has only been partially successful. Alternatives include Hexamoll, DINCH, Pevalen, Diethylhexylterephthalate (strictly speaking also a phthalate as an ester of terephthalic acid), alkyl sulfonic acid ester, acetyl tributyl citrate, acetylated castor oil derivative or epoxidized soybean oil. Nevertheless, phthalates are still present in large quantities in all other applications listed in Table 2.29. Even in medical products such as blood bags, infusion bags, dialysis bags, urine bags, catheters, intubation tubes, gloves, contact lenses and many other PVC-containing medical products, DEHP is still used as a plasticizer (Rosado-Berrios et al. 2011). Although there are now alternative plasticizers in medicine (Lagerberg et al. 2015), DEHP remains the only plasticizer with FDA approval (Choi et al. 2012) in medical devices. Its high toxic effectson reproductive systems (BMG 2005) are offset by the good protection of red blood cells from hemolysis in blood preservation containers. The cells maintain their morphology over a long period of time, osmotic stabilityis ensured and the cells show an improved recovery rate 24 hours after a transfusion, when the blood is preserved in DEHP blood bags (Lagerberg et al. 2015).

Spread in Water and Health Risks

What does the release of plasticizers into surface water look like? This was investigated using the example of a shower hose made of soft PVC. On the one hand, drinking water flows through it in our showers before it leaves the house as greywater in the direction of the sewage treatment plant and becomes surface water. On the other hand, the bathtub hose comes into contact with the foamy bath water. In both cases, plasticizers can be released into the water, which would be a problem especially if the water from the shower head is actually used and drunk as drinking water.

Most shower hoses are made up of several layers of PVC: an inner, middle and outer hose. A nylon fiber mesh around the inner liner improves the mechanical properties such as tensile strength, while the middle hose serves as a carrier material for a metal embossing foil (Fig. 2.117). The outer hose protects this foil and serves, for example, as a reservoir for antibacterial additives, which can also be contained in the inner hose.

The base material PVC is without "plasticizer" a rigid pipe (Fig. 2.113), which would be completely unsuitable for the application of a flexible shower hose for the wet room. As plasticizers, the aforementioned phthalates (esters of phthalic acid) such as the DINP (Diisononyl phthalate) shown in Fig. 2.115 are used.

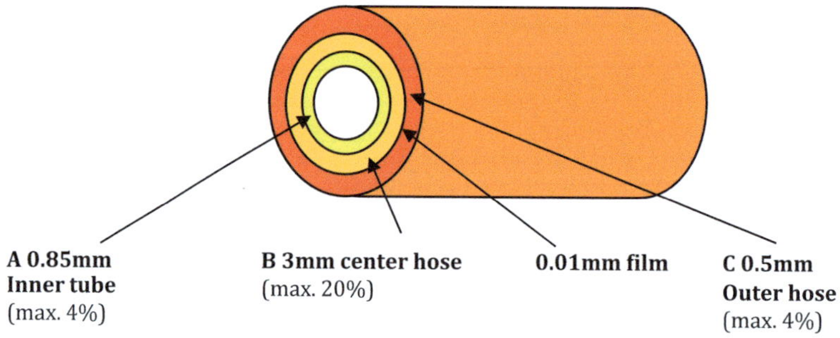

Fig. 2.117 Layer structure of a PVC shower hose

The question of whether and in what quantity the used plasticizers are released into the water is then discussed and examined in detail. In principle, the phthalic acid diesters are to be described as insoluble in water depending on the length of the alkyl chains. This is shown by the clear phase boundary between DINP and water in Fig. 2.118, which also forms again after vigorous shaking. The diester phase is colored yellow for contrast optimization with an organic dye.

Products used in the drinking water sector in Germany must comply with the Drinking Water Ordinance. For plastics, this requires compliance with the standards of two testing procedures. Firstly, the DVGW-270 test, which examines the growth of microorganisms on plastic surfaces, and secondly, the so-called KTW (Plastics in Drinking Water) test, which primarily evaluates the parameters of clarity, color, odor, taste, and the release of organically bound carbon (TOC).

For approval, the recipe of the product to be tested must be provided to the testing center. Table 2.30 lists the recipe of a hose.

The KTW test certificate for the shower hoses made from the granulate listed in Table 2.30 shows that the product complies with the KTW recommendations of the Federal Health Office and therefore does not pose a health risk to humans.

To answer the question about the release of plasticizers, the C(carbon) release = TOC value *(Total Organic Carbon)*indicated in the test certificate can be used over a period of 9 days. The cold water value of 0.4 mg C/m^2d is used for an assessment. Assuming that the TOC value results 100% from the plasticizer used, then a person who drinks the shower hose completely filled in one day would have ingested a plasticizer amount of 0.016 mg. With a body weight of 70 kg, this

Fig. 2.118 Test tube with DINP (yellow phase below) and water (above colorless)

Table 2.30 Composition of a PVC granulate for hose production

Substance	Percentage
PVC	>60
Phthalate Plasticizer (DINP)	<40
Tin-based stabilizer mix	<2
Lubricant (amide wax)	<1
UV stabilizer (based on benzotriazole)	<0.5
Refining pigment (bluing pigment)	<0.5

Table 2.31 NOAEL values, TDI values *(tolerable daily intakes)* and RfD (EPA reference dose) for some phthalates. (taken from CSTEE 1998)

Phthalate	NOAEL mg/kg/day	*Tolerable daily intake* µg/kg/day	RfD (Reference dose) µg/kg/day
DINP	15	150	n. a.
DNOP	37	370	n. a.
DEHP	3.7	37	20
DIDP	25	250	n. a.
BBP	20	200	200
DBP	52	100	100

n. a. = no information, EPA = Environmental Protection Agency, NOAEL = No Observed Adverse Effect Level (toxicological endpoint in toxicity determination)

would be 0.0002 mg/kg (0.2 µg/kg). It is very unlikely that a consumer would do this. But even if so, the intake amount of 0.2 µg/kg is significantly below the values listed in Table 2.31. Thus, a health hazard from the plasticizer DINP in the hose is excluded.

Since a huge amount of greywater comes together from each household, which flows through PVC hoses towards the sewage treatment plant, the pollution of surface water by plasticizers was also investigated to obtain a statement on the ecological hazard.

A study by the State Environmental Agency (Braun et al. 2001) on the "Occurrence of phthalates in surface water and wastewater" has shown that wastewater pipes do not make a significant contribution to the pollution of surface waters, as the phthalates are almost completely retained in the wastewater treatment plants. Phthalates are mainly released through the leachate from landfills and are mainly found in sediments near their release source.

Since the sewage sludge is incinerated due to the PFT problem and is no longer used as fertilizer on fields and meadows, phthalates from wastewater no longer enter the environment in this way.

A new source has emerged through the microplastic contamination of our waters, through which we ingest plasticizers, adsorbed to microplastic particles, either through the consumption of fish or other seafood (Stryer 1990; Torre et al. 2016; EFSA 2016; Lart 2018; Collard et al. 2017; Van Cauwenberghe and Janssen 2014; Catarino et al. 2018; Lusher et al. 2017) or through the consumption of marine salt, which is contaminated with microplastic particles (Karami et al. 2017; Jander 2017). That DEHP not only adheres to sediments (Skrzypek 2003), but also to polyethylene particles in high concentration, has already been proven.

Due to their endocrine (hormone-like) and reproductive damaging effects, phthalates have a very negative image. However, in order to have an effect, they must enter the human bloodstream. As already mentioned, this is excluded within the limits through drinking water.

To avoid the discussion under which circumstances the discredited phthalates are harmful to health and at what concentration they are still released into various aqueous solutions (tap water, bath water, mouth saliva, nutrient solutions), phthalate-free plasticizers are on the rise, especially in products with water contact. To maintain the flexibility of PVC, DINCH (1,2-Cyclohexandicarbonsäureisononylester), a healthily harmless compound, is used. The structure is hardly distinguishable from the phthalate DINP (see Fig. 2.119 with Fig. 2.115), but there is a significant structural feature that determines how strong the migrationfrom the plastic matrix is and whether the compound is potentially genotoxic.

If you rotate the molecules 90° around their longitudinal axis, the difference is more clearly recognizable (Fig. 2.120). The spatial extension perpendicular to the aromatic ring of the phthalate corresponds to about one atomic diameter, i.e., approximately one angstrom or 0.1 nm.

In the chair form of the plasticizer DINCH based on cyclohexane, the spatial extension in the same direction is 4 times higher, at about 0.4 nm.

Phthalates are flat due to the aromatic ring (benzene ring + carbonyl group) and capable of intercalating between the hydrogen-bonded base pairs, which hold the DNA strands together in a double helix, due to their hydrophobicity. For DNA replication (duplication), an enzyme (DNA helicase) opens the base pairs and a second enzyme (DNA polymerase) copies the single strands. The disruption during the duplication of DNA, which is important for cell division to pass on the genetic information, can be well imagined by the insertion of foreign compounds into the double helix (Chi et al. 2016).

The plasticizer DINCH does not contain a flat aromatic structural component, but instead a bulky cyclohexane framework (chair conformation; Fig. 2.108

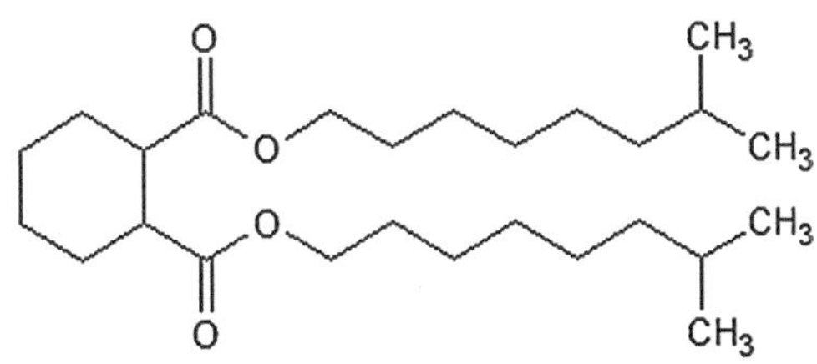

Fig. 2.119 Structural formula of DINCH

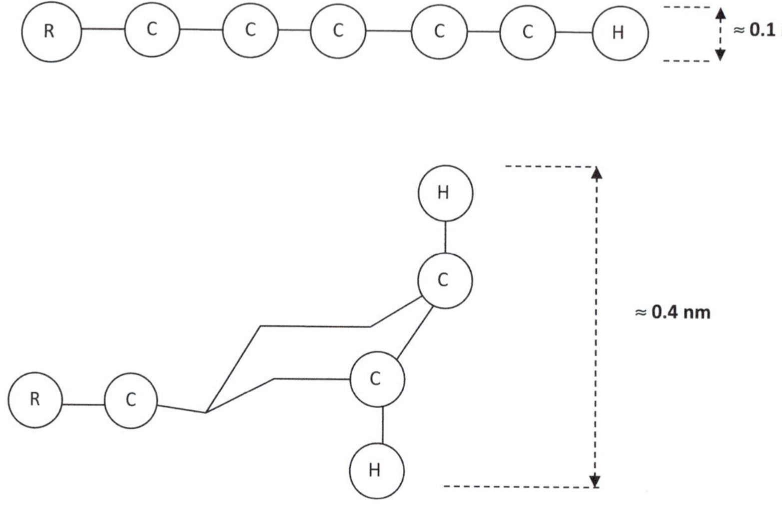

Fig. 2.120 Spatial extension of the dicarboxylic acid portion of the plasticizers DINP (above) and DINCH (below)

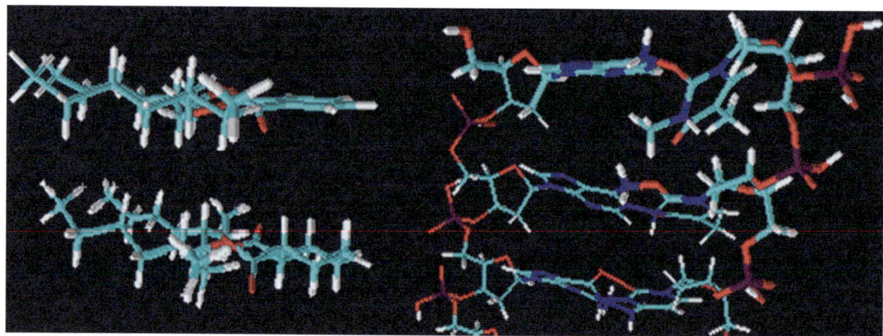

Fig. 2.121 Left: Plasticizer molecules DINP (top) and DINCH (bottom). Right: Section of the DNA

below), which is not capable of inserting itself between the DNA strands. The insertion into the DNA double helix is illustrated in Fig. 2.121. The distances of the base pairings from each other within the double helix vary depending on the helix type, at 0.23 (A-type), 0.34 (B-type) or 0.38 nm (Z-type) (Stryer 1990).

The time-dependent release of the phthalate-containing plasticizer DINP and the phthalate-free plasticizer DINCH in shower hoses, filled with pure water, was investigated at the Fraunhofer IPA Institute by Dipl.-Biol. Markus Keller. Fig. 2.122 shows the temporal course of the plasticizer release.

In both shower hoses, equipped with DINP and DINCH as plasticizers, no temporal migration of the plasticizer into the pure water in the hose could be detected.

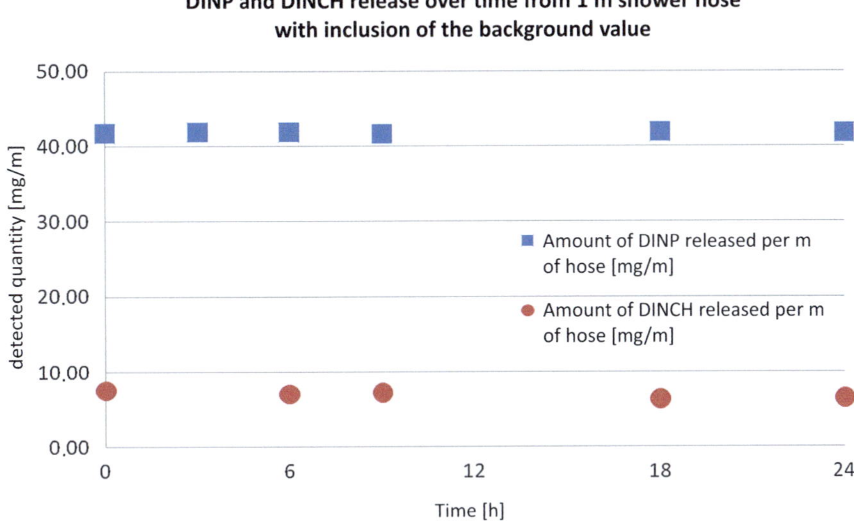

Fig. 2.122 Temporal DINP and DINCH release from the shower hoses after incubation with pure water with the background load

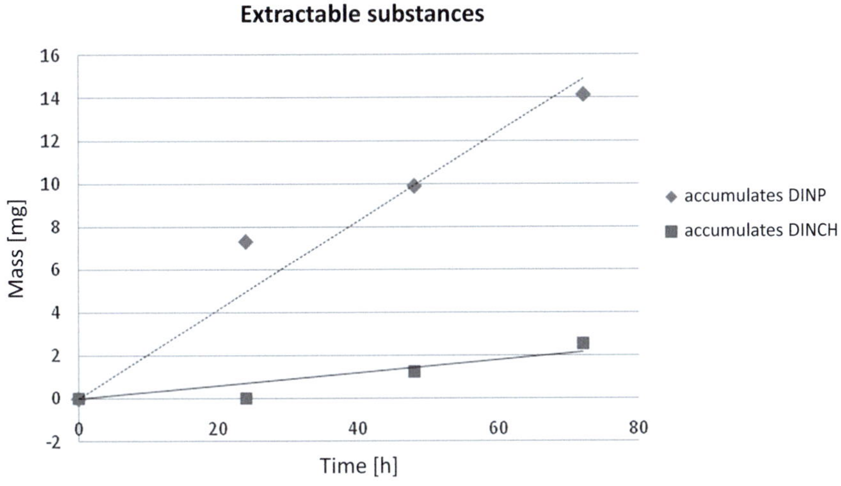

Fig. 2.123 Temporal course of the plasticizer elution with hexane from shower hoses

Only a higher background load could be detected in the non-rinsed DINP hoses. Once the surface-near plasticizer is detached, no further plasticizer migrates from the PVC matrix.

With a nonpolar solvent, both plasticizers can be eluted from the plastic matrix and differentiated in their behavior (Fig. 2.123).

In the case of the DINP hose, a significantly stronger temporal release of the plasticizer can be demonstrated compared to the DINCH hose. This result was to be expected due to the different structures of the two plasticizers and is also confirmed by a study which examined the release of DEHP and DINCH in nutrient solutions (Welle et al. 2004 2005). The bulkier cyclohexane framework can pass less unhindered past the linear PVC chains and remains more firmly anchored in the matrix.

The replacement of the plasticizer DINP with DINCH leads, based on the present results, to a lower temporal release of the plasticizer and thus a lower burden on the conveyed water. Due to the almost non-existent solubility in water, the plasticizer is not additionally extracted from the hose matrix by flowing tap water. In addition, due to current discussions and the recommendation of the Federal Environment Agency, a replacement of DINP as a plasticizer is highly recommended. The Federal Environment Agency explicitly mentions DINCH as a possible alternative (Federal Environment Agency 2007). To determine which plasticizers are contained in a plastic product or in microplastic particles, a closer look at the ATR-IR spectra is worthwhile.

Identification of Plasticizers in Microplastics

An example of plasticizer identification is the question of whether the plasticizer DEHP is still contained in blood bags (erythrocyte bags; Fig. 2.124) today.

Fig. 2.124 PVC blood bag

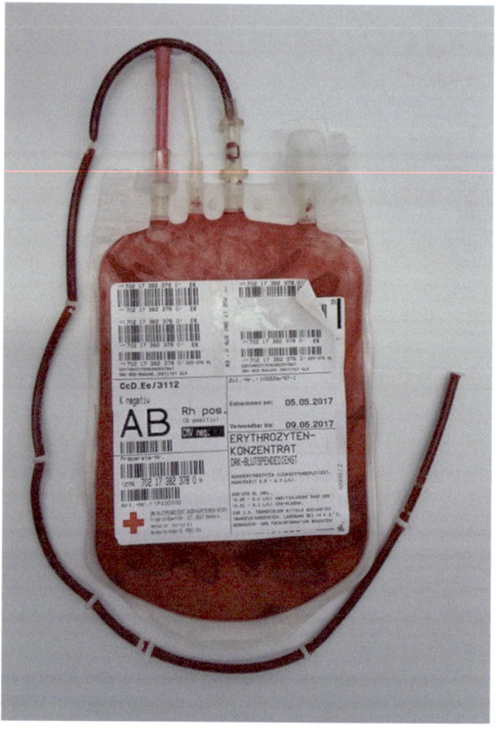

This question is not easy to answer due to the structural similarities of the plasticizers. However, it is possible to determine whether a soft-PVC material contains DEHP as a plasticizer by carefully studying the infrared spectra. The decision is only possible by precisely comparing the signals with other similar plasticizers as a reference. Fig. 2.125 shows the IR spectrum of the blood bag (erythrocyte bag; gray spectrum) shown in Fig. 2.124 over the entire MIR range from 4000 to 400 wavenumbers.

The fact that the bag material contains plasticizers can be recognized by the absorption at 1725 wavenumbers, which is a typical characteristic of an ester group. The spectra of a possible selection of plasticizer esters can also be seen in the spectrum. The DEHP, the DINP, and the DINCH spectra, which all show similar signals, are used for comparison. If you look at certain sections enlarged (Fig. 2.126), you will find that DINCH as a plasticizer is not an option.

In spectrum 1, the corresponding bands in the deformation area are missing for DINCH, which are present in both the PVC bag and the plasticizers DEHP and DINP. In addition, due to the lack of an aromatic system, DINCH lacks the C=C stretch and deformation vibrations in the range 1500–1600 cm^{-1} (Fig. 2.114, spectrum 2). The decisive difference in the IR spectrum is found in the C-H stretching vibration range between 2800 and 3000 cm^{-1}. The four signals from the alkyl

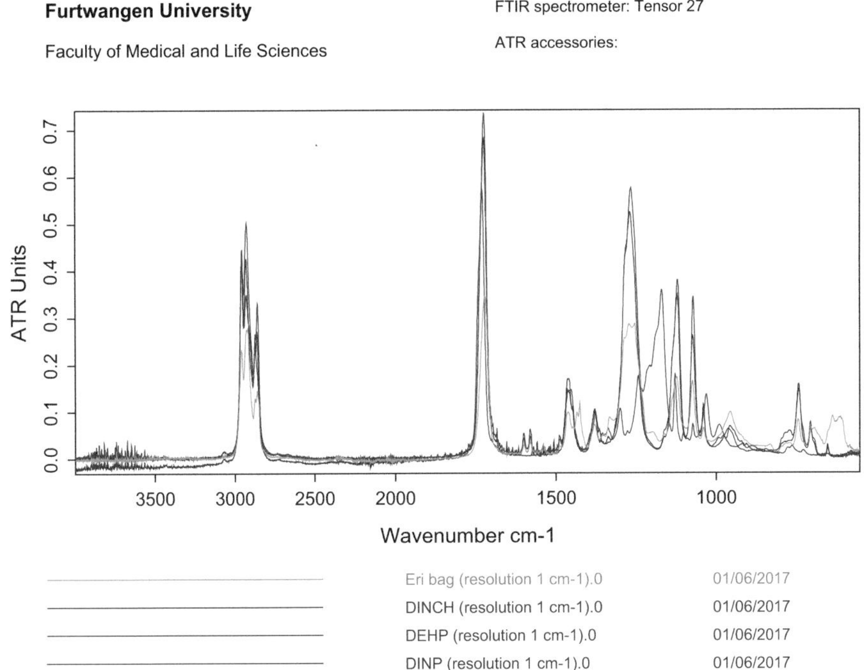

Fig. 2.125 Comparison of the infrared spectra of a soft-PVC blood bag with the three plasticizers DEHP, DINCH and DINP

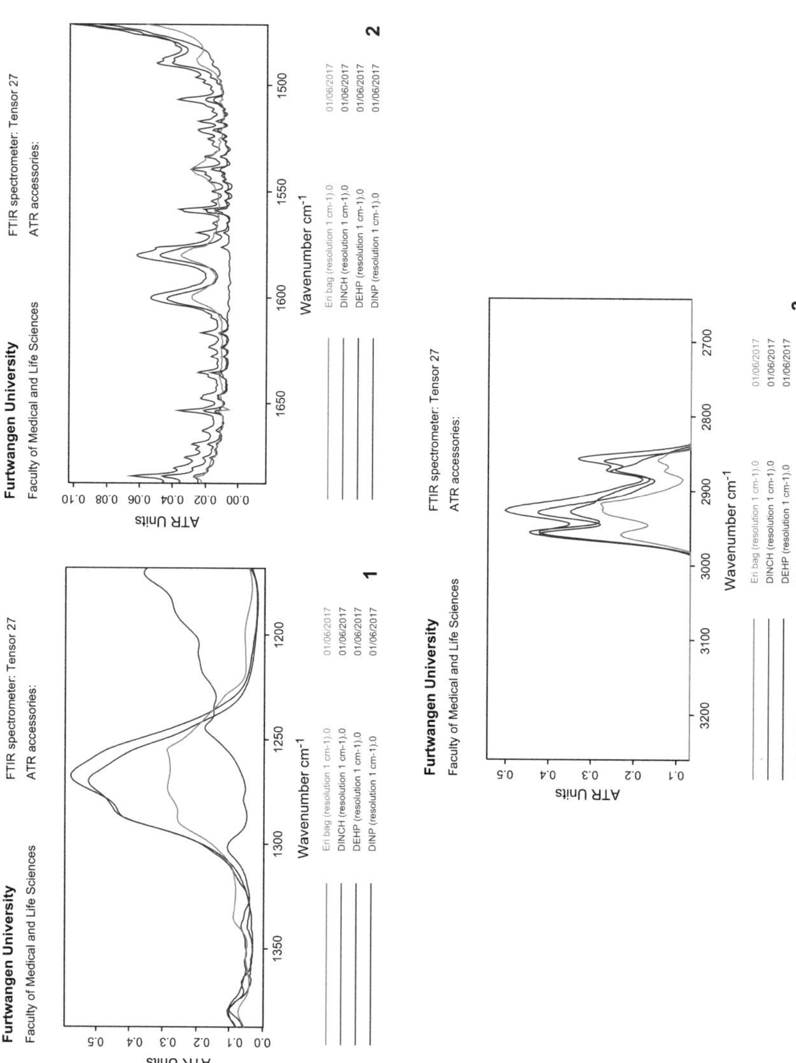

Fig. 2.126 Comparison of the infrared spectra of a soft-PVC blood bag with the three plasticizers DEHP, DINCH and DINP in different wavenumber ranges

groups best match the signals of the pure plasticizer DEHP. Although DINCH also shows the 4 signals, based on the alkyl groups of the side chains and the cyclohexane framework, DINCH can be excluded as a component due to the aforementioned discrepancies in other energy ranges. Therefore, it can be concluded that the ATR-IR spectroscopy can clarify whether DEHP is still contained in soft-PVC. The blood bag was used in a hospital in 2017.

2.5.2 Lubricants

Among the other additives, the so-called amide wax is presented here, which is also contained in PVC granulate (Table 2.30). Essentially, amide wax is distearylethylenediamide (Fig. 2.127). It is a semi-synthetic substance, as it is synthesized from naturally occurring stearic acid from plant or animal fats through a condensation reaction with synthetic ethylenediamine.

Distearylethylenediamide is used as a lubricant, release agent, as an additive for coatings, or as an aid to disperse pigments. The carboxylic acid diamide is listed in the HSDB (Hazardous Substances Data Bank), a toxicological database, and is therefore considered not harmless for use in commercial products (Bibra Toxicology AdviceabdConsulting 2005).

Whether the lubricant is contained in microplastic particles or in granulate, which is also classified as microplastic, can only be determined by IR spectroscopy after a 1–2 hour thermal exposure, if the concentration is low. In plastic granulate, the amide wax serves to prevent the granulate from clumping and thus to keep it conveyable or free-flowing until it reaches the extruder or the injection molding machine. In PVC hose production, the wax prevents the cut hoses from sticking together, which must be mechanically fed for further assembly.

The amide wax, which is contained in the PVC granulate (Table 2.29) at less than 1%, is not visible in the IR spectrum (Fig. 2.116; black line) of the extruded inner hose. Only a 1–2 hour tempering causes the wax to diffuse from the plastic matrix to the PVC surface and agglomerate there. This increases the concentration on the plastic surface that lies on the ATR crystal. The amide wax becomes detectable on the PVC surface due to this accelerated diffusion by the characteristic amide vibrations. The N–H stretching vibration becomes visible in the red spectrum in Fig. 2.128 at 3300 cm^{-1}, as does the N−H deformation vibration at about 1600 cm^{-1}.

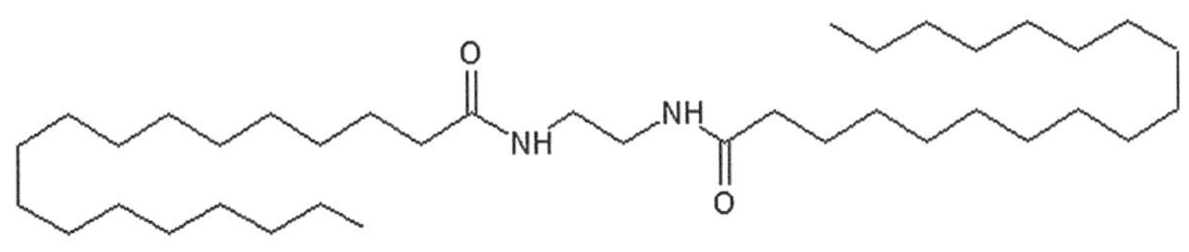

Fig. 2.127 Structural formula of distearylethylenediamide

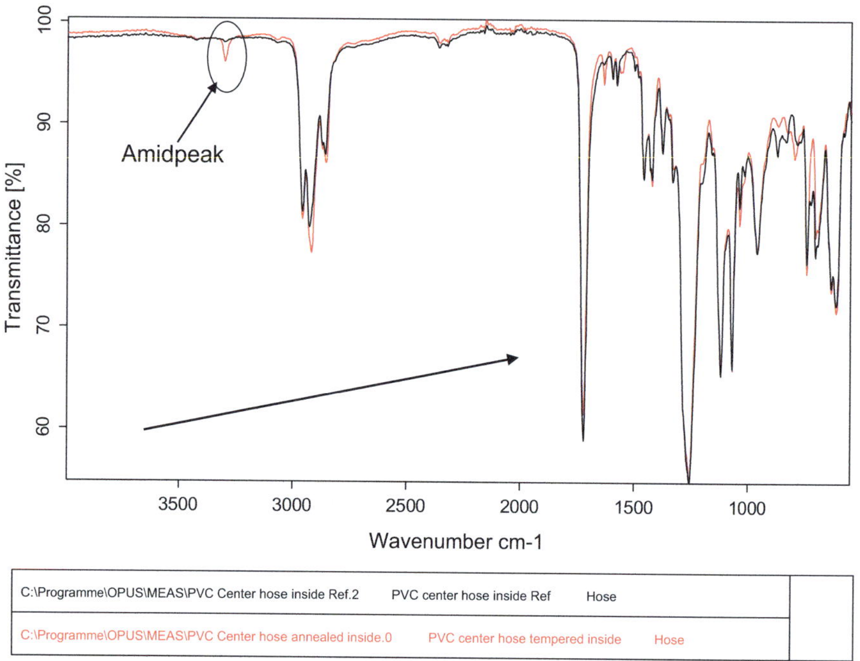

Fig. 2.128 IR spectra of the tempered (red) and untempered inner hose with amide wax additive

If the amide wax surfaces between the inner and outer hose, it can cause adhesion problems between the inner and outer hose on the PVC shower hoses, leading to the bilateral detachment of the embossed metal foil between the inner and outer hose and thus to bubble formation. This effect occurs particularly at higher temperatures. This quality problem could be solved with the help of IR spectroscopy. The bubble formation problem did not occur with a PVC granulate without amide wax. The example also shows that for plastic samples with a low amide wax content or when checking whether amide wax is present, the sample must be heated before recording a spectrum.

To illustrate the matter, the characteristic peak of the amide at a wavenumber of 3300 cm^{-1} is shown enlarged again. In addition, other sample forms are depicted (Fig. 2.129).

The samples without amide wax (yellow, black, blue, green spectrum) show a negligibly small peak despite heat treatment, which suggests a very low concentration of the lubricant in the plastic. This fact is confirmed visually, as no bubbles form on the modified tube anymore. The change in the composition of the plastic granulate has thus led to a quality improvement of the product through the use of the IR spectrometer. The example shows that in the case of plastic samples with a low amide wax content or when checking whether amide wax is contained in plastic particles, the sample must be heated before recording a spectrum.

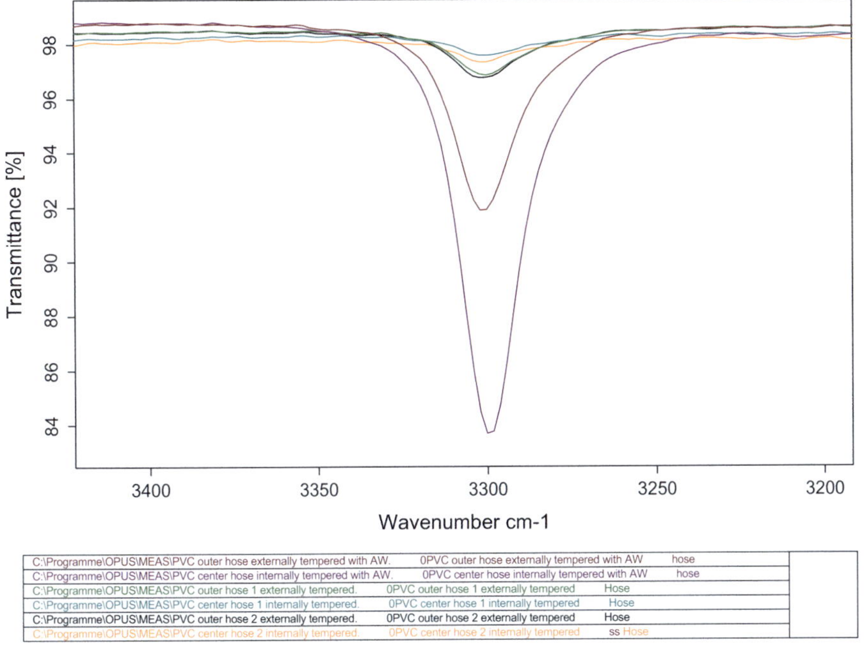

Fig. 2.129 Characteristic peak of the amide at different hose parts

2.5.3   Stabilizers

The majority of plastic products are exposed to sunlight, especially in outdoor areas. The UV component of this radiation, which is not absorbed in the atmosphere by the ozone layer, the so-called UV-A radiation in the wavelength range of 315–380 nm, is sufficiently energy-rich to homolytically cleave carbon-carbon bonds with a binding energy of about 300 kJ/mol according to $E = h\nu$. Due to the decrease in the ozone layer caused by propellants, an increasingly larger proportion of the higher energetic UV-B radiation (280–315 nm) reaches the earth. The UV radiation-induced decomposition of the hydrocarbon polymer framework via radical intermediates and recombinations with the oxygen diradical leads to changes in the material. The photolytic breakup of the main chains and the peroxide formation leads to color changes in the plastic, which becomes increasingly yellowish the longer it is exposed to sunlight. This is referred to as an increasing yellowing of the plastic. Continuous irradiation not only changes the color of the polymeric materials, but the new molecular structures also negatively affect the mechanical properties such as elongation at break, modulus of elasticity and ductility. The optical defects (yellowing) together with the material breakage or crack formation on the surface contribute to material failure. To improve the so-called weathering stability of plastics, UV absorbers are added to the polymer as an additive. The effectiveness of different UV absorbers is determined by the so-called yellowing induction time. This is the time it takes for a plastic with a certain UV absorber concentration to measure the same yellowing after irradiation as without the use of a UV absorber. Depending on the concentration, over 3000 hours can be achieved by using UV absorbers (MaierabdSchiller 2016). To ensure reproducible weathering or irradiation conditions, the standardized Sun-Test is used. This uses a xenon gas discharge lamp that emits the same radiation spectrum as daylight, but at a higher intensity. With this device, the UV resistance of materials is tested in quality assurance. The most commonly used UV stabilizers are based on the structural unit of benzotriazole or benzophenone (Fig. 2.130).

Phenol-benzotriazoles and hydroxybenzophenones, such as benzophenone 3(2-hydroxy-4-methoxybenzophenone or oxybenzone, are added not only to plastics but also to sunscreens and cosmetics. UV absorption in both classes of compounds can occur when the UV energy is sufficient for an isomerization of the compounds and the return of the higher energetic isomer to the more stable favorable state is not associated with radiation emission, but with heat release. The mechanism of the described reversible structural change in the tautomeric equilibrium is shown in Fig. 2.131.

The UV absorption is dependent on the concentration and the layer thickness of the substrate according to the Lambert-Beer's Law. Either the absorber substance is continuously integrated into the plastic at a lower concentration <1% or an additional coating with a high absorber concentration>1% protects the underlying plastic.

In contrast to the unsubstituted benzotriazole in Fig. 2.118, the hydroxyphenyl-derivatized benzotriazoles (UV-326, UV-320, UV-329, UV-350, UV-328, UV-327, UV-928, UV-234 and UV-360) used as UV stabilizers are very poorly water-soluble. As lipophilic substances, they dissolve well in plastics and varnishes, but are also released from these or from cosmetic articles into the aquatic environment. Due to their lipophilic properties, they have a high sorption and bioaccumulation potential. The aforementioned nine UV stabilizers have all been detected in both sedimentsand suspended solids. The phenol-benzotriazole UV-360 (Fig. 2.132) proved to be one of the dominant substances in sediments and suspended solids, reaching maximum concentrations of about 60 ng/g dry weight (Wick et al. 2016).

Degradation attempts have shown that the mentioned phenol-benzotriazoles in the aquatic environment are almost exclusively sorbed and are very persistent (Wick et al. 2016). With microplastics, a new artificial type of suspended solids with large surfaces has now entered our waters, whose adsorption potential towards most environmental pollutants, including the UV absorbers, has not yet

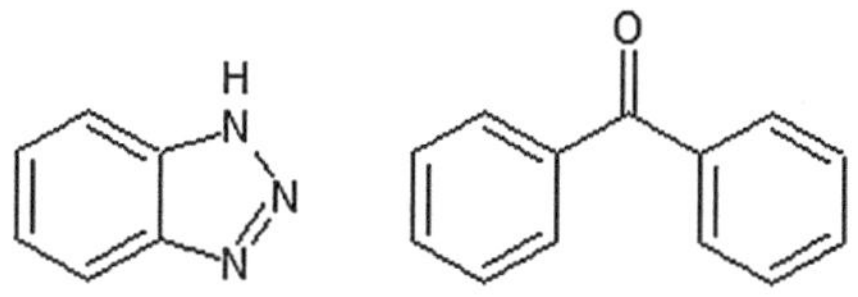

Fig. 2.130 Structure of benzotriazole (left) and benzophenone (right)

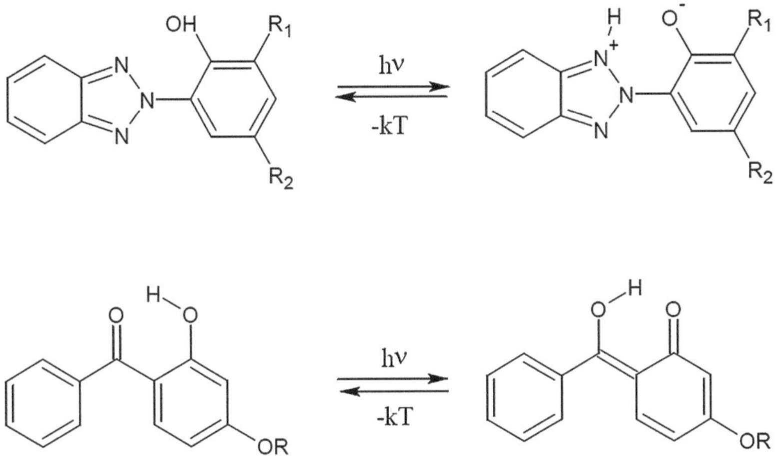

Fig. 2.131 Functioning of the UV stabilizers using the example of a hydroxyphenylbenzotriazole (top) and a 2-hydroxybenzophenone (bottom)

Fig. 2.132 Structural formula of the UV stabilizer UV-360

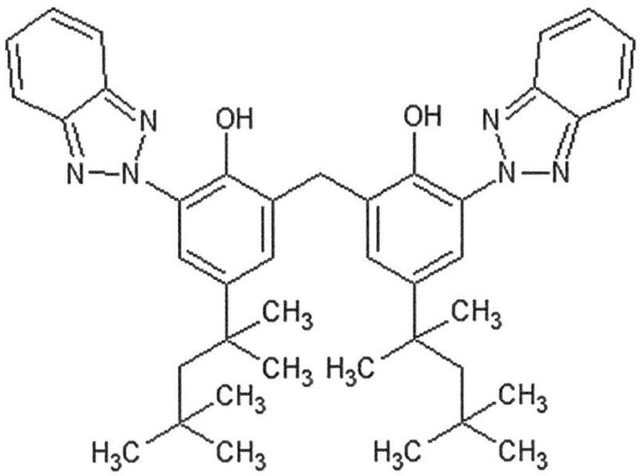

been tapped. Microplastic particles can act as "sorption islands" for UV absorbers due to their unipolar properties and low surface tension, just as they do for DDT and DEHP (Bakir et al. 2014). The persistence or stability and the property of accumulation in organisms (bioaccumulability) together with the partly proven toxic properties make some UV absorber substances those with so-called pbt properties. They are SVHC substances and are gradually included in the REACH candidate list for substances of very high concern. Meanwhile, with UV-327, UV-350, UV-320 and UV-328, four phenol-benzotriazoles are on the candidate list. This results in an obligation to inform for the distributor towards his customers, if, for example, the UV absorber concentration in a plastic exceeds the 0.1 mass percent mark. If the ECHA decides on a licensing requirement, the substances end up on the prohibition list in Annex XIV of the REACH Regulation and can therefore only be produced or used from a certain expiry date (SunSet day) by means of a temporary authorization.

Benzophenones, such as the carcinogenic 4,4'-Bis(dimethylamino)benzophenone (Michler's Ketone), are also listed as SVHC substance on the candidate list (CIRS 2008).

While lipophilic hydroxybenzotriazoles and benzophenones sorb onto sewage sludge if wastewater streams pass through a wastewater treatment plant, the water-soluble unsubstituted benzotriazoles are only adsorbed to a small extent in wastewater treatment plants onto the sewage sludge. The larger proportion is directed into rivers with the treated wastewater, as the microorganisms of the activated sludge basin cannot metabolize the stable benzotriazole (Hinterbuchner 2006), leading to comparatively high concentrations of these trace substances in our waters (Hinterbuchner 2006). In the Rhine, the concentration course of benzotriazole from the source to the mouth was determined in August 2014 (Fath 2016). Figure 2.121 shows the continuous increase up to the mouth. With a discharge volume of 2500 m^3/s (WSV n.d.) (measured in Emmerich), this corresponds to a load of about 40 tons of benzotriazole flowing into the North Sea. The Rhine is a drinking water source for 22 million people. The polar 1H-benzotriazoles and tolylbenzotriazoles can penetrate into the raw water of drinking water extraction due to their water solubility, their poor biological degradation by microorganisms during bank filtration, and the low removal efficiency in wastewater treatment plants (Kurzenberger 2010). There, they must be eliminated from the raw water by special drinking water treatment methods such as ozonation or activated carbon treatment.

The main cause of the benzotriazole entry into wastewater or waters is the use of the substance as a corrosion inhibitor and not the cleavage of the hydroxyphenyl groups from the UV absorbers. Whether such fragmentation is possible at all within the conditions in wastewater treatment stages is the subject of current investigations.

Due to the complex-forming properties with non-ferrous metals such as copper, brass or silver, benzotriazoles and the tolyltriazoles methylated on the phenyl ring are also used as corrosion inhibitors and are found in aircraft de-icing agents, in antifreeze and cooling fluids for power plants and motor vehicles, in brake and cutting fluids, and in dishwasher cleaning agents (dishwasher powder and tabs) as tarnish protection.

The unsubstituted bezotriazoles seem to be harmless to human health, as the clean cutlery from the dishwasher is put back into the mouth. The largest proportion of the benzotriazoles has left the dishwasher due to the good water solubility (20g/L) and is on its way to the wastewater treatment plant as wastewater. There, the benzotriazoles are only removed from the wastewater to a small extent and thus flow into the adjacent river via the inlet (Fig. 2.133). 1H-Benzotriazoles and the benzotriazoles methylated on the aromatic compound show no mutagenic

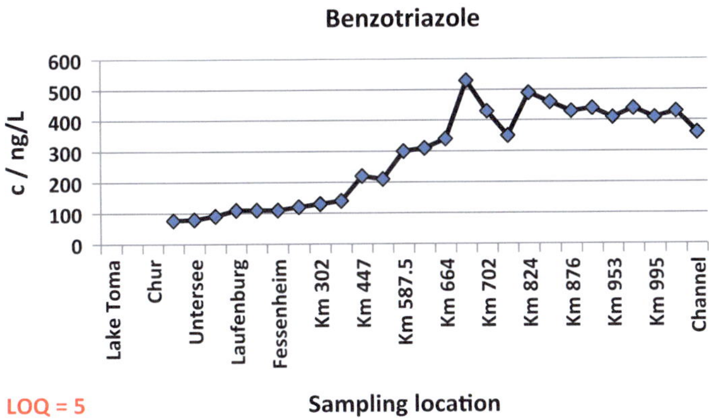

Fig. 2.133   Concentration course of benzotriazole along the Rhine in August 2014. (Fath 2016)

properties and are not capable of causing chromosome breaks. Nevertheless, they show a toxic influence on three different water organisms such as *Vibrio fischeri* (bioluminescent bacterium), *Ceriodaphnia dubia* (water flea) and *Pimephales promelas* (fathead minnow), which can be determined by the LC_{50} value. The LC_{50} value indicates at which concentration 50% of the test organisms used die. While a concentration of 22 mg/l 5-methyl-benzotriazole caused 50% of the fathead minnows to die after an exposure time of 96 hours, three times the amount of 1H-benzotriazole was needed for the same result (Reemtsma et al. 2010). The methyl groups on the aromatic compound of the bezotriazole increase the toxicity towards the test organisms (Pillard et al. 2001). Due to the lower health-damaging effect of benzotriazoles, an effective treatment of benzotriazole-derivatized UV stabilizers, which are capable of cleaving the aromatic and thus stable benzotriazole group with maximum yield, would be desirable. To what extent an electrochemical or a UV(AOP) treatment in real wastewaters can achieve this is the subject of current research work.

The extent to which the adsorption or absorption of benzotriazole UV absorbers can be determined by recording the IR spectrum of a microplastic particle is shown by a comparison of the plastic IR spectrum with that of pure crystalline benzotriazole. At a surface concentration >1%, the characteristic and sharp peaks of benzotriazole (Fig. 2.134) can be recognized within the plastic spectrum depending on the type. Specifically, the N-H valence vibration at 3400 cm^{-1}, the aromatic C-H vibrations at 3100 cm^{-1}, the aromatic conjugated C=C stretching vibrations at 1623 cm^{-1} and the N=N stretching vibration at 1575 cm^{-1} and the aromatic *out-of-plane* vibration for 1–2 substituted aromatics at 737 cm^{-1} should be mentioned.

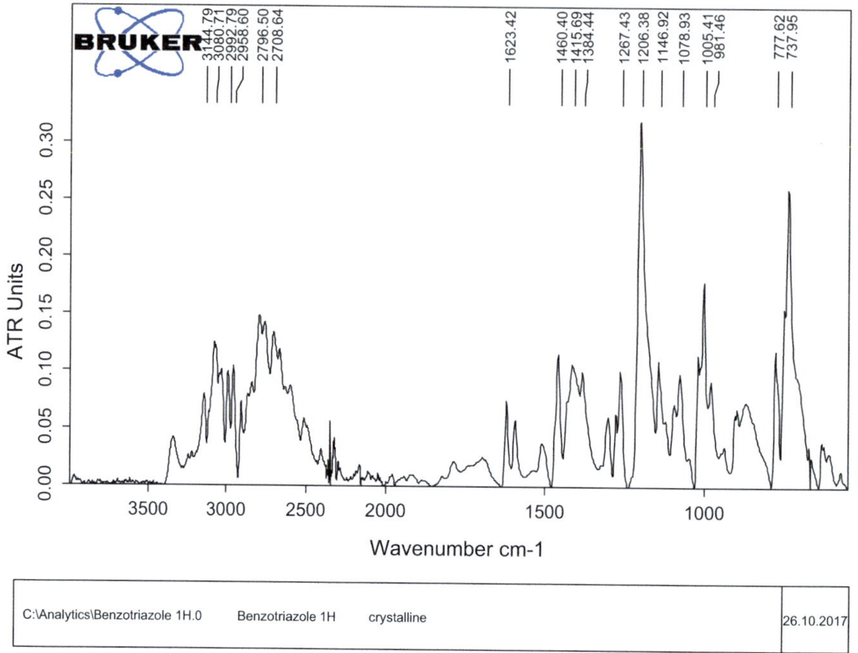

Fig. 2.134 ATR-IR spectrum of crystalline 1H-benzotriazole

2.5.4   Flame Retardants

Flame retardants are substances that are added to flammable materials such as plastics to delay or prevent the spread of flames in the event of a fire. They are used in textiles, furniture, carpets, facades, insulation materials as well as in electronic devices, where ignition sources can lead to a fire (GächterabdMüller 1993). The delay time until a fire spreads to flame-retarded objects can save lives. Different chemical compounds are used as flame retardants. In addition to inorganic flame retardants based on aluminium hydroxide, brominated organic flame retardants represent the largest group, even before chlorinated and organophosphorus compounds. Fig. 2.135 presents some common representatives of aromatic and aliphatic brominated flame retardants. The halogen functionalization is crucial for the flame retardant function, regardless of the compound class. Thus, in addition to cyclohexanes, biphenylmethanes, biphenyls, dibenzofurans or dibenzodioxins can also be used as brominated or polybrominated compounds (Fig. 2.135).

Flame retardants work through a combination of chemical and physical processes during combustion. During the pyrolysis of the flame retardants, halogen radicals are formed in the gas phase, which bind the oxygen that can no longer promote combustion. The flame retardant, which chars under oxygen deficiency, forms a protective layer on the burning material and virtually suffocates the fire,

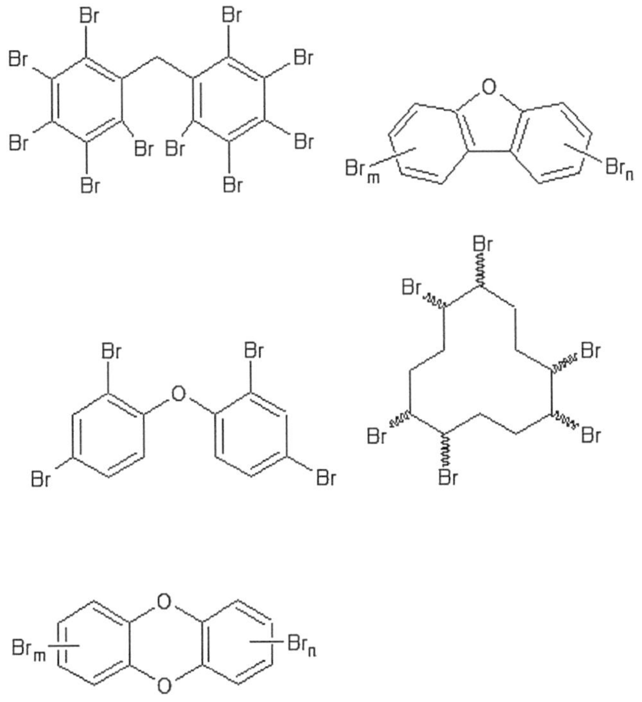

Fig. 2.135 Selection of some brominated flame retardants

as the oxygen supply is interrupted. Endothermic reactions of the flame retardant remove energy from the fire, causing the temperature to drop. This cooling slows down the exothermic combustion process. The dilution of the flammable gases by inert gases such as HBr, which is formed during the pyrolysis of the brominated flame retardants, also reduces the reaction speed.

The effect of halogenated flame retardants can be described by the following reactions in the gas space. The pyrolysis of the halogen hydrocarbon flame retardants initiates a radical chain reaction (Federal Environment Agency 2008; https://de.wikipedia.org/wiki/Flammschutzmittel). The homolysis of R–X generates an alkyl and a halogen radical (Eq. 2.50).

$$R-X \rightarrow R\cdot + X\cdot \tag{2.50}$$

The formed halogen radicals react with the hydrocarbons of the burning material and form non-flammable hydrogen halide (Eq. 2.51) and they react with the dioxygen diradical from the air and bind it over several endothermic intermediate stages (Eq. 2.52 to 2.54).

$$X\cdot + R-H \rightarrow R\cdot + HX \tag{2.51}$$

$$X\cdot + \cdot O-O\cdot \rightarrow X-O\cdot + \cdot O\cdot \tag{2.52}$$

$$\cdot O\cdot + H-X \rightarrow OH\cdot + X\cdot \tag{2.53}$$

$$X{-}O \cdot + HX \rightarrow 2X \cdot + OH \cdot \qquad (2.54)$$

At the end of the reaction cascade, the resulting reactive radicals neutralize themselves through various recombination reactions to thermodynamically stable compounds (Eq. 2.55 to 2.57).

$$H{-}X + OH \cdot \rightarrow H_2O + X \qquad (2.55)$$

$$R \cdot + OH \cdot \rightarrow ROH \qquad (2.56)$$

$$R \cdot + R \cdot \rightarrow R{-}R \qquad (2.57)$$

Brominated flame retardants are inexpensive and mix well with plastics. Many compounds of this group are persistent, i.e., difficult to degrade in the environment, and accumulate in living organisms—they are therefore bioaccumulative. Due to the pbt (persistent, bioaccumulative, and toxic) properties and the risk to infants and the environment, the flame retardants hexabromocyclododecane, pentabromodiphenyl ether (Penta-BDE) and (Octa-BDE) are classified as SVHC substances and are listed in Annex XIV of the REACH list, which is equivalent to a ban on use without authorization. Worldwide, around 2 million tons of flame retardants are produced (2012) and processed into the corresponding household products (Troitzsch 2012). It is therefore no surprise that they reach everywhere through evaporation and leaching. The lipophilic flame retardants can be found in household dust, in the blood serum of animals and humans, in breast milk, and in sediments all the way to the polar caps, where they are carried by air currents, bound to microplastics or dust (Lunder et al. 2008; Sjödin et al. 1999; Fromme et al. 2016). An even greater danger arises in plastic fires, where highly toxic dioxins are released during the pyrolysis of structurally similar polybrominated and polychlorinated flame retardants. A well-known example of this is the Seveso poison, which is 2,3,7,8-tetrachlorodibenzodioxin (TCDD), which was released in a fire in Seveso (Italy) near Milan. During plastic combustion in waste incineration plants, the resulting dioxins are retained by appropriate filter systems before emission. Flame retardants from plastic articles can escape during the use phase, the production and disposal phase of the articles either on a landfill or during thermal recovery (Kemmlein et al. 2003), as can other additives such as the previously described plasticizers. Depending on their vapor pressure, the additives can initially pass into the gas phase and deposit elsewhere or be extracted by liquid media, mainly aqueous media. Through abrasion during the use phase or also in production, where sprues or reject parts are shredded to recycle the resulting granulate (Fig. 2.136), microplastics are also produced in plastic combustion processes. The resulting microplastics naturally still contain all the additives used or adsorb, when microplastics are released into the environment, other pollutants in the water or air. In microscopic form, the plastic with its health-endangering additives, such as flame retardants (Birnbaum and Staskal 2004) or adsorbed SVHCs, poses a higher risk potential not only due to its lung penetration. Nutritional sources such as sea salt (Karami et al. 2017) and drinking water

Fig. 2.136 POM recyclate in a shredder. The dimensions of the particles and the plastic powder meet the definition of microplastics in the size distribution

(Carrington 2017) can be contaminated with microplastic particles and thus enter our bodies, where they are exposed to entirely different conditions in our digestive tract, where a higher desorption and extraction rate is expected (Bakir et al. 2014). Microplastics as carriers or trojans can thus introduce toxins into our organism. A quick and cost-effective method to determine whether plastic parts or microplastic particles contain brominated flame retardants is also provided by ATR infrared spectroscopy.

Figure 2.137 shows the IR spectrum of the brominated aromatic flame retardant BDE 47. This is 2,2′,4,4′-tetrabromodiphenyl ether. Characteristic for the aromatic system are the C_{sp2}-H valence vibrations at 3080 cm^{-1}. Since some plastics also contain aromatic systems, such as polycarbonate or polystyrene, this aromatic vibration absorption does not conclusively determine whether these plastics contain, for example, BDE 47 or other bromodiphenyl ethers. A more characteristic criterion to pursue this question is provided by the C_{sp2}-Br valence vibrations, which absorb at wavenumbers of 1028–1073 cm^{-1}. The corresponding peaks for this vibration are circled in the spectrum (Fig. 2.137).

Since aliphatic brominated flame retardants of the compound class polybrominated or -halogenated cycloalkanes are also used in plastics, there is also the possibility to quickly find out by means of ATR-IR spectroscopy whether these types of flame retardants are present in microplastic particles. Of course, it is important to compare with the IR spectrum database, where the spectrum of the suspected flame retardant is also stored. Figures 2.138 and 2.139 shows the IR spectrum of the brominated aliphatic flame retardant HBCD. This is 1,2,5,6,9,10-hexabromocyclododecane. If this substance is present in the examined plastic, this could be recognized, for example, by the presence of the C-Br valence vibrations in the low

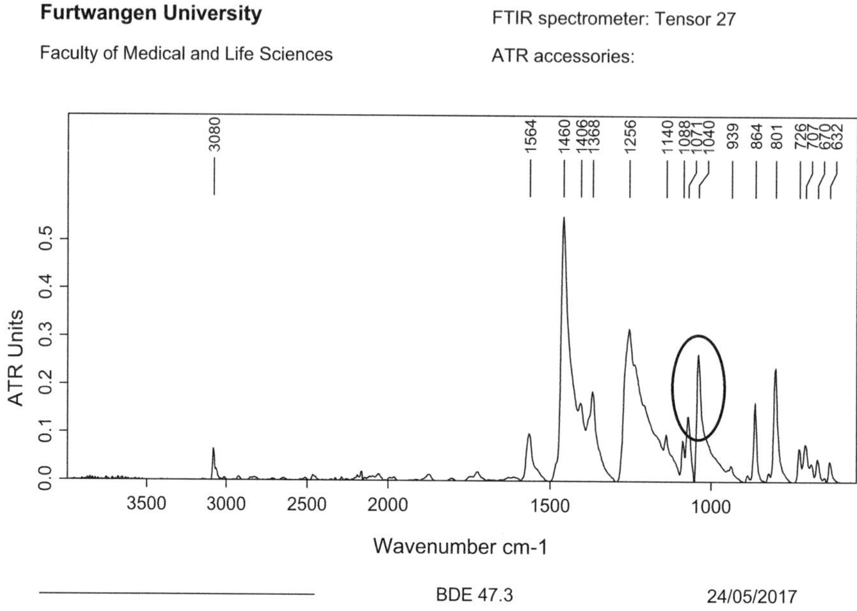

Fig. 2.137 ATR infrared spectrum of BDE 47

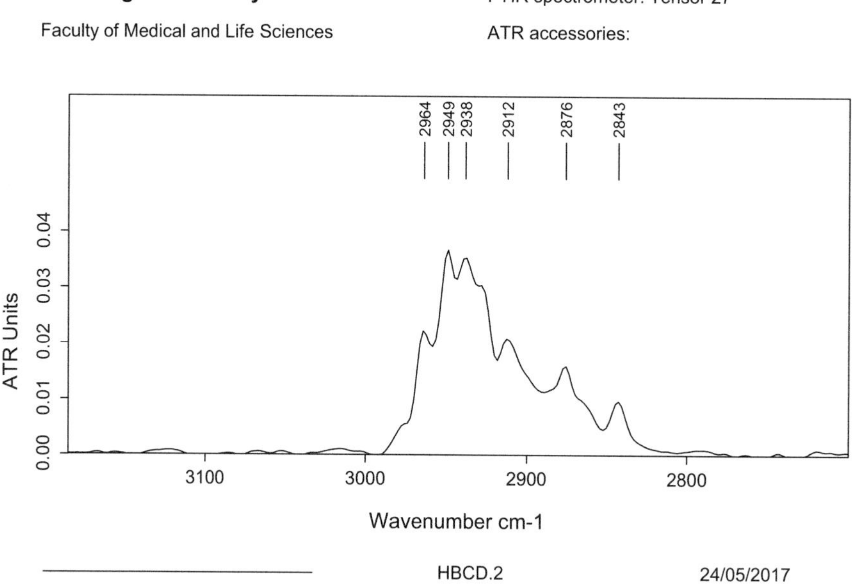

Fig. 2.138 ATR infrared spectrum of HBCD in the wavenumber range 2700–3200 cm^{-1}

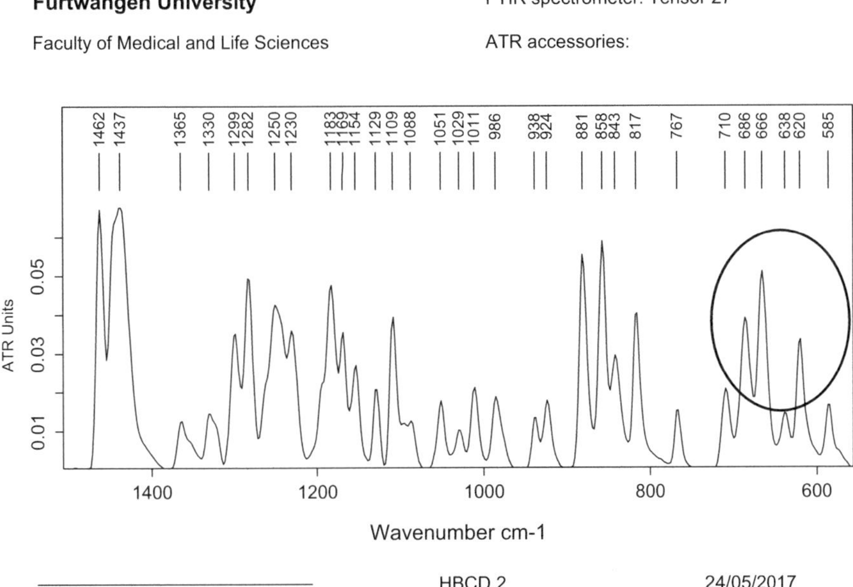

Fig. 2.139   ATR infrared spectrum of HBCD in the wavenumber range 550–1500 cm^{-1}

energy range between 515 and 680 cm^{-1} (see circled peaks in Fig. 2.139). In addition, the C-H valence vibrations in the aliphatic hydrocarbon are split by the influence of the six heavy bromine substituents in the wavenumber range of 2800–3000 cm^{-1} (Fig. 2.138).

2.5.5   Pigments

Plastics are colored by adding pigments. These can be of inorganic or organic nature and are not soluble in the plastic matrix, but are dispersed. In addition to the colored heavy metal compounds, which are more or less toxic depending on the heavy metal, soots are used for black coloring, so-called Carbon Black. Seals and black plastic pipes in production facilities or hydraulic lines in automobiles or tires contain soot as a black pigment. Since soot is an incomplete combustion product of hydrocarbons, it mainly contains carbon, contaminated with other stable combustion products such as polycyclic aromatic hydrocarbons, briefly called PAH. This class of compounds consists of at least two condensed ring systems. They are found in mineral oils, bitumen, tar, pitch, soot and products made from them. Until the 1980s, PAHs were used as wood preservatives or as binders in road asphalt. When disposing of these old products from the construction industry, the PAH-containing waste must be declared as hazardous waste. PAHs can be found in all products made of rubber and plastic, in the rubber-coated hammer handle as well as in the rubber duck and the bath shoes. The background paper of the Federal Environment Agency (2016) provides information on which PAHs occur in which concentrations in the different household and consumer items, including children's toys, and what current limit values are set for the respective products in Europe.

According to the REACH regulation, products containing more than 1 mg/kg of one of the eight carcinogenic PAHs have been banned since December 27, 2015. These are Benzo[a]anthracene, Benzo[a]pyrene, Benzo[b]fluoranthene, Benzo[e] pyrene, Benzo[j]fluoranthene, Benzo[k]fluoranthene, Chrysene, and Dibenzo[a,h] anthracene. For toys and baby items, the limit is 0.5 mg/kg. This also applies to imported items. Since 2015, manufacturers and importers have had to ensure that the new limits are adhered to. So far, 16 different PAHs have been routinely examined and quantified in products, sediments, sewage sludge, or wastewater using standardized sampling, processing, and analysis methods by GC-MS or HPLC. In products, Benzo[a]pyrene is often analyzed as a representative for all 16 PAHs, as most other PAHs are usually associated with this indicator substance. Benzo[a] pyrene was chosen as the indicator substance because it is one of the particularly strong carcinogens. Simple photometric determination methods, based on a Friedel-Crafts alkylation on the aromatic system using chloroform and aluminum trichloride as a catalyst, are possible in certain concentration ranges (LAGA 2013). We cannot get rid of PAHs, as they are ubiquitously distributed in our environment through combustion processes of natural or industrial origin. Even though the PAH limits in products have been reduced in recent years, they are still used as plasticizer oils in winter tires (see IR spectrum in Fig. 1.1). Until 2009, PAH-containing plasticizer oils were legally used in car tires. Since January 1, 2010, an EU-wide limit has been in place for PAH-containing plasticizer oils in car tires. According to the UBA, the average PAH content in old tires is currently 40 mg/ kg (Federal Environment Agency 2016). Even in lower concentrations in car tires, which wear out by the ton every year, they can unleash their environmental toxic potential via microplastics. The realization that the annual tire wear, in addition to the fibers of artificial textiles, is the main cause of microplastic distribution in aquatic systems (BoucherabdFriot 2017), means that the danger from PAHs is not averted. On the one hand, the rubber microplastic particles enter the environment through tire wear (Kole et al. 2017) and now for several years additionally as a coating of artificial turf fields (Fig. 1.2). The granulate used as a coating is nothing more than the recyclate of old tires, which contain PAHs. A statement by the RAL Quality Association for Plastic Coatings in Outdoor Sports Facilities e.V. RAL Quality Association (http://www.ral-ggk.eu/ral-standards-news/news/49-news/198-ral-news-pak) states that a PAH limit has been set for these granulates and international studies have concluded that no significant PAH exposure for users of artificial turf fields has been demonstrated and that rubber granulate from old tires does not pose a specific health risk (BAG 2017). At first glance, it seems like a good thing that not all old tires are thermally recycled, but are instead put to meaningful reuse as an elastic filler for artificial turf coatings. On second glance, one realizes from personal experience that the recyclate does not remain on the

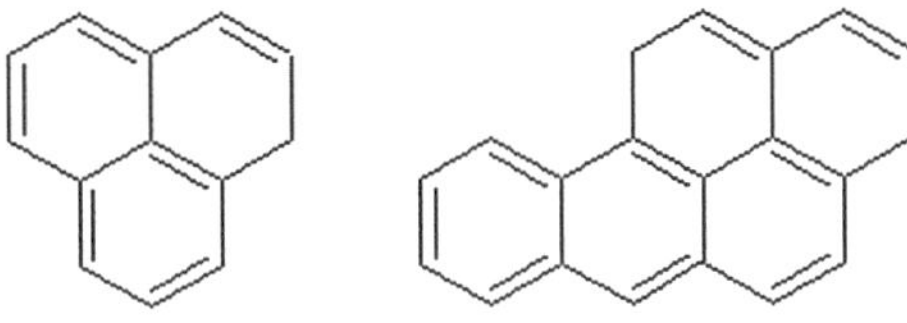

Fig. 2.140 Structural formula of phenalene (left) and benzo(a)pyrene (right)

sports field. It is carried onto streets, into sewer channels, and into households via the clothing and cleated shoes of artificial turf users, or it is carried into the environment and into the bodies of water adjacent to the sports field during extreme weather conditions by wind and rain. Even though the PAHs are not water-soluble due to their hydrophobicity, they can be ingested in the granulate as food by fish and other water organisms. In the digestive tract, environmental conditions change, so that, for example, phenanthrene sorbed to polyethylene desorbs about twelve times faster at pH 4 and 37 C in a sodium taurocholate solution, which simulates the in-vitro conditions in the digestive tract, than in water (Bakir et al. 2014).

PAH are multi-core benzoic hydrocarbons that derive from benzene as a basic structural element. The chemical structure of all PAHs consists of at least 2 or more unalkylated and unsubstituted, condensed (annelated or edge-attached) benzene rings. In addition to the σ-electron framework, there is a delocalized π-electron system that covers the entire molecular framework. As a result, all carbon atoms are sp^2-hybridized. As a consequence, the PAHs are planar and resonance-stabilized like benzene. When two annelated benzene rings are combined, naphthalene is formed as the simplest PAH compound. If more aromatic rings are added to the two-ring system, linear addition (ortho-annelated) results in the so-called acenes (e.g., anthracene). With an angled linkage of a so-called *peri*-condensation, phenanthrene is obtained as the simplest representative of this group. If the third annelated ring is bound to both rings of a two-ring system, the result is phenalene, shown in Fig. 2.140, with an internal carbon atom. All PAHs are components of coal tar, including naphthalene (10%), phenanthrene (5%), and fluoranthene (3.3%) as the most abundant percentage-wise (Rippen 1993).

To assess the behavior of the compound class of PAHs in the environment, the parameters water solubility, vapor pressure, Henry's coefficient (distribution coefficient between gas and liquid phase) and the K_{OW} value are important. The K_{OW} value indicates the distribution coefficient of a substance in a mixture of *n*-octanol and water as a logarithmic value. It is one of the most important physicochemical parameters for environmental research. In this context, *n*-octanol is the model substance for animal fat tissue. By comparing the K_{OW} values, one can therefore make a statement about whether a substance accumulates more strongly in an organism than another. The K_{OW} value correlates well with the so-called bioconcentration factor (BCF) in fish, mussels, and other aquatic organisms. The BCF describes the accumulation of substances from the water phase in the fat tissue. Due to this good correlation, *n*-octanol has established itself as a model solvent in environmental research (Rippen 1993).

Regarding water solubility, it is the case with PAHs that it decreases with the increase of annelated benzene rings. While 30 mg of naphthalene still dissolve in one liter of water, it is only 0.2 mg/l for fluoranthene (Rippen 1993). The same applies to the decrease in the air/water distribution coefficient (Henry's coefficient). Naphthalene, with a Henry's coefficient of 1.71×10^{-2}, is three orders of magnitude more volatile in water than, for example, benzo[a]pyrene (Fig. 2.140) with 2.13×10^{-5} (Rippen 1993).

There are also large differences regarding the adsorbability of the substances. The larger the molar mass, the better the respective component is adsorbed and the more lipophilic is the compound (Skrzypek 2003). This is also confirmed by the logarithm of the distribution coefficient of the substances in the n-octanol/water system (K_{OW}), as the suspended solids in waters are hydrophobic non-water-soluble materials. The smallest K_{OW} value of 3.6 is possessed by naphthalene, while higher molecular PAHs like fluoranthene have a value of 5.1 (Rippen 1993). Suspended solids can be of geogenic mineral/inorganic nature and consist of silts, clays, and sands or be made of biogenic organic material such as phytoplankton, diatoms, fungi, bacteria, algal blooms, humic substances, etc.

The adsorption of chemicals to suspended solids is described by the corresponding temperature-dependent adsorption coefficient K (T), which indicates the ratio of the adsorbed substance concentration to the dissolved substance concentration.

The current water pollution with plastic waste has expanded the list of suspended solids in waters by an additional organic material, microplastics. Relatively little is known about the adsorption properties of this new artificial type of suspended solids with regard to the adsorption of different pollutants to different types of plastics, depending on the surface texture. For the POPs *(persistent organic pollutants)* PFOA, phenanthrene, DDT and DEHP, there are significant differences in the adsorption coefficient depending on the plastic material PE or PVC (Bakir et al. 2014). How strongly and in what concentration hormones, such as the birth control pill hormone ethinylestradiol, sorb to different types of microplastics with different particle sizes and surface structures, is the subject of current scientific investigations (Skrzypek 2003).

To gain insights into potential ecotoxic effects of pollutants, the K_{OW} value and the bioconcentration factor provide important clues. Since it is known that aquatic organisms ingest microplastic particles with their food (Hummel 2017), an additional parameter for assessing the effects of pollutants is added. The distribution equilibrium between sorbed pollutants on the plastic (also additives) and the n-octanol as a model substance for organic tissue. In the metabolic process, e.g. in the digestive tract of a fish, pollutants adsorbed and absorbed on microplastic particles from the environment or additives of the plastic can desorb or be released and be taken up into the fatty tissue. The pollutants introduced into an organism via microplastic particles as "Trojans" may have a greater toxic potential than the substance dissolved or dispersed in water, depending on the high adsorption concentration or additive concentration.

Toxicology of PAHs

Carcinogenic, teratogenic and mutagenic properties have been attributed to numerous representatives of the PAHs PAHs. Benzo(a)pyrene (Fig. 2.129) has the highest carcinogenic potential. The carcinogenic effect of polycyclic aromatic hydrocarbons is based on the formation of reactive metabolic products, which are capable of reacting with the DNA and thus compromising its function. In the detoxification process of our organism, foreign substances can be oxidized to water-soluble carboxylic acids in the liver, filtered from the blood in the kidney and excreted in the urine. The oxidation of the methyl group in toluene, for example, results in the excretion of benzoic acid as a degradation product. In benzene and the polycondensed aromatics, there is no functional group that could be oxidized, so the degradation takes place in a different way. With the help of the cytochrome P-450 coenzyme and the epoxide hydrolase, a double bond of the conjugated system is first epoxidized in the first step and then hydrolyzed to the *cis*-diol (Bravo et al. 2012). After the second epoxidation step, the reactive epoxide remains, as the adequate attachment of the epoxide hydroxylase is not possible as in the first step due to steric hindrance. The result is that nucleophiles, such as the exocyclic amino group of guanine, attack the diolepoxide in an Sn_2 reaction and break open the strained three-ring. The mechanism described is shown in Fig. 2.129 using the example of the double epoxidation of benzo[a]pyrene. The exocyclic amino group of the purine base is now no longer available for the formation of the hydrogen bond with adenine of the complementary strand of the DNA to form the double helix. This explains the carcinogenic effect of some representatives of the substance class of PAHs. The planar PAHs reach the nucleophilic base pairs as reaction partners by intercalation into the DNA double helix. Such a significant structural change due to the reaction influence on base pairs can lead to errors during DNA replication.

Since PAHs are unavoidable due to natural and industrial combustion processes of hydrocarbons and these substances have a high persistence, they agglomerate in our environment. Although biological degradation with the help of microorganisms is considered the most important process for eliminating PAH concentrations in soils, the kinetics of degradation is slow (Klöpffer 2012). The microbial degradation of PAHs also leads to the formation of a diol (Fig. 2.141) and subsequent ring cleavage to water-soluble carboxylic acids (Klöpffer 2012). For the different degradation mechanisms of environmental chemicals in air (photochemical), water (hydrolytic) and soil (microbial), reference is made here to the further literature (Rippen 1993).

Artificial Turf Granulate

In Car tires, the PAH content was reduced by European legislation by setting a limit value. This is a positive development in light of recent findings that the annual Tire wear (DekantabdVamvakas 1994) is a major cause of microplastic

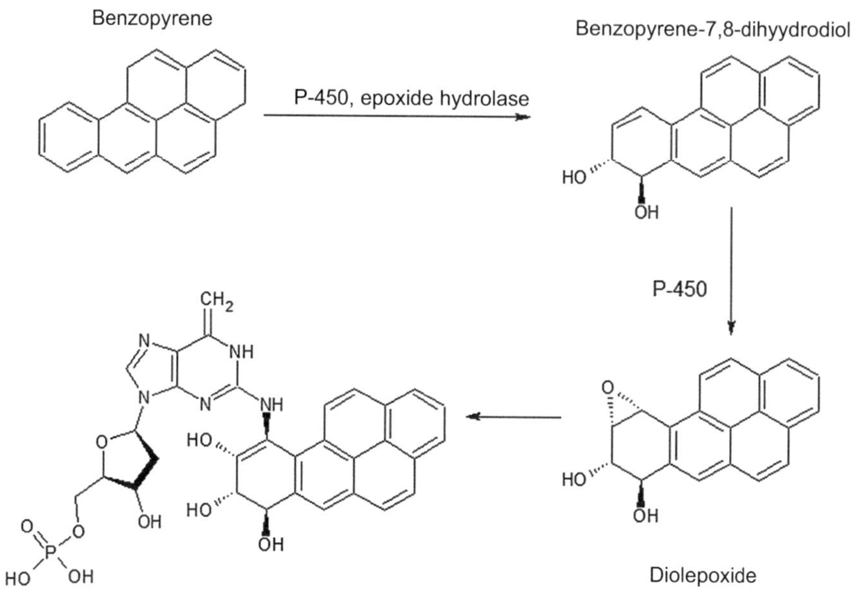

Benzopyrene

Benzopyrene-7,8-dihyydrodiol

P-450, epoxide hydrolase

P-450

Diolepoxide

Fig. 2.141 In-vivo degradation mechanism of benzo[a]pyrene with DNA adduct formation. (Bravo et al. 2012)

pollution in waters. It is not known whether the rubber particles found actually only enter our aquatic environment through tire wear, or whether the rubber granulate used on artificial turf fields (Fig. 2.142) from Old tires is also to blame. Some environmentally conscious club representatives have already switched to cork particles as an alternative to rubber granulate.

The IR spectra in Fig. 2.131 show by the matching of the signals that the artificial turf granulate in Fig. 2.131 is indeed the tire material NBR (Kumar et al. 2014). Natural rubber (latex) is no longer used for the production of car tires. The main component of natural rubber is isoprene (2-methyl-1.3 butadiene), which is replaced by 1,3-butadiene in synthetic rubber. By polymerizing the monomer, BR (butadiene rubber) is synthesized, or by copolymerization of butadiene with styrene or acrylonitrile, SBR (styrene-butadiene rubber) or NBR (nitrile-butadiene rubber) is synthesized. To harden this plastic material during shaping, the synthetic rubber is vulcanized by adding sulfur. This results in a ladder-like cross-linking of the thread-like linear macromolecules with each other. Other additives such as lime and soot and other fillers influence the abrasion properties and the color of the vulcanized tire.

To ensure good adhesion of rubber tires to the road surface at low temperatures in winter, a winter tire compound must be softer than a summer tire compound. This is achieved by so-called plasticizer oils, which contain PAHs. The hardness

Fig. 2.142 Old tire granulate on artificial turf field

of a tire can also be altered with fillers such as lime. The lime content is significantly higher in the examined summer tire (see red IR spectrum in Fig. 2.143). This is evident from the strong absorption of the yellow spectrum in the wavenumber range 1000–1300 cm⁻¹. In the fingerprint region of the IR spectrum, in addition to the valence vibrations of the polymer framework between 1600 and 1000 cm⁻¹, the vulcanization can be recognized by the broad peaks between 570 and 710 cm⁻¹, which can be attributed to the C-S valence vibration (von Moos et al. 2012).

The excerpt of the IR spectra in the C–H valence vibration range (Fig. 2.144) shows that aromatic C–H vibrations are present in all three samples, including the artificial turf granulate, suggesting a small proportion of PAH. In this wavenumber range >3000 cm⁻¹, winter tires differ from summer tires. If we compare this range together with the low-energy IR spectrum, it can be determined that the examined artificial turf granulate is mainly the recyclate of summer tires, but still contains PAH.

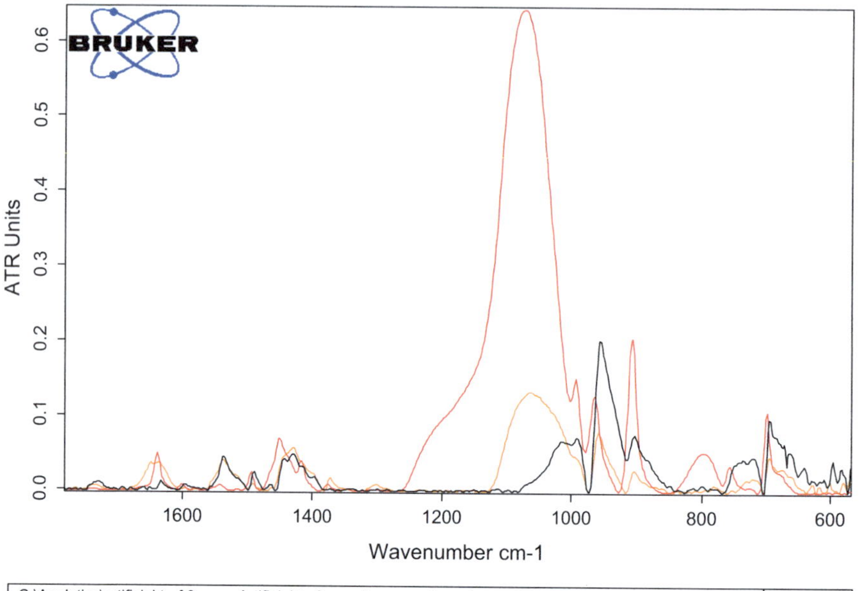

C:\Analytics\artificial turf.0	Artificial turf granules	Granules	15.11.2017
C:\Analytics\Summer tires.0	Summer tires	Piece	06.11.2017
C:\Analytics\Winter tires.0	Winter tires	Piece	06.11.2017

Fig. 2.143 Comparison of the IR spectra of the artificial turf granulate from Fig. 2.142 and two samples from winter and summer tires in the fingerprint region

The small selection of plastic additives is far from complete, it is only intended to show that organisms, including humans, not only ingest the polymeric material with the microplastic, but also the sometimes significantly more harmful additives. The list of these additives from the categories "antioxidants, metal deactivators, lubricants, plasticizers, light stabilizers, reinforcing agents, optical brighteners, flame retardants, colorants, blowing agents, crosslinking agents, biostabilizers, antistatics" etc. can be continued at will, but is not the content of this reference book. The interested reader can find information about this in the handbook of plastic additives (Lusher et al. 2013).

The influence of microplastics on humans is not yet known. However, it is known that they regularly ingest some of it with food and drinking water. If microplastic particles are detected, nanoparticles are most likely also present, which we cannot analytically capture. As soon as plastic particles reach nanometer dimensions, they are able to penetrate cells with all their pathogenic chemicals they harbor, and thus reach every organ. The rule of Paracelsus also applies to micro and nanoplastics, according to which "the dose makes the poison". This simple rule provides a convincing reason to improve plastic waste management worldwide by massively reducing the use of plastic, reusing plastic products or reworking them into new products, instead of disposing them in nature.

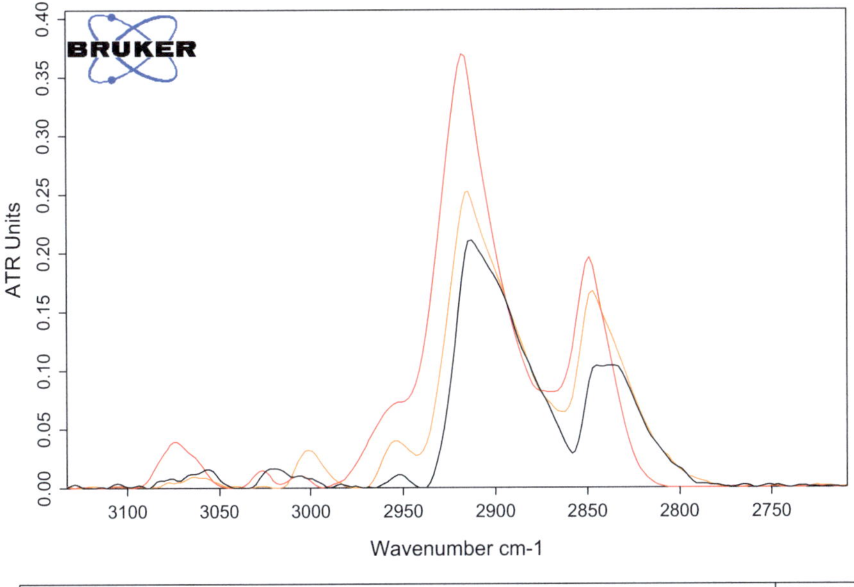

C:\Analytics\artificial turf.0	Artificial turf granules	Granules	15.11.2017
C:\Analytics\Summer tires.0	Summer tires	Piece	06.11.2017
C:\Analytics\Winter tires.0	Winter tires	Piece	06.11.2017

Fig. 2.144 Comparison of the IR spectra of the artificial turf granulate from Fig. 2.142 and two samples from winter and summer tires in the C–H valence vibration range

2.6   Previous Results of Microplastics in Waters

Industrial plastic production began in the 1950s with 1.7 million tons of plastic per year worldwide (PlasticsEurope 2013). By 2012, production had already risen to 288 million tons, with 57 million tons produced in Europe (PlasticsEurope 2013). Reasons for this rapid increase are the various advantages of this material over other materials. These include good chemical resistance (Cole et al. 2011), high strength at low weight (Andrady 2011) and cost-effective production (Derraik 2002).

Based on the amount of production, polypropylene (PP), polyethylene (PE), polystyrene (PS), polyvinyl chloride (PVC), and polyethylene terephthalate (PET) are the most important types of plastic (PlasticsEurope 2013).

Humans come into direct contact with substances such as stabilizers, antioxidants, pigments or coupling reagents like bisphenol A (BPA), which are added during the manufacturing process (WagnerabdOehlmann 2009), either directly through handling or indirectly through food packaging (Thompson et al. 2009).

Since the 1960s (Thompson et al. 2004), the amount of plastic waste in marine habitats has been continuously increasing (Barnes 2005; Derraik 2002) and the aforementioned positive properties of plastic, such as durability, due to chemical resistance, make this material with all its additives a massive problem when it ends up as waste in marine habitats.

75% of marine pollution is based on plastic (Galgani et al. 2013), with macroplastic input occurring either sea-based or land-based. Sea-based, for example, through ship waste disposal or land-based through pollution of beaches and coasts (Derraik 2002).

If the particle sizeof the plastic waste is less than 5 mm, it is referred to as microplastic (Browne et al. 2007).

The entry of microplastics can occur either via the mainland or directly at sea. On the mainland, the particles are introduced via sewage systems or drifting waste. At sea, plastic is directly introduced as waste. However, although it is known that plastic waste is introduced into the marine environment via the mainland, there are so far only a few publications on microplastics in river systems.

In Sect. 2.6.1, the results of the Rhine investigation in the summer of 2014 are presented (Fath 2016). The Rhine was examined for microplastics in a unique sampling from the source (Tomasee) to the mouth in Hoeck van Holland in cooperation with the Alfred Wegener Institute on Helgoland. Using a method developed there (Löder et al. 2015a), the concentration of microplastics and the polymer types were examined by infrared spectroscopy (IR) in Rhine water samples at the HFU and on Helgoland. To prepare the water samples for the measurements, they were enzymatically cleaned with SDS, protease, cellulase, chitinase and also with H_2O_2 and divided into two size fractions. For the quantification and qualification of the microplastics, two different IR spectrometers were used. For the analysis of particles >500 μm, the *attenuated-total-reflection* (ATR)-IR spectrometer was used and for the analysis of particles <500 μm, the *micro-Fourier-transformed-infrared* (FTIR) spectrometer was used.

Our environment is increasingly being polluted with plastic waste. Plastic waste can be found worldwide on beaches, in surface waters, in river beds (Fig. 2.134) and even in the deep sea (Fath 2016). Our plastic waste reaches places where we do not live, whether it be at Lake Toma (Fath 2016; Taylor et al. 2016) or in the deep sea (Taylor et al. 2016). This is a worrying state of affairs, as aquatic organisms ingest microplastic particles (<5 mm) with negative consequences for their survival, health and reproduction (Kershaw 2014; von Moos et al. 2012; FarrellabdNelson 2013). The particles are transferred along the food chain from lower biological systems to higher ones (Eerkes-Medrano et al. 2015).

In addition to intrinsic pollutants such as antioxidants, processing aids, UV light, stabilizers, flame retardants, dyes and plasticizers, microplastic particles also adsorb hydrophobic pollutants from their environment (HüfferabdHofmann 2016). Microplastic contamination was first discovered in marine ecosystems (Thompson et al. 2004). An estimated 80% of the plastic found in the oceans comes from the mainland and is carried into the seas via rivers and coasts (Jambeck et al. 2015). The North Sea is heavily polluted with microplastics, and major rivers such as the Thames or the Rhine contribute to this (Morritt et al. 2014; Klein et al. 2015). The contamination of the Rhine with industrial chemicals has been investigated by several Rhine monitoring stations since the Sandoz scandal (Süddeutsche Zeitung

2011) (RuffabdSinger 2013). The microplastic load along the entire course of the Rhine was only investigated in 2015 by two independent working groups. In one case, the Manta Trawler was used between Basel and Rotterdam (see results in Sect. 2.6.1) (Mani et al. 2015). While in the project "Rheines Wasser" of the HFU a filter pump, as shown in Fig. 2.149, was used (Fath 2016). With a load of 8–10 tons of microplastics per year, which the Rhine washes into the North Sea, comparable results were obtained with both different methods. The basis for the load calculation is the water discharge of the Rhine during the sampling period. In August 2014, it was about 2500 m^3/s (Fig. 2.145).

The calculation of the load from the analysis results referred in both studies to the near-surface distribution of microplastics. Additional depth profile and sediment investigations would certainly further increase the tonnage of the microplastic load per year, because in addition to plastics with high density, lighter plastic fragments also sink over time due to growing biofilm (Barnes et al. 2009; Browne et al. 2010; Lobelle and Cunliffe 2011; Dris et al. 2015).

The result of the sampling at the source of the Rhine (Lake Toma) was striking and surprising. Here, 270 particles/m^3 were filtered from the water. A reference of this amount to adjacent industrial plants or the population density (Mani et al. 2015) cannot be established at this altitude. At Lake Toma in the Grisons Alps at an altitude of 2345 meters, there is neither industry, agriculture nor private households. The microplastic input there has a different origin. It is conceivable that the input is via precipitation and thus via the melting snowfields that feed the mountain lake.

So far, the Rhine is the only river of its size that has been examined for microplastics along its entire length. The topic of "microplastics in an aquatic

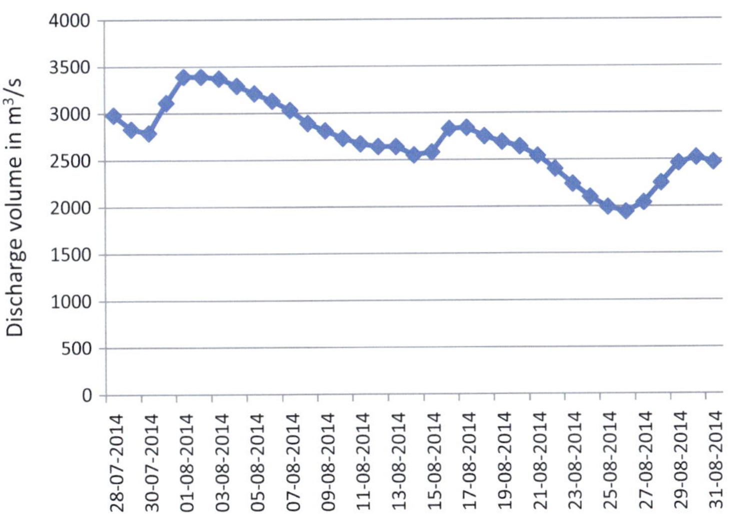

Fig. 2.145 Water discharge of the Rhine at Emmerich in August 2014. (Klein et al. 2015)

environment" has seen a significant increase in both public and scientific interest in recent years.

Mass production of plastic began in the 1950s with 1.7 million tons worldwide. Nowadays, about 75% of marine waste is made up of plastic. Microplastics are plastic particles that are smaller than 5 mm. Microplastics are either produced directly at this size (primary source) or they are formed through the decay of larger pieces (secondary source). The ingestion of microplastics by marine organisms can lead to suffocation or starvation. Another problem is the interaction between microplastics and toxic chemicals. The entry of microplastics can occur either via the mainland or directly at sea. On the mainland, the particles are introduced via sewage systems or drifting waste. At sea, plastic is directly introduced as waste. However, although it is known that plastic waste is introduced into the marine environment via the mainland, there are so far only a few publications on microplastics in river systems.

In this research project, the concentrations of microplastics and the polymer types in the water of the Tennessee River are to be investigated using infrared spectroscopy (IR). To prepare the water samples for the measurements, they must be enzymatically cleaned with SDS, protease, cellulase, chitinase and also with H_2O_2 and divided into two size fractions. For the quantification and identification of the microplastics, two different IR spectrometers were used. Both with the ATR and with the help of the micro-FTIR measurements, different polymer types can be identified and particle sizes can be measured.

As already mentioned above, mass production of plastics began in the 1950s (Barnes et al. 2009) with 1.7 million tons of plastic worldwide (PlasticsEurope 2013). In 2012, plastic production rose to 288 million tons worldwide, with 57 million tons of it produced in Europe (PlasticsEurope 2013). The reason for the rapid increase in production are on the one hand the material advantages of plastics such as durability (Cole et al. 2011), light weight and stability (Andray 2011) as well as the cost-effective production (Derraik 2002). Depending on the use, different types of plastics are synthesized. With various additives, the range of applications of these materials becomes even more extensive. The most frequently produced plastics are polypropylene (PP), polyethylene (PE), polystyrene (PS), polyvinyl chloride (PVC) and polyethylene terephthalate (PET) (PlasticsEurope 2013). Additives that are added to the polymers are stabilizers, antioxidants and coupling reagents (WagnerabdOehlmann 2009). Monomers, such as Bisphenol A(BPA), plasticizers such as phthalates or flame retardants such as polybrominated diphenyl ethers (PBDE) can be absorbed by humans directly through plastic contact or indirectly through food packaging (Thompson et al. 2009).

Since the 1960s, the amount of plastic waste in marine habitats has been continuously increasing (Derrai 2002; Barnes 2005), and the aforementioned advantage of the durability of the plastic material thus becomes a problem in the marine environment. About 75% of all marine waste is made up of plastic (Galgani et al. 2013). The pollution of the world's oceans by large plastic items occurs at sea and from the mainland, at sea through the intentional or unintentional waste entry from ships. Pollution of seawater and beaches originating from the mainland is mainly caused by municipal sewage systems and lying and drifting plastic waste (Derraik 2002).

Not only does this waste contain large pieces of plastic, but also smaller parts, so-called microplastics. In the formation or origin of microplastics, two sources are distinguished, whereby primary microplastics are those that are already industrially produced and used in microscopic particle size, while secondary microplastics are formed by the decay of large plastic pieces (macroplastics) (Thompson et al. 2004). Primary microplastics are used in body care and cosmetic products (Barnes et al. 2009), as granules for injection molding or powder for 3-D printers, in medication (drug capsules) and much more (Browne et al. 2011). The decomposition of large plastic fragments mainly occurs photochemically (Cozar et al. 2014). Underwater, decomposition cannot take place due to lack of UV radiation. In the sediment, decomposition therefore mainly occurs mechanically or bacterially. Due to UV radiation and the extraction medium water, the plastic becomes brittle and due to the wave movement, which rasps the plastic over sand and stones (mechanical decomposition), large pieces gradually fragment into microplastics (Ivar do SulabdCosta 2014).

A main problem of plastic waste is its uptake by marine biota (Barnes et al. 2009). Depending on the size and density of the plastics, they can be found along the entire water column, with impacts on the marine food web. Floating and suspended microplastics are ingested by corals, reef clams, sponges, tube worms and other organisms that obtain their food by filtering out microorganisms and plankton from the water. On the other hand, plastic particles with high density, which over time become covered with a biofilm that makes the particles even heavier, slowly sink and thus become accessible as a "food source" for organisms living in or on the sediment of the seabed (Browne et al. 2007).

Through the ingestion of plastic waste, the digestive tract can be blocked or the stomach lining damaged, causing the animals to starve, or the respiratory tract is blocked, and the animals suffocate on microplastics.

Environmental organizations have published a multitude of images of dead seabirds on different beaches on social media (Halang n.d). During the decomposition process, the stomach contents of the carcass gradually become visible, shockingly proving the cause of death, as plastic is the last thing to rot and remains (ZarflabdMatthies 2010).

Microplastics pose another danger. Plastics can transport toxins in two ways. Plastics contain toxic chemicals that are added during their manufacturing process and can be released into the environment. These toxins have an ecotoxicological impact on the habitat of animals and humans. An example of this is the mentioned estrogenic effect of bisphenol-A and alkylphenol additives or the reduction of testosterone production by phthalate-containing plasticizers (Teuten et al. 2009).

In addition, due to their large surface area (Sect. 3.2.4), microplastic particles can adsorb toxins from their aquatic environment.

The pbt (persistent, bioaccumulative and toxic) substances settled on the plastic particles can be ingested as food by various organisms through the above-mentioned uptake possibilities. The microplastic particle thus becomes a transport vehicle for a toxic load (Trojan horse). During the attempt to metabolize the plastic, the carried toxins can be released, stored and accumulated in the organism (Engler 2012).

Although plastic pollution of the world's oceans originates from the mainland and it is assumed that rivers play a major role as a transport medium for microplastics, the uptake of microplastics by freshwater organisms has so far been little researched. Freshwater studies, in which the uptake of microplastics by freshwater fauna was investigated, have so far only been carried out in sewage treatment plants, river mouths and large lakes (Imhof et al. 2013). It was shown that the microplastic contamination is of the same order of magnitude as that of marine habitats (Imhof et al. 2013). This inevitably leads to the assumption that river systems are not only pathways for microplastic entry into our seas, but also act as a sink for microplastics.

There are two directives for the protection against pollution in maritime habitats. Firstly, the WFD (Water Framework Directive), which focuses on a good ecological status of all organisms dependent on surface waters, and secondly, the MSFD (Marine Strategy Framework Directive), which works to achieve or maintain a good state of the marine environment.

The uptake of microplastics by limnic organisms has so far been little researched. Both in the investigation of rivers regarding the microplastic load and the different types, as well as in the investigation of freshwater organisms, there is a scientific gap.

The Rhine investigation has made an incomplete approach to this, but has provided a first contribution that needs to be expanded and compared with other rivers in order to obtain further information about the contamination and the accumulation of microplastics in rivers and thus to be able to better assess the potential danger for flora and fauna.

2.6.1 Microplastics in Inland Waters Using the Example of the Rhine

The Rhine originates in Switzerland and flows into the North Sea in the Netherlands after 1328 kilometers. About 50 million people live in the catchment area of the Rhine, the most important and most diversely used river in Europe, of whom about 22 million people are supplied with drinking water from the Rhine. The Rhine has great economic and demographic significance, and in the past, the river formed the basis for successful settlement and a functioning economy. However, the increasing industrial use of the Rhine as a trade route and transport route has led to increasing pollution of the water—unfortunately, the river has also been misused as a disposal route for all kinds of waste.

Before the discovery of the industrial synthesis of plastics, the waste still consisted of organic components. These can be decomposed into harmless degradation products by natural processes. However, the introduction of long-lasting synthetic plastics has disrupted the natural degradation process (Hanser Customer Center 2017). Once plastic is in the waters, it breaks down into ever smaller particles and poses a great danger to the ecosystem. Numerous studies on marine organisms demonstrate the consequences of microplastics. It is ingested by animals and blocks their digestive tract, or they succumb to the pollutants mixed into the plastic. About half of the plastic waste in the sea comes from land. Rivers transport large amounts of plastic waste seawards. During an investigation of the Los Angeles River and San Gabriel River in the southwest of the USA, an enormous amount of plastic particles was found within 24 hours. Extrapolations resulted in 2.3 billion plastic fragments weighing 30 tons per year (ntv 2017).

The following sections present the methods for determining the pollution status of an inland water body like the Rhine by microplastic particles in the surface water.

Filtration

To investigate the microplastic load in waters, three methods have been used so far:

1. *Manta-Trawler*: A Manta-Trawler has already been used in investigations of the microplastic load in the Weser and Elbe. The shape of the construction resembles a manta ray. The main element is a rectangular, buoyant body, with a wide opening, at which a fine-meshed net made of plastic is located. The Manta is towed behind a water vehicle through the water for sampling, in this way possible plastic particles are fished from the surface of the water and collect in a collection container at the end of the plastic net.
2. *Candle filter*: In this method, a defined volume is pumped through special candle filters. Solids remain on the fine-meshed fabric of the filter and can be examined in the further course.
3. *Sediment samples*: In this method, sediment samples are taken at different points of a water body. The samples are then prepared and examined in the laboratory.

The most suitable method for sampling within the entire course of the Rhine is filtration with stainless steel mesh candles. A Manta-Trawler would have been a comparable alternative, but its use, especially in the Vorderrhein with its narrow and quite wild passages, is not possible due to its enormous size. Also, a water vehicle would be necessary to tow the Manta-Trawler through the water. In most care and hygiene products, microplastic particles made of polyethylene or polypropylene are used (BUND 2014), these have a lower density than water and float on the water surface of a water body, instead of settling on the bottom or in the shore area in the sediment. Furthermore, the later purification of the sediment samples is more time-consuming compared to the candle filter, as it contains a lot of sand, rock fragments, and organic material.

Therefore, the use of candle filters is the best choice for sampling the Rhine. With the help of a diaphragm pump, a precisely defined volume can be filtered. In addition, the few components can be mounted on a compact and space-saving frame.

The filter apparatus was based on a system from AWI on Helgoland. However, this was expanded with certain components and adapted to the special conditions on the Rhine.

To transport the water through the candle filters, a four-chamber diaphragm pump was used, which is operated with 12 V direct current. This allows a 12 V battery to serve as a power source, for example, on a boat. If no 12 V source is available, a generator can be used in combination with a power supply unit. The results of a particle size analysis of selected hygiene and care products, which contain primary microplastics, led to the selection of a filter fabric of 10 μm.

The following parts were used to build the filter apparatus:

- Four-chamber diaphragm pump Series 5050 from Shurflo 12 V
- Filter housing made of polypropylene with stainless steel mesh filter candle from Wolftechnik; pore size 10 μm; length 124 mm
- Control unit of the diaphragm pump with speed control; On/Off switch with additional 20-A fuses; wiring
- Power supply unit—laboratory power supply ea-ps 3016-20b from EA Elektro-Automatik
- Water meter type 08188-20 from Gardena
- Generator type EU20i from Honda
- Stainless steel hose olive (self-made)
- PMMA plate (Plexiglas) for the frame of the system

Couplings and connectors:

- Quick couplings (Insert & Body) with check valve made of polysulfone from Colder Products Company; ½ inch AG
- Laboratory hose made of PVC from Carl Roth; ½ inch with additional hose clamps
- Hose coupling from Gardena; ½ inch
- Brass reducing piece ½ inch internal thread to ¾ external thread
- Tap piece ½ inch to ½ inch internal thread from Gardena
- Teflon sealing tape

Figure 2.146 shows the schematic structure of the filter apparatus used, including all components.

1. Stainless steel hose olive (Fig. 2.147)
2. Laboratory hose made of PVC; ½ inch; length: approx. 2 m (2.1: 0.6 m, 2.2: 2 m)
3. Hose coupling from Gardena; ½ inch

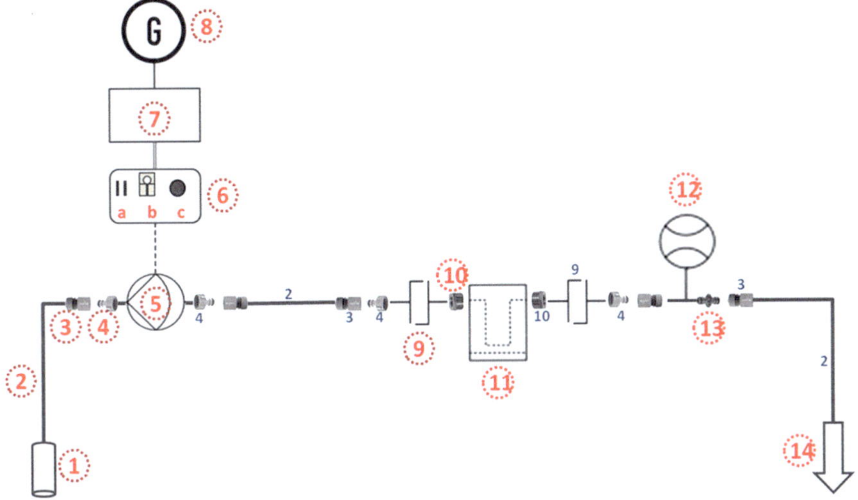

Fig. 2.146 Sketch of the filter apparatus with all components used

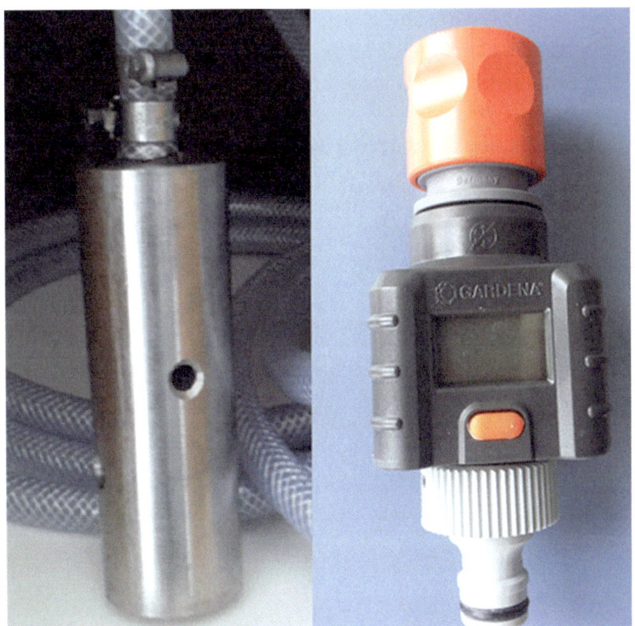

Fig. 2.147 Stainless steel hose olive (left), Gardena water meter (right)

4. Faucet piece ½ inch to ½ inch internal thread from Gardena
5. Four-chamber diaphragm pump Series 5050 from Shurflo
6. Control unit of the diaphragm pump
    a. 2 × fuses 12 V, 20 A
    b. On/Off switch
    c. Speed controller
7. Power supply unit—laboratory power supply ea-ps 3016-20b from EA
8. Power source (EU20i generator from Honda)
9. Quick couplings (Insert & Body) with non-return coupling from Colder Products Company; ½ inch
10. Brass reducing piece ½ inch internal thread to ¾ external thread
11. Filter housing with stainless steel mesh filter cartridge from Wolftechnik; Pore size: 10 µm
12. Water meter from Gardena
13. Hose adapter ½ inch from Gardena
14. Wastewater drain

The water is sucked in by the diaphragm pump through a suction olive made of stainless steel (Fig. 2.147) and then conveyed through the candle filter. The system is connected by laboratory hoses made of PVC (½ inch) and special quick couplings with integrated check valve.

The core element of the system is a filter housing made of polypropylene (Fig. 2.148) with a diameter of 14 cm. The lid of the housing is removable, inside is a stainless steel mesh candle with a length of 12.4 cm and a diameter of 5.8 cm. The pore size of the mesh is 10 µm.

A four-chamber diaphragm pump (12-V direct current) draws in the water via the hose olive and pumps it with a maximum delivery rate of 20 l/min through the candle filter. A generator serves as the power source for the power supply unit, which provides direct current at 12 V. A water meter was installed directly after

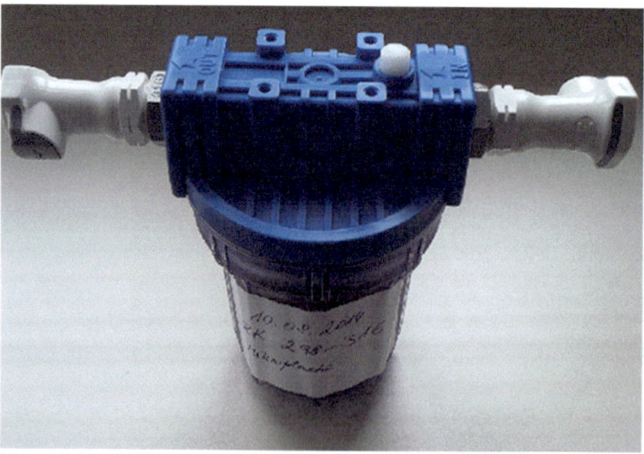

Fig. 2.148 Filter housing made of polypropylene from Wolftechnik, 660 ml. Right: Stainless steel mesh filter candle, pore size: 10 µm

the candle filter and allows control of the already promoted sample volume. The control of the diaphragm pump is built into a waterproof switch housing. A speed controller allows stepless control of the delivery rate. The filter and the pump are connected with a laboratory hose made of PVC with a diameter of ½ inch, quick couplings with check valve on the filter housing allow a quick exchange of the filter elements. All components are mounted on a Plexiglas plate with dimensions 60 cm × 45 cm. For transport, two brackets were additionally mounted on the sides.

As part of the "Rheines Wasser" project, the investigation of microplastics in the surface water of the Rhine began on 28.07.2014. With the help of the filter system (Fig. 2.149), 11 samples were taken in duplicate determination. The installed diaphragm pump draws in Rhine water and promotes it through the stainless steel mesh candle filter. Suspended solids and possibly existing plastic particles settle on the surface of the filter. All drivable sections of the Rhine were sampled from the boat, at the remaining stations (at Tomasee, in Chur and in Laufenburg) the filtration was carried out from the shore. The sample volume was controlled using a water meter (Fig. 2.138). 1000 liters of Rhine water were filtered per filter. For the filtration, the suction hose was placed just or a maximum of 15 cm below the water surface. Each sampling point met certain requirements: It had to be freely accessible and ideally located at the main current and could not be in a groyne or

Fig. 2.149 Complete filter apparatus from above: 1. Filter housing with stainless steel mesh filter candle from Wolftechnik; pore size: 10 µm, 2. Four-chamber diaphragm pump Series 5050 from Shurflo, 3. Control unit of the diaphragm pump, 4. Water meter from Gardena

Fig. 2.150 Sampling at the Alpine Rhine near Chur

the like (Fig. 2.150). To place the end of the suction hose as much as possible in the main current, the hose was placed in the current from the shore using an extension of approx. 2 m.

The starting point for the sampling was Lake Toma in the canton of Graubünden (CH). The lake is considered the source of the Rhine and was therefore the first to be sampled. The first sample served as a reference for later evaluation. Table 2.32 lists all the sampled locations along the Rhine. After Lake Constance, the sampling

Table 2.32 Sampling locations along the course of the Rhine

Sampling location	Date of sampling	Sample volume in liters
Source of the Rhine, Lake Toma	28.07.2014	1000
Chur (CH)	29.07.2014	800
Lake Constance, Rorschach	01.08.2014	1000
Stein am Rhein (CH)	03.08.2014	1000
Laufenburg (CH)	05.08.2014	1000
Rhine kilometer 319–332	10.08.2014	1000
Mainz, Rhine kilometer 503–521	13.08.2014	1000
Rhine kilometer 635–648	15.08.2014	1025
Rhine kilometer 775–795	19.08.2014	1000
Wageningen (NL), Rhine kilometer 887–901	21.08.2014	1000
Rhine kilometer 963–973, side channel (Lek)	23.08.2014	1000

was carried out from a boat, with the corresponding sections indicated by Rhine kilometers.

The suction hose, with an inner diameter of 127 mm, was held directly in the current, thus larger particles could also be sucked in, which could lead to pump blockage when accumulated. After a thorough cleaning of the components, the sampling can continue. An improvement can be made by attaching a simple sieve to the end of the hose, with, for example, a 5 mm pore size, to avoid the intake of "larger" particles and still filter microplastics according to definition. It is also important to ensure that the diaphragm pump does not suck in air before and during operation, as this reduces the pump performance, as no vacuum can be created by the pump. On average, it takes about 90 minutes to filter 1 cubic meter of water. The diaphragm pump delivers between 10 and 14 liters per minute. In the shore area and at shallow points of the river, ground contact of the suction hose should also be avoided. It can also happen that the water meters used switch off after a certain time. Therefore, one should also keep an eye on the total time and the average delivery volume in order to be able to maintain the set total volume of one cubic meter per filter unit in the event of a failure of the water meter.

Overall, the filtration system has proven itself in its use on the Rhine. Due to its compact design, it takes up little space during transport and on the boat and can be carried without much effort to the corresponding sampling points (Table 2.32). When choosing the sampling points, some details should be taken into account. They should ideally be located at fast river passages near the main current and not at inlets from factories and sewage treatment plants. Since a single point cannot be representative of the entire river section, sampling the main current is most sensible in order to obtain a realistic picture of the respective river section. In groynes or in backwaters, the concentration of microplastics can be significantly higher compared to the main current, as well as in the outlet of sewage treatment plants. In advance, the total volume was set at 1000 liters per filter. The number was adopted in consultation with the Alfred Wegener Institute, as the institute has already conducted similar investigations in the past. This allows the results to be compared with other projects later on. It is not always possible to filter the planned 1000 liters, as the proportion of suspended matter can sometimes be so large that the filter clogs prematurely and there is a risk of damaging it. This must be taken into account in the evaluation. The use of Teflon tape to seal the connections often cannot be avoided, so special attention must be paid to this in the evaluation of the samples. It is very likely that plastic fibers from Teflon will detach and settle on the surface of the filter. The advantage of Teflon over other sealing materials lies in its density (2.20 g/cm^3). In the later density separation, Teflon sinks in the zinc chloride solution (density $= 1.65$ g/cm^3) and can thus be easily separated. In summary, it can be stated that the filtration system is suitable for microplastic filtration in flowing waters, using the example of the Rhine, and that concentrations of microplastics in a flowing water body can constantly vary, as floods or changing weather conditions cause fluctuations.

Selection

During the sampling on the Rhine, a total of eleven samples were taken at different points along the entire course of the Rhine. After the sample collection, the samples were cleaned. In this process, microplastic particles were freed from suspended solids and isolated. The suspended solids are undissolved, inorganic minerals or organic particles. In surface water, the majority of the organic particles consist of plankton, which includes algae, bacteria and other microorganisms. The inorganic portion mainly consists of soil particles such as sand or clay compounds. Insects and tiny creatures such as crabs or larvae are also sucked into the filter. These components of the sample need to be removed in order to isolate the microplastics.

A selection or isolation of the microplastic particles is achieved through a multi-stage enzymatic treatment, in which all organic components are removed. A subsequent density separation with zinc chloride separates inorganic particles. The following listed materials and devices are used to process the filter residues.

Devices:
- Filter systemfor microplastics
- Water meter and various couplings
- Shaking incubator
- Vacuum controller
- Drying oven
- Ultrasonic bath
- Stainless steel mesh candles (Pore size 10 µm) with filter housing
- Laboratory scale
- Bottle tops for filtration at the vacuum controller (Bottletop); 250 ml
- Round filters made of stainless steel mesh (500 µm and 10 µm)
- Anodisc membrane filters (0.2 µm pore size; diameter: 25 mm) Whatman
- Erlenmeyer flasks and beakers in various sizes
- Glass funnel
- Separating funnel 250 ml
- Spray bottles
- Small wire brush
- Tweezers
- Pipette (1000 µl)
- Compressed air connection
- Schott bottles 250 ml and 1000 ml

Consumables:
- Aluminum foil
- Nitrile gloves

Enzymes:
- CellulaseTXL; 30 U/ml
- ProteaseA-01; 1,100 U/ml
- Chitinase; 40 U/m, Manufacturer: ASA Special Enzymes

Chemicals and buffers:
- Pure water (Milli-Q)
- Ethanol (25%), Sigma
- Hydrogen peroxide (H_2O_2) 35%, Sigma
- Zinc chloride($ZnCl_2$); Density 1.65 g/cm^3, Carl Roth
- PBS buffer solution; Solubility: 150 g/l, Merck
- Sodium dodecyl sulfate (SDS), Merck

A time-consuming procedure must be carried out to process the filter samples. The diagram in Fig. 2.151 schematically shows all the successive purification steps of the method.

First, a sufficient amount of tap water must be filtered using the filter system and the associated stainless steel mesh filter candle (pore size: 10 µm). The filtered

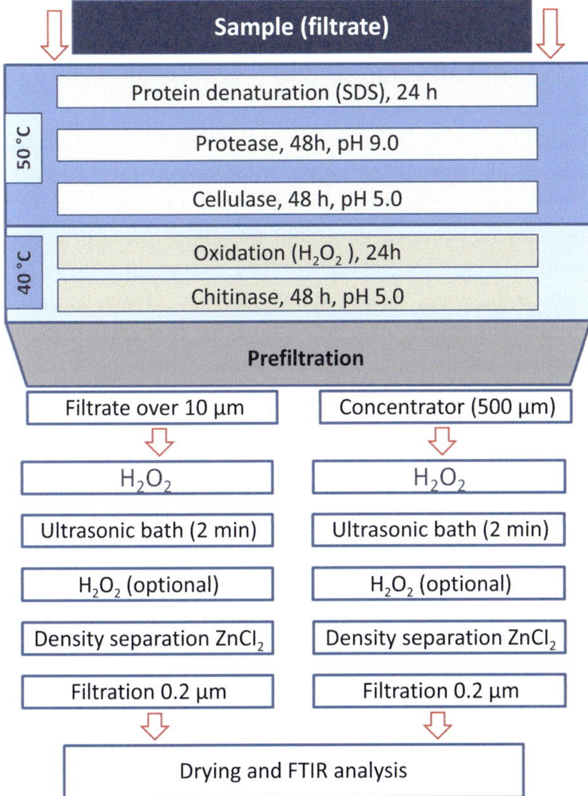

Fig. 2.151 Flowchart of the method for the isolation of microplastic File: Preparation scheme Microplastic sample.png

tap water is used to prepare the enzyme and buffer solutions. The enzymes and enzyme solutions are prepared according to the various data sheets and stored cool. Subsequently, the sodium dodecyl sulfate solution and different PBS buffer solutions are prepared. The correct pH value must be set for all solutions, the pH optima of the enzymes are taken from the respective data sheets. The entire treatment takes place in the filter housing. The housing remains closed during the entire enzymatic treatment and is only opened for pre-filtration. The filter housings are flushed with about 10 liters of the filtered tap water using the membrane pump, remaining water is removed with compressed air. Once all preparations have been made, the actual enzymatic treatment can begin.

The first step of purification is a protein denaturation by a sodium dodecyl sulfate solution (150 g/l), also referred to as SDS *(sodium dodecyl sulfate)*. SDS is an ester compound that is formed from sulfuric acid by the elimination of water. The anionic surfactant destroys non-covalent bonds of proteins. For this purpose, the prepared solution is conveyed into the filter housing with a volume of 660 ml using the filter system. Incubation takes place at 50 °C in the shaking incubator. Sodium dodecyl sulfate dissolves larger protein complexes by eliminating the hydrophobic effect. This allows organic suspended solids, mostly plankton, to be better digested.

After an incubation period of 24 h, the sodium dodecyl sulfate solution is emptied by compressed air and rinsed again with filtered tap water. The process is repeated with the enzymes protease and cellulase. The samples are treated for 48 hours each. The enzyme protease continues the digestion of proteins by hydrolyzing covalent protein bonds (Stryer 1990). The treatment with cellulase degrades plant components in the water. The enzyme causes a 1,4-glycosidic cleavage of the cellulose, the main component in plant cell walls, through a hydrolysis reaction. The enzyme chitinase cleaves and degrades chitin. The polysaccharide is the main component of the exoskeleton of insects and is also found in cell walls of yeasts, fungi, and algae.

After the enzyme treatment with cellulase, the sample components are oxidized with hydrogen peroxide (35) for 24 h and 40 °C in the incubator. An alternative would be the use of caustic soda or hydrochloric acid. However, tests have shown that plastics are not or only relatively slightly attacked by hydrogen peroxide compared to acids and alkalis.

Finally, chitin residues from insects and crustaceans are removed by treatment with the enzyme chitinase in 48 h at 40 °C.

After the enzyme treatment, pre-filtration is carried out. To facilitate later evaluation, the samples are separated according to particle size. For the pre-filtration, the filter housing is opened for the first time. The filter candle is removed and placed in a cleaned beaker. It is important to note that only nitrile gloves are used and all required items such as beakers or tweezers are cleaned with pure water and then with ethanol (25%) before use. The filter cake, which has built up on the surface of the candle filter, is carefully scraped into the beaker with a small wire brush and rinsed with pure water and ethanol. Any residues in the filter housing are also placed in the beaker, in addition, the housing itself and all involved components such as seals and the lid are rinsed with pure water and ethanol and collected in the beaker. This is followed by treatment of the filter candle in an ultrasonic bath for a maximum of 3 minutes, after which the cleaning process of the filter candle is repeated and all used items (wire brush, tweezers, etc.) are rinsed again and the rinse liquid is collected in the beaker. The contents of the beaker are then passed through the 500-μm round filter. The filtrate is collected. The residues on the filter are rinsed with pure water and stored in a Schott bottle until evaluation. Larger plastic parts are separated by the filtration through a round filter with a pore size of 500 μm. Particles larger than 500 μm are optically selected under the binocular. Particles of the smaller fraction can no longer be manually selected and are therefore analyzed spectroscopically.

For this purpose, the collected filtrate is sucked off in the next step over a stainless steel mesh filter (10 μm) at the vacuum controller. The round filter is placed in the bottle cap for this purpose, which is mounted on a 1-L Schott bottle. After vacuum filtration, the stainless steel mesh filter is placed in a 250-ml Schott bottle and again treated with hydrogen peroxide for 24 hours. To avoid contamination, a new bottle cap is used for each sample, Schott bottles are not covered with the corresponding lid, but with aluminum foil. Before the final step of the isolation, the round filter is removed and cleaned with pure water and ethanol. The residues are placed in a separating funnel for density separation with zinc chloride. The density separation can take up to 48 hours. Inorganic material, such as sand or clay particles, are thus removed in the final step of purification. This is done by a density separation, using a zinc chloride solution as a separating medium.

Through the classic float/sink method, all plastics with low density accumulate on the surface. Plastics with higher densities, such as polyvinylidene fluoride (1.78 g/cm^3) or polytetrafluoroethylene (2.15 g/cm^3) sink to the bottom in the zinc chloride solution (1.65 g/cm^3), making separation impossible. For this reason, a further density separation of the lower phase would be necessary. In addition to the zinc chloride solution, concentrated salt solution or sodium iodide with correspondingly higher densities can theoretically also be used for the separation. This allows the separation of specifically heavier plastics. Difficulties could arise when draining the lower phase, as it is not exactly known when only microplastics are left in the separating funnel. The timing is difficult to standardize and several preliminary tests are needed to optimize the timing. The zinc chloride solution, including the contained microplastic particles, is again vacuum-filtered onto membrane filters with a pore size of 0.2 µm. The membrane filters are finally dried for 24 hours at 40 °C. This handling allows the analysis of the entire filter.

During the purification process, some details must be observed: The aluminum foil is used to cover Schott bottles, as the lid of the bottles, especially the inner sealing ring, represents a source of contamination. Likewise, all open vessels must be covered, as plastic fibers from the ambient air can contaminate the samples. To control the individual purification steps, three checks are carried out (see below). Primarily, this is to find errors in the setup of the apparatus and to secure the later evaluation. The recovery control is intended to rule out that plastic particles are lost during the filtering process. This control mainly refers to the filter system itself. Microplastic particles can settle during the filtering process, which means that not all particles are detected and thus the result is falsified. To determine the efficiency of the enzymatic treatment, a dry mass determination is carried out before and after the treatment. Furthermore, a blank sample is carried along and analyzed in the same way during the entire purification process. If contaminations are found in the subsequent evaluation, these must be taken into account and deducted in the analysis of the results. Plastic fragments from the air and fibers from clothing are very likely. During the filtration process, plastic parts from the filter apparatus can detach (hose connections, couplings, filter housing, membrane pump) and thus contaminate samples. Since it was not possible to completely avoid plastic in the construction of the filter system, a list of used materials and devices made of plastic must also be created. If necessary, an FTIR database of the materials must be created.

The isolation of microplastics is a very time-consuming process, in addition, work must be done very carefully and cleanly to avoid contamination. Therefore, all materials used should be as plastic-free as possible.

Controls

In order to avoid misinterpretations in the quantification of microplastic particles, the following controls are carried out in the applied method:

1. *Dry mass determination*: The efficiency of the enzymatic treatment is to be determined. For this purpose, the dry mass of the filter cake on the candle filter is determined before and after purification.
2. *Recovery control:* In the recovery control, so-called *PE beads* with a diameter of 200 μm are pumped through the filter system with water. Afterwards, the plastic particles are picked and counted under the microscope.
3. *Blank sample*: In the blank sample, an unused candle filter including housing is used. The housing is filled with filtered water and also incubated during purification. After the treatment, the water is filtered and the filter is dried.

Detection

A specialized method of Fourier Transform Infrared Spectroscopy is used to evaluate the samples. A FPA detector is used for this purpose. The so-called Focal-Plane-Array detector allows spectroscopic evaluation of larger areas. With the help of FTIR-FPA technology, complete filters can be analyzed. The filter is divided into individual sections, and one section is detected at 4096 points. An interferogram is created for each detection and compared with existing plastic databases. This technique allows each individual particle on the surface of the filter to be qualitatively evaluated. This method can analyze particles down to a minimum size of 20 μm.

The measurement takes 10–12 hours and the FPA detector must be cooled down with liquid nitrogen every three hours. To identify plastics, the integration of four wavenumber ranges (2980–2780 cm^{-1}, 1800–1740 cm^{-1}, 1760–1670 cm^{-1}, 1480–1400 cm^{-1}) is necessary. Depending on the type of plastic, specific absorptions or transmissions occur in these ranges. Figure 2.152 shows the procedure for identifying individual plastic particles on the dried filter.

Results and Interpretation

In August 2014, the author Prof. Dr. Andreas Fath swam downstream the Rhine, from the source at Tomasee, to the North Sea at Hoek van Holland. This *Sports-meets-Science-* project "Rheines Wasser" was accompanied by students from Furtwangen University (HFU) and several international scientific institutes (EAWAG/TZW/AWI/Wetsus). A variety of analyses were carried out to investigate the water quality of the Rhine: from rapid tests for nitrates, phosphates, COD, pH value, O_2 and turbidity, to the instrumental analysis of heavy metals, pharmaceuticals, industrial chemicals, X-ray contrast media and perfluorinated surfactants, to

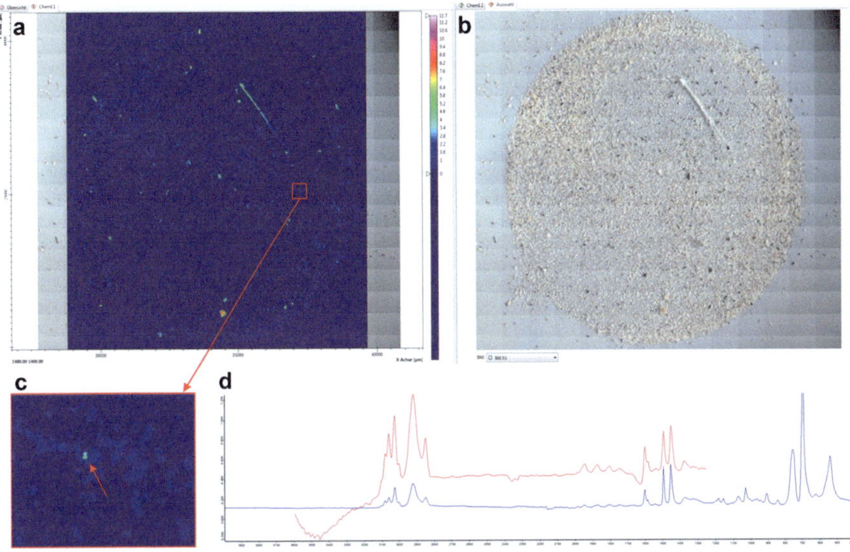

Fig. 2.152 Procedure for identifying a single microplastic particle. A particle (C) is selected on the false color plot (A). Its characteristic IR spectrum can be seen in (D) (red spectrum). This is compared in the overlay with a spectrum from the plastic database (blue spectrum). If they match, the plastic is identified. The image (B) of the filter surface provides an overview of the entire filter and serves as a guide

microorganisms and microplastics. A main goal was and is to draw the public's attention to water protection with a spectacular action.

For the first time within this research project, the Rhine as an inland water was investigated for microplastics along its entire course. We are already informed about the plastic garbage vortex in the North Pacific and plastic in Lake Constance as well as on Germany's coasts. But what about the Rhine? Where the view to the bottom is possible, such as at the weirs in the Upper Rhine, numerous PET bottles and tin cans can be found Fig. 2.153. The rock transported in the Rhine not only grinds itself but also the softer plastic. The Rhine is a huge plastic mill Fig. 2.154.

On the 1200 km long route, every 100 km, 1000 liters were filtered through a 10-µm metal sieve using a portable filter pump manufactured at HFU. At the Alfred Wegner Institute in Helgoland (Birte Beyer; Dr. Gunnar Gerdts, Dr. Martin Löder) and at HFU (Jonas Loritz, Helga Weinschrott; Dr. Andreas Fath), the filters were processed in a multi-stage, time-consuming process in which all accompanying materials such as insects, stalks, bark pieces, sand, shells, etc. were eliminated by the use of enzymes and hydrogen peroxide. Figure 2.155 shows the work at Tomasee, the source of the Rhine.

With the help of infrared spectroscopy, an identification and quantification of the microplastic particles in the surface water of the Rhine could be determined (Figs. 2.156, 2.157 and 2.158).

Fig. 2.153 Macroplastic fragments on the Atlantic beach in northern Spain (left) and macroplastic with tin cans on the bottom of the Upper Rhine near Rheinau (right)

Fig. 2.154 The Rhine is a plastic mill

In the surface water of the Rhine, ten different plastics were detected in August 2014 when filtering 1000 liters of water about 15 cm below the water surface. The plastics are listed with their abbreviations. These are the ones that we use and consume daily. An overview of the most common plastics and their use can be found in Table 2.12 and about their density in Table 2.33.

Polypropylene accounts for the largest share with around 80%, followed by Polyethylene with around 10%. The remaining 10% is mainly shared by polystyrene, PVC, and ABS. Due to the determined frequency of the plastics, no mass balance can be drawn about the entire plastic load in the Rhine, because 15 cm below the water surface, you will primarily find those plastic particles that float on the water surface due to their low density, which in the case of PP and PE is even

Fig. 2.155 Microplastic filtration at the source of the Rhine

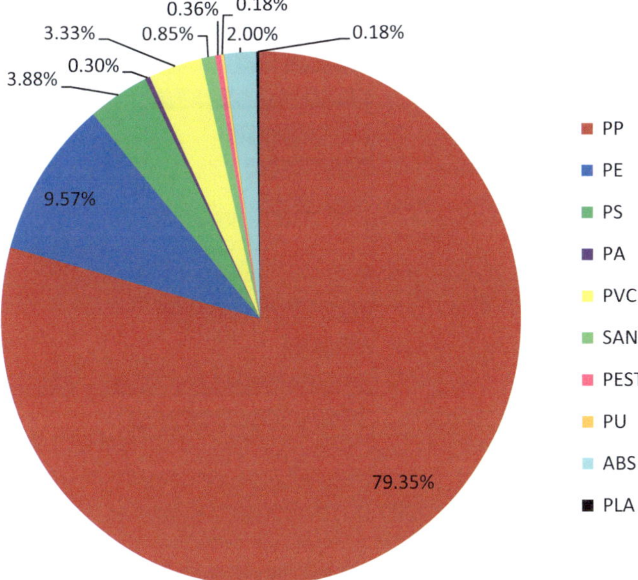

Fig. 2.156 Plastic type share of all microplastic samples <0.5 mm particle size

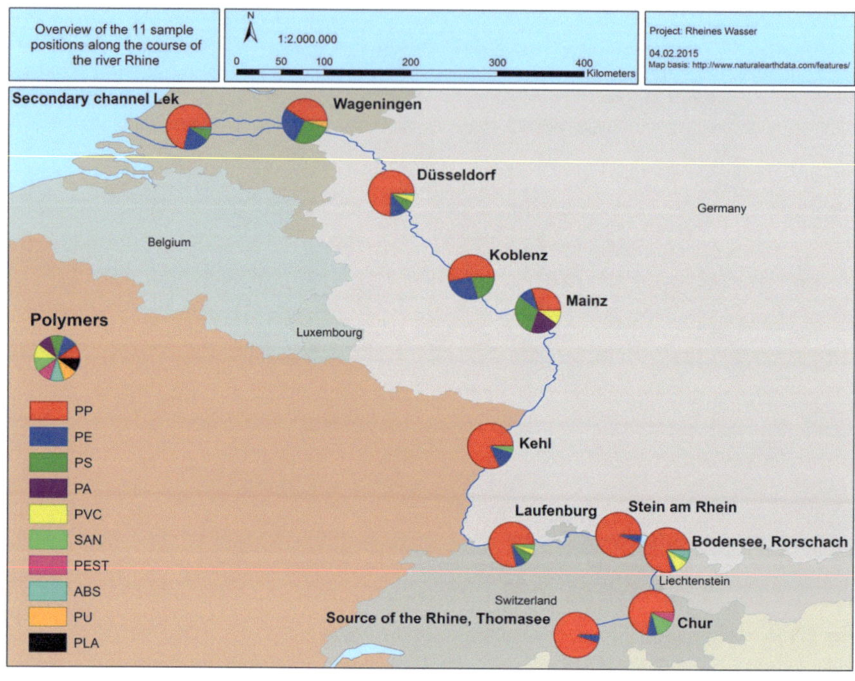

Fig. 2.157   Microplastic distribution along the Rhine

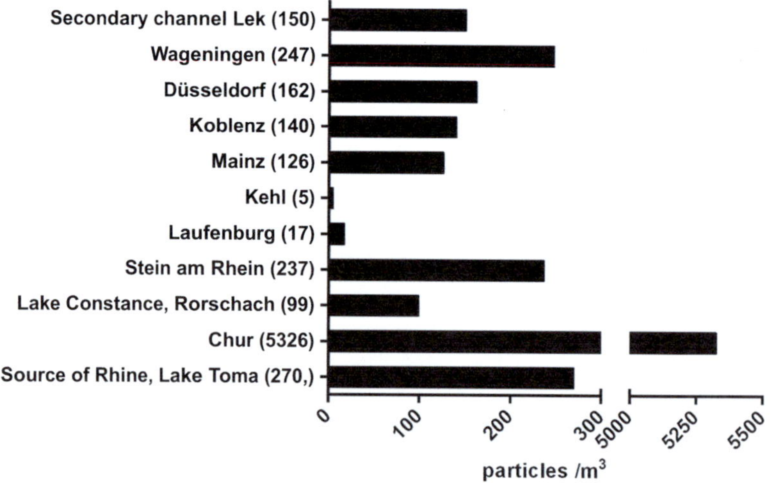

Fig. 2.158   Number of microplastic particles <0.5 mm along the Rhine

Table 2.33 Density of commonly used plastics in g/cm^3		
	Polypropylene (PP)	0.85–0.92
	Polyethylene (PE)	0.88–0.98
	Polyamide (PA)	1.01–1.16
	Polystyrene (PS)	1.05–1.08
	Styrene Acrylonitrile (SAN)	1.06–1.10
	Polymethyl Methacrylate (PMMA = Plexiglas)	1.16–1.20
	Polyester Epoxy (PEST)	1.10–1.40
	Polyvinyl Chloride (PVC)	1.19–1.41
	Polycarbonate (PC)	1.20–1.22
	Polyurethane (PUR)	1.20–1.26
	Polyvinyl Alcohol	1.21–1.31
	Polyethylene Terephthalate (PET)	1.33–1.41
	Polyoxymethylene (POM)	1.41–1.43

smaller than that of water. Sampling at greater depth or from the sediment would result in a different plastic distribution, especially since the density of PET, known from the PET bottles, which are stuck in the bottom of the river at the weirs of the Upper Rhine, is significantly higher at 1.4 g/cm^3 than PP with only 0.9 g/cm^3. The smaller the density of the microplastic particles, the more often they are found on the surface in the area where a swimmer stays (see the pie chart in Fig. 2.144 and Table 2.33).

The proportion of polyamides is smaller than expected despite their low density. Since polyamides mainly enter the wastewater via textile fibers and these fiber structures get caught everywhere, e.g. on fish scales, gills, in fish stomachs, on driftwood and water plants, this is an explanation for the small proportion in the surface water of the Rhine.

The initially unexpected result of the frequency of plastic particles along the entire course of the Rhine, shown in Fig. 2.157, and the location-specific distribution in Fig. 2.158 can be better interpreted through the close experience of the Rhine, if you have swum the 1231 km yourself and have felt all vertical and horizontal currents.

Just as with drugs, such as Diclofenac, Sulfamethoxazole, Metoprolol or sweeteners like Acesulfame and Sucralose, one would have expected a continuous increase in the number of microplastic particles parallel to the increase in the population with each kilometer of the Rhine. However, as Fig. 2.158 shows, the course is partly different.

From the last lock in Iffezheim, the gradient remains almost constant between 0.1 and 0.2 ‰ (exception St. Goar 0.4 ‰, no sampling) until the mouth. On this stretch, a continuous increase from 126 to 247 microplastic particles <500 μm is recorded. The last measurement in a side channel of the Lower Rhine is not connected to the current and is therefore to be considered as a single measurement. It is also noticeable that from Mainz onwards, the proportion of other types of plastic besides PP and PE increases. This can have two reasons. Either the input of these types of plastic has increased due to wastewater discharges, or the grinding effect of the Rhine has transformed macroplastic into microplastic. In addition, in this section of the Middle Rhine, the current is stronger compared to the canal between Basel and Karlsruhe and is not interrupted by locks. Turbulences caused by high shipping traffic also create a stirring effect and do not allow the heavier plastic particles time to settle.

In the canal from Basel to the Iffeszheim lock, the smallest value of five particles per cubic meter is measured. This is due to the very weak to non-existent current, which allows even plastic particles with lower density to settle (the high PP content is an indicator of this).

In the Vorderrhein and the Alpine Rhine, no plastic particles remain on the bottom or in the sediment. Everything that can be washed away by the whitewater masses is entered into Lake Constance as a "standing water" with 99 particles per cubic meter. By the time the Untersee flows out at Stein am Rhein, the current and turbulence increase sharply, and so does the particle count to 237. The extremely high value at Chur clearly shows the dependence of the measurement result on the flow behavior of the Rhine. In whitewater, plastic particles do not undergo density separation, so that, in terms of the number and type of plastic, the complete microplastic content is recorded due to the mixing. The high particle count of $5326/m^3$ may indicate that in calmer sections of the Rhine, only a smaller proportion of microplastic particles stay near the surface, while the rest is distributed in depth and on the bottom.

Another unexpected and surprising result is the relatively high number of 270 plastic particles per cubic meter in Tomasee, the source of the Rhine in the Grisons Alps at 2345 m altitude. Up there, there is neither industry nor agriculture. Apart from the goats and marmots that can be found there, there are also occasional hikers. However, they could at most be held responsible for macroplastic contamination through irresponsible disposal of their food packaging, but not for the microplastic contamination in Tomasee, which should actually be the zero sample of our series of analyses. So how does microplastic get into Tomasee? First of all, it can be stated that Tomasee is fed by meltwater and has a very low discharge rate, which allows microplastic that has been introduced to accumulate in this sink over a longer period of time. The melting snowfields around the lake already contain the microplastic, which rises into the atmosphere with the hot exhaust air and other dusts due to incomplete and improper burning of plastics (diaper pants in the fireplace etc.) and is deposited elsewhere with the precipitation, e.g. in the Alps. This also explains why perfluorinated surfactants (e.g. due to the use of PFT-containing fire-fighting foam in major fires) can already be detected in Tomasee.

If one wanted to get a rough idea of the microplastic load flowing into the North Sea each year based solely on the surface-near Rhine water layer from the present study, one would arrive at a minimum of 8 tons. This assumes an average particle count of 200 per cubic meter. With a discharge volume of 2500 m^3/s in

August 2014 and a particle diameter of 500 µm (density about 1 g/cm^{3}and spherical shape), one gets approximately 8 tons of microplastic per year. This would be the lower limit, because the result of the sample in Chur shows that it can indeed be a multiple of this. In the flow conditions there in the Alpine Rhine, there are no water layers due to the strong turbulence, so that by sucking off 1000 liters of water, the particle content of the complete water depth cross-section is recorded. This results in a higher particle count and particle diversity.

This microplastic particle load, however many tons per year it may ultimately definitely be, represents an underestimated potential hazard for flora and fauna. It is known that microplastic particles act like a magnet for organic pollutants such as PFTs (reference PFT scandal Düsseldorf) and adsorb these adsorbable substances. Thus, pollutants bypass the analytics, including those of the Rhine monitoring stations, because the water samples are filtered before they are injected into the HPLC *(high performance* or *high pressure liquid chromatography)* and analyzed by mass spectrometry, to prevent clogging of the chromatographic separation columns (columns). The pollutants that stick to the plastic particles are therefore not detected.

This means that the load of PFTs and also that of other trace substances, which were not completely eliminated in sewage treatment plants, are higher than the analytics reveal to us. Depending on the type, quantity, and surface structure of the microplastic load and the distribution equilibrium, the discrepancy can vary in size.

A look through the microscope at microplastic particles reveals very large fascinating surface structures, which equally increase the adsorption capacity (Figs. 2.159 and 2.160). The already lively looking transparent "jellyfish", comparable to the beauty and diversity of seawater jellyfish, almost make us forget the toxic potential of both species.

Comparison of Results From Two Independent Studies on the Rhine

In a MP study commissioned by the LUBW at the end of 2014 in southern and western German inland waters, which includes the Rhine and the Danube as well as their tributaries, the collected water samples were also pretreated with the methods already described and microplastics were quantified non-destructively using infrared spectroscopic methods.

The analysis results obtained within the study, which were recently published (LUBW 2018), provide a comprehensive comparison between the selected waters and provide insights into the type and quantity of plastic particles and partly also provide information about their origin. The study covers the Rhine, starting from Lake Constance to the Dutch border.

Since the project "Rheines Wasser" 2014 (Fath 2016) already examined the Rhine from the source to the mouth for microplastics, a comparison with the LUBW study on the Rhine is offered, especially since both studies complement each other. In both studies, the samples were taken 15 cm below the water surface. This is a crucial criterion for comparability due to the different densities of

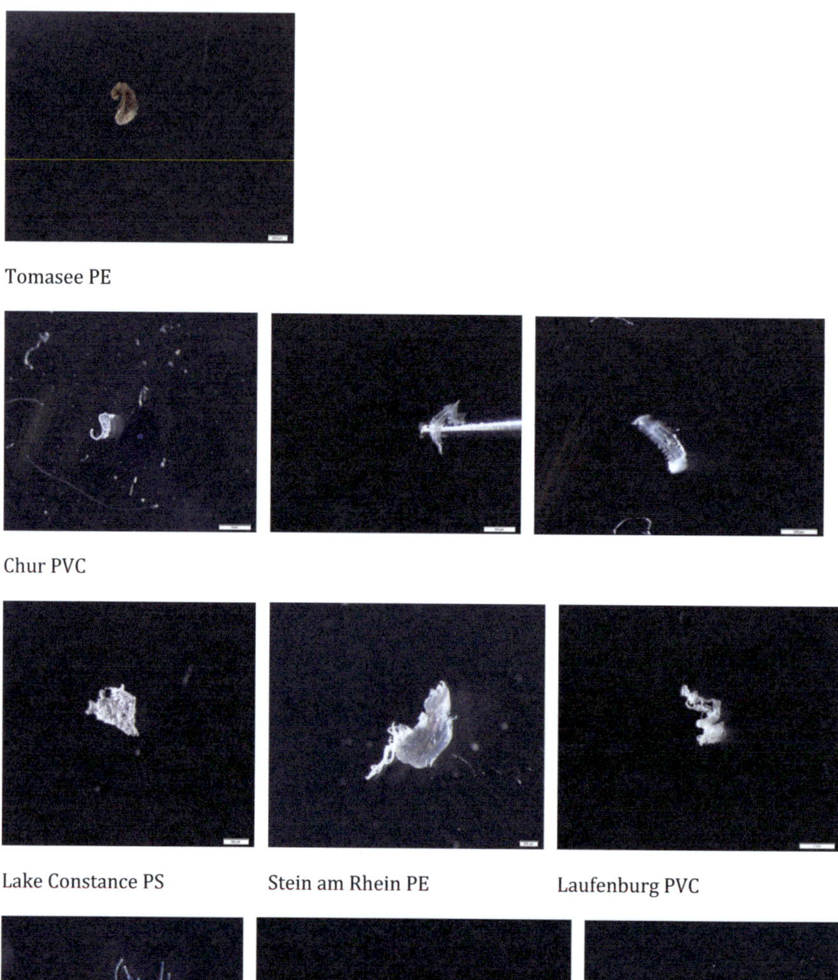

Fig. 2.159 Microscope images of microplastic particles >500 μm in the Rhine from the Tomasee

plastics. In both cases, the lighter plastics PP, PE, and PS predominate at almost all locations. In the tributaries of the rivers (Rhine and Danube), the micro particle numbers are significantly higher, sometimes by a factor of 10 (Emscher estuary).

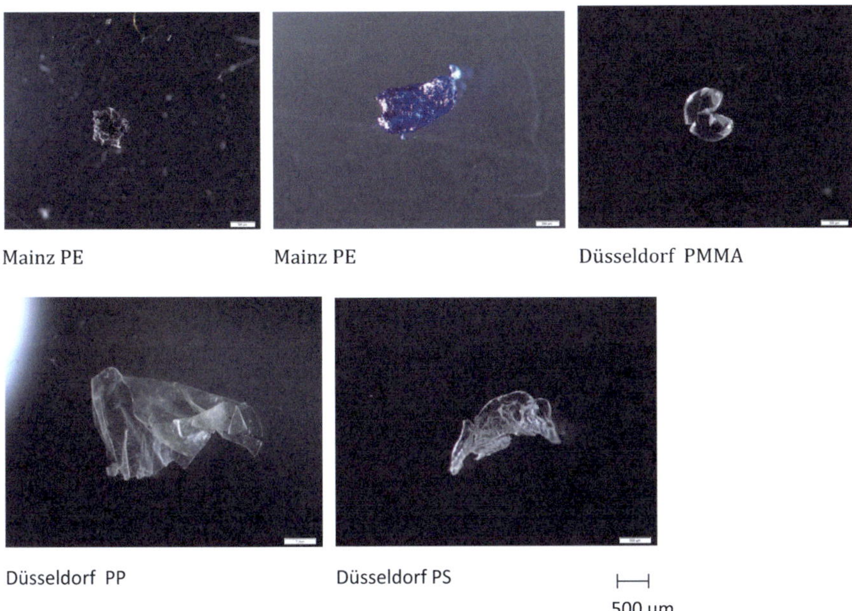

Fig. 2.160 Microscope images of microplastic particles >500 µm in the Rhine from the Tomasee

A significant difference, however, can be observed in terms of the number of particles. While the number of larger microplastic particles >300 µm in both studies is in the single-digit range, in the LUBW study on the Rhine, "only" low double-digit numbers of particles between 20 and 300 µm in diameter are counted, whereas in the "Rheines Wasser" project, three-digit numbers are counted (Table 2.34). The difference can be explained by the fact that in the LUBW study, all samples were collected with a trawl net with a mesh size of 300 µm, while in the "Rheines Wasser" project, a candle filter with a pore size of 10 µm was used. In both cases, 1000 liters of water were filtered each time. The comparison shows that more than 90% of the smallest examined MP category (20–300 µm) is missing in the LUBW inventory.

The fact that it is still possible to capture particles <300 µm with the trawl net is due to the net filling with all sorts of "bycatch" (leaves, algae, wood fibers, insects, etc.), which clogs the pores. Since the smallest particles are 20–300 µm, which occur in large numbers and are also more easily absorbed by marine habitats, it is all the more important to include these in an inventory. With the trawl net used, this is not possible quantitatively, and the comparison in Table 2.34 confirms the statement of the LUBW study that the number of very small particles in the water is actually much higher than recorded in this study, despite the relatively high number of smallest MP particles in the Rhine of 100–200 particles. If the sample from Chur (Fath) is included in the average calculation, 200 particles/m^3 were

Table 2.34 Comparison of particle numbers in the Rhine, based on two different samplings

Measurement point	>300 µm (LUBW)	>500 µm (Fath)	20–300 µm (LUBW)	25–500 µm (Fath)
Lake Constance	1	1	4	99
Mainz	2	4	1	126
Koblenz	9	0	13	140
Düsseldorf	5	3	2	162

determined in the Rhine in August 2014 (Fath 2016). In an international river comparison (Table 2.35) with an American (Tennessee River August 2017; Fath et al. 2019) and an Asian river (Yangtze; Wang et al. 2017), the values in the Rhine are low. As in the LUBW study, it is also shown here, including the Kinzig ($777/m^3$) (Jander 2017), a tributary of the Rhine, that the MP particle count from the tributaries is significantly higher than in the main stream.

In the LUBW study, the highest values for microplastic particles of the order of 300–1000 µm are determined at the Ruhr estuary with 117 particles. These are mainly PP and PE. The measuring point was located below a sewage treatment plant (sample Ruh 03 NW). In the comparison sample above the sewage treatment plant (Ruh 02 NW), only 8 particles were counted. These two data provide a clear indication that microplastic particles from the sewage treatment plant effluent are introduced into the Ruhr and that these are primary microplastics. This is suggested on the one hand by the particle size of the PP and PE particles, which are mainly used in cosmetics and peelings, and on the other hand by the fact that macroplastic was already eliminated in the first cleaning stage of the sewage treatment plant. In the biological treatment stage, after the completion of nitrification and denitrification, the sinking sludge is returned to the digester. The light microplastic particles with a density less than 1, including PE and PP, flow into the river together with the treated wastewater over the outlet.

Interpretation of Particle Distribution on Different Types of Plastic

In the distribution of the counted microplastic particles on the different types of plastic, as already mentioned, it is not surprising due to the sampling near the water surface that mainly the light plastics polypropylene and polyethylene are found and detectable. Nevertheless, there are significant differences both between

Table 2.35 Surface-near microplastic concentrations in different rivers

River	MP/m³	Discharge volume m³/s
Kinzig	800	5–10
Tennessee	16000	2000
Rhine	200	2500
Yangtze[a]	9000	31.900

[a] Wang et al. 2017 (50–500 µm)

the two studies and within the studies regarding the polymer type distribution (Tables 2.36 and 2.37).

The reasons for this are multifaceted. On the one hand, random events have a greater influence on statistical surveys with smaller amounts of data. While the distribution calculation in the case of the LUBW study is based on very low MP particle numbers of 5, 10, 3 or 7 (Table 2.37, Lake Constance, Grenzach-Wylen, and Düsseldorf), the distribution calculation in the comparative study (Fath et al. 2019; Table 2.36) is based on three-digit particle numbers. A simple numerical example illustrates the influence of the amount of data on the distribution calculation. With three MP particles (1 PP, 1 PE and 1 PS) per 1000 liters of river water, a distribution of 33.3% for each of the three types of plastic would result. If you randomly catch another plastic particle in the trawl net, for example a polyamide

Table 2.36 Distribution of microplastic types along the Rhine. (Fath et al. 2019)

Sample	MP Σ	PP	PE	PS	PA	SAN	ABS	PU	PEST	PVC
Tomasee	270	94.7	5.3							
Chur	5326	72.8	6.9			13.3		7		
Lake Constance	99	77.8	3.1			2.5	6.9			9.7
Stein am Rhein	237	94.4	5.6							
Laufenburg	17	78.7	6.9	7.2		3.6				3.6
Kehl	5	81.4	13.9			4.7				
Mainz	126	28.4	10	30	20.6					11
Koblenz	140	53	27	20						
Düsseldorf	162	73.3	12.5	7.2		2				5
Wageningen	247	41.7	27	28				3.3		
Side Channel Lek	150	70.7	19	8.9		1.4				

Table 2.37 Distribution of microplastic types along the Rhine. (LUBW 2018)

Sample	MP Σ	PP	PE	PS	PA	SAN	PU	PEST/PET
Lake Constance 1	18	53.3	23.7	3	1			12.6
Lake Constance 2	5	37	27.8	7.4			7.9	14.3
Grenzach W.	10	37.5	18.8					43.8
Weisweil	21	35.8	22.3	11.8			19	9.6
Hügelsheim	12	47.5	29.3	3.2			5.5	13.2
Nackenheim	3	9.4	76.6	4.7		4.7		
Lahnstein	22	37.9	43.8	1.6	7.8		0.4	7.4
Bad Honnef	15	51.8	44.3	1.9				
Düsseldorf	7	24.9	63.7	8.6			1.9	

particle (PA) stirred up from deeper layers by a ship's propeller, a new, significantly deviating distribution with 25% for each of the four polymers results. If the above hypothetical distribution calculation is based on 300 particles, then a randomly caught PA particle has a negligible influence on the distribution. The example shows that more reliable, statistically secured statements regarding the polymer distribution can be made with larger particle numbers.

Furthermore, a direct data comparison of the two studies is only possible to a limited extent. Although the same sampling locations can be compared from Tables 2.36 and 2.37, distances of a few kilometers between the sampling locations can lead to different results. In this context, the distance from tributary mouths plays a role, as well as whether the sampling took place before or after a confluence.

Physical, hydrodynamic and weather-dependent factors, as well as local factors, also play a role in the MP type distribution, and these factors are never the same at any given time.

Both the flow velocity, discharge volume and turbulence (ship traffic, tides) of a body of water, as well as the water temperature, water depth, width of the shipping channel and the proportion of dissolved and dispersed substances, as well as rainfall events and sewage plant inlets, influence the degree of distribution of the plastics due to their different densities. Turbulence can transport heavier plastics towards the water surface and their sinking speed depends on the viscosity of the water, which in turn is influenced by many other factors. Since the analyzed plastics differ relatively little in their density (Table 2.33), the hydrodynamic and physical factors of the water body are not negligible. This becomes clear using the example of the sediment content in the water at the water surface. The main component of river sediment is sand with a density of 1.5–2 g/cm^3 and is larger than that of the plastics and depending on the flow conditions and rainfall events, it can be found more or less at the water surface. No one can judge this better than the crawl swimmer, who dips his hand into the water at the same distance from his eyes before the start of a pull phase. In some swimming sections in the Rhine, the hand is clearly visible and in others it is no longer visible due to the stirred up or introduced sediment. Similarly, the swimmer, who himself has a density of <1 g/cm^3, feels the water's upward, downward and propulsive forces in the turbulence of the current and perceives the water depth visually. Taking this perspective into account makes an interpretation of the Rhine results, which are shown in Table 2.36, more plausible. The inhomogeneous distribution of plastic types at the water surface can have many causes. In calm water (Tomasee), plastic particles with a density greater than 1 g/cm^3 settle in deeper layers or in the sediment, and in the filter, besides PP and PE, no other heavier plastic is found. The same result is observed in laminar flow, where the lake forms into a river (Stein am Rhein). Suspended solids on the bottom do move with the current, but remain in the deeper water layers near the bottom.

In the Alpine Rhine near Chur, the particle count is above average with over 5000 MP particles/m^3, even though the Rhine has only covered a few kilometers from the Tomasee and comparatively few municipal wastewaters have been discharged. Such a high concentration, based on primary microplastics from cosmetic articles, is therefore unlikely. The plastic spectrum depicted in Fig. 2.161 also includes plastics with higher density such as styrene acrylonitrile (SAN, 1.1 g/cm^3) and polyester(PEST, 1.2 g/cm^3).

Due to the low water depth of less than one meter and strong turbulence before and after the rock in the whitewater, neither introduced plastic particles nor the sand can settle. The result is that all occurring types of plastic are captured by the filter system and the filter becomes clogged before 1000 liters are filtered. The amount of water transported by the Rhine at this point is comparatively low, even with heavy rainfall of 100 m^3/s at the end of July 2014, so that inputs from local chemical companies and polymer producers weigh much more heavily than elsewhere on the Rhine (Lower Rhine near Emmerich 2500 m^3/s in August 2014).

On the High Rhine, the flow of the Rhine is interrupted by a multitude of dams and weirs and highly dammed up by 11 hydroelectric power plants. The foaming water cascading down from the high dams does not calm down from the Albbruck Dogen power plant to the Laufenburg dam over the few kilometers. At the historic bridge near Laufenburg, there is also a dangerous bottleneck in the river, with strong rapids and pressurized waters, where the water pushes up from below like in a cooking pot at a water depth of less than two meters. Just like in Chur, microplastic particles of all types dispersed in the water have no chance to settle in the sediment due to the flow conditions, so that the entire occurring plastic spectrum can be captured by a sample taken at the surface. Thus, plastics with a higher density, such as SAN and PVC, are also found in the filter candle.

The Lake Constance sample was taken at the height of Rohrschach, not far from the mouth of the flood-bearing Old Rhine, which carries 12 °C cold water along with its microplastic load into the 19 °C warm Lake Constance. The cold water falls off in the lake and stirs up the sediment and the bottom water layers, so that a wider spectrum of plastics can be detected there, in addition to the standard representatives PP and PE, with SAN, ABS and PVC, than at the end of the Untersee. Whether the local plastic processors in Rohrschach also have an influence on the quantity and type distribution is speculation. In general, however, it can be stated that the amount of sand in the filter correlates with the proportion of heavier plastics. In a filter where hardly any sand is found because clear surface water was filtered, the polymers PE and PE are the only MP representatives.

The SAN input at Chur extends over Lake Constance into the High Rhine at Laufenburg. The fact that SAN does not appear in the diagram in Stein am Rhein is because it was not captured during the sampling. The SAN is transported with the laminar depth flow into the High Rhine.

Since the measuring station in Mainz was located between Rhine kilometer 495 and 505 in the immediate influence area of the Main, the same phenomenon as with the Lake Constance sample can be observed here. However, the high proportion of polyamide is striking. No other sample on the Rhine, if any, shows such high values for polyamide, except in Lahnstein with 7.8% (LUBW study). This can be explained by the fact that the polyamide microplastic input is mainly due to the washing of textiles (Boucher and Friot 2017), i.e., the municipal wastewater

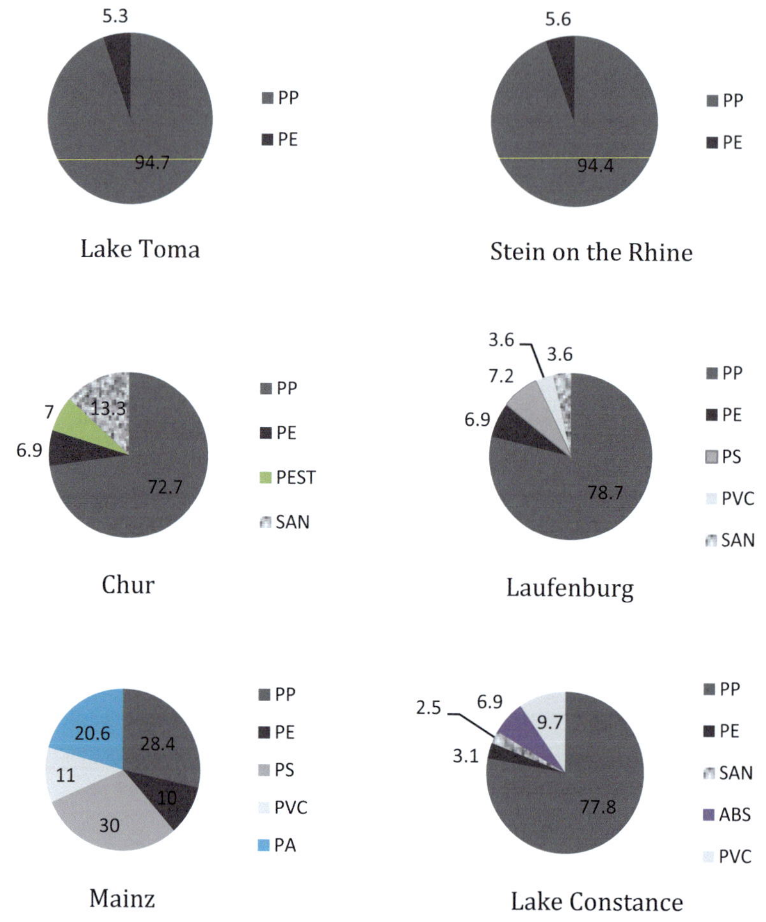

Fig. 2.161 Type distribution at selected positions in the Rhine in August 2014

from private households. In the still young Rhine, the input quantity is relatively low due to the low population density. By the time it reaches the mouth in Hoek van Holland, the Rhine has collected the wastewater from 56 million people. This means that already in Mainz, over half of the flow distance, a high load of washing machine wastewater has indirectly flowed into the Rhine, and with it also the proportion of polyamide fibers that were not retained in the sewage sludge of the sewage treatment plants, either because the fibers are too small and thus remain in the filtrate of the dry pressing of the sewage sludge, or because the municipal wastewater cannot be treated in the sewage treatment plant due to heavy rainfall events. Due to the combination of domestic wastewater with rainwater (combined sewer overflow), the sewage treatment plant's wastewater retention basins are too small during heavy rainfall events, so that the excess water is discharged untreated into rivers. In such cases, the concrete sewer pipes are flooded to such an extent that a removal of the biofilm formed during dry periods occurs. How high the MP concentrations in these biofilms are and what their distribution looks like is the subject of current investigations. In Mainz, the polyamide that has sunk into deeper water regions with a density >1.1 g/cm3, as well as the even heavier PVC (11%; $\rho =$ 1.2–1.4 g/cm^3), is stirred up by the entry of the Main and can be captured in the filter.

In Koblenz, before the entry of the Moselle at the height of Lahnstein (km 587), 1000 liters of Rhine water were filtered at the same location in both studies, with a different result (Tables 2.36 and 2.37). While in August 2014, besides PP, PE, and PS, no heavier plastics were detected, the LUBW study additionally found PA and PET and little PU. The discrepancy shows that in addition to the described influence of the flow conditions, there are other temporary factors that can significantly influence the result. These include, for example, shipping traffic and water level. Both factors influence both the concentration of the MP particles and their distribution. The effects of an upstream traveling, fully loaded freighter with draft are clearly audible and noticeable for a swimmer in the Rhine despite some meters distance. An impairment of the distribution of plastic suspended matter in the water due to different traffic volumes on the water is quite conceivable and provides a plausible explanation for the deviation in the quantification of microplastic particles at the same location, but at different times.

The values in the Lower Rhine at Wageningen and in the Rhine side canal at Lek are not directly comparable. Due to the missing connection of the canal to the main stream, the microplastic concentrations and the distribution in the canal are regionally influenced, while the Lower Rhine is influenced over a larger area. Since it brings the water, despite the diversion into the much more water-bearing Waal at the German-Dutch border, from all riparian countries, not only is the MP particle number with 247/m^3 above average, but also the concentration of all other anthropogenic organic trace substances such as pharmaceuticals, insecticides, sweeteners, corrosion inhibitors, industrial chemicals etc. (Fath 2016).

In Breisach, the Old Rhine meets the Rhine Side Canal, and the Upper Rhine is then interrupted by eight locks up to Iffezheim (last lock before the Dutch border). The flow rate of the approximately 100 km long straightened stretch from Breisach to Iffezheim is greatly slowed down, and the water splits before each lock into the Rhine and the lock side canal. Here, a comparison of microplastic samples is difficult, as the conditions in the water change abruptly. Due to the low gradient, the current stagnates several kilometers before the lock, and swimming in the warmed, standing water is tough. Directly after the lock, it continues with fresh, cool, oxygen-rich water with a stronger current into the subsequent storage basin, with the flow rate and water temperature decreasing continuously. The microplastic results in Kehl (Fath), Weisweil near Rust (LUBW), and Hügelsheim near Rastatt (LUBW) cannot be compared due to the changing conditions. Weisweil is under the influence of the Schönau lock and Hügelsheim under the influence of the Iffezheim lock, while the sample at Kehl, where the water is distributed into a multitude of harbor basins, and the Strasbourg and Gambsheim locks are about 10 km away from the sample collection in both directions. This explains the very low (lowest concentration in the Rhine in August 2014; Fath 2016) concentration of only 5 particles/m^3 (Table 2.36) and that, despite the high input of the Kinzig with 777 particles/m^3 of various types of plastic, as the pie chart in Fig. 2.162 shows.

The Rhine tributaries usually carry less water compared to the main stream (the exception is the Aare) and have a steeper gradient and thus a stronger, more turbulent current, with less water depth. As a result, the microplastic concentrations are higher, and a broader spectrum of various types of plastic can be detected. This issue was already discussed using the sample in Chur as an example. At the Kinzig mouth into the Rhine, on the one hand, a more than 100-fold dilution effect comes into play (discharge volume Kinzig 10 m^3/s; discharge volume >1000 m^3/s) and on the other hand, the entry of the shallow Kinzig into the deep and calm Kehler "harbor basin" causes the heavier types of plastic to sink.

Compared to the approximately equally long Tennessee River in the USA, which also has a similar water discharge volume (Table 2.38), the microplastic particle numbers are relatively low. With just under 18,000 particles/m^3(Fath et al. 2019), the Tennessee River is extremely heavily polluted with microplastics. The concentrations are even higher than in the Yangtze (Table 2.38) (Wang et al. 2017), although it should be noted that the particle sizes are not directly comparable. The lower limit at Wang et al. (2017) was 50 μm, while it was 25 μm at Fath et al. (2019). Due to the tendency for the particle number to increase sharply inversely proportional to the particle size, this difference in the size spectrum should not be neglected.

Fig. 2.163 shows the distribution of microplastic particles from 25–500 μm in the Tennessee River around river mile 219 at the height of Waterloo in Alabama. The pie chart includes the typical plastics that can also be detected in the Rhine, also in a similar distribution. Only the PSU share and the silicones are noticeable. At Waterloo, about 20 miles before the Pickwick Dam, the water is anything but calm. Current in the fairway, rain, strong wind, and shallow water depth (2–4 m) are the boundary conditions for sampling and type distribution. The high

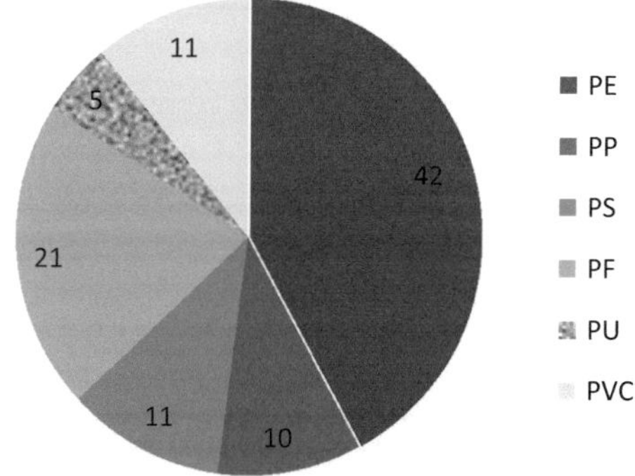

Fig. 2.162 MP type distribution of the 777 particles from the Kinzig water sample in percent. (Fath et al. 2019)

Table 2.38 Comparison of two rivers (Rhine vs. Tennessee River) according to selected parameters

	Rhine	Tennessee River
Length	1231 km	1049 km
Swimming stages	25	33
Start	July 28, 2014 at Tomasee (CH)	July 27, 2017 in Knoxville
End	24th Aug. 2014 in Hoek van Holland (NL)	29th Aug. 2017 in Paducah (Kentucky)
Catchment area	56 million inhabitants 185,000 km^2	4.8 million inhabitants 105,868 km^2
Countries	Switzerland, Germany, France, Netherlands	USA (four states: Tennessee, Kentucky, Alabama, Mississippi)
Found MP/Number Types	Between 5 and 5326 particles/m^3 Ø 200/m^3 PP, PE, PS, PA, SAN, ABS, PU, PEST, PVC, PLA	Between 14140–17674 particles/m^3 Ø 16,000/m^3 PP, PE, PS, PA, ABS, PSU, PVC, Silicones
Discharge	2500 m^3/s (in August 2014 in Emmerich) the most water-rich river in Germany	1998 m^3/s
Flow speed	between 0.7 and 2.9 m/s	between 0.4 and 1.5 m/s
Tributaries >40 m^3/h	14	6
Lakes	1 Lake Constance	9 Fort Loudoun Lake Watts Bar Lake Chichamauga Lake Nickajack Lake Guntersville Lake Wheeler Lake Wilson Lake Pickwick Lake Kentucky Lake
Total gradient	0.05–0.89 ‰ 2345 m from Tomasee to the North Sea For comparison: 248 m from Basel to Rotterdam	156 m from Knoxville to Paducah
Monitoring stations	7	5
Locks and weirs	25	9
Hydropower plants	24	29
Nuclear power plants	3	3
Freight transport	over 200 million tons per year	over 50 million tons per year
Name	The name "Rhein" originates from the Indo-European root word reih- for "flow". This root also gave rise to the verb "to run"	Tennessee comes from the Cherokee village Tanasi

proportion of sand and plant suspended matter throughout the Tennessee River meant that only 500 liters of water could be filtered until the pores were clogged. The qualitative statement that the higher the proportion of sand in the filter, the higher the proportion of microplastic particles of higher density, was also confirmed in the filter T8 evaluated in Fig. 2.150.

Of course, the question arises as to why the microplastic concentration in the Tennessee River, in whose catchment area relatively few people live with 4.8 million (compared to Rhine = 56 million), is so high. One explanation would be that neither in Tennessee nor in Mississippi nor in Alabama or Kentucky is waste separated. Plastic waste is neither collected separately nor recycled, nor is it subjected to thermal recovery. It goes to the landfill together with the residual waste or, to put it in the local language, in "Landfill". The consumption of plastic in the beverage and food sector (shopping bags, cutlery, fast food; straws etc.) is significantly higher in the aforementioned US states than here and environmental protection is, if at all, only treated stepmotherly. Nothing will change in this regard during Donald Trump's legislative period, who commented on the topic of "environmental protection" with: *"we can leave a little* bit". So plastic waste will continue to fill the land and the soils for the time being.

The decomposition processes already mentioned, which turn macroplastic, mesoplastic and finally microplastic, continue in the soil, where tiny creatures such as ants, insects, worms, caterpillars, beetles etc. carry plastic parts into deeper layers. Rain events thus carry microplastic particles from landfills either via surface water into rivers and lakes or they are also carried into larger water reservoirs via underground waterways, even into the groundwater (LUBW 2018; Rillig 2012; DRZE 2018).

All results presented in this chapter regarding the quantification and qualification of microplastic particles from filtered and processed water samples were obtained with the inclusion of infrared spectroscopy. In order to be able to analyze even the smallest particles, the recording was not made in reflection mode, as the ATR crystal to be placed on the sample shifts, displaces or even misses the sample, but in transmission mode.

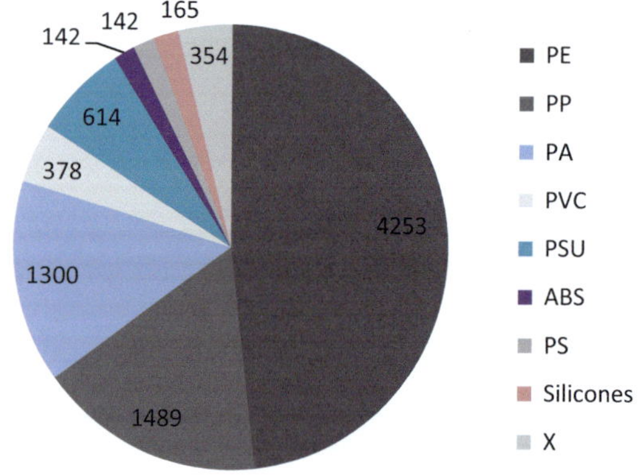

Fig. 2.163 MP type distribution of the 17,674 particles from the Tennessee filter sample T8 (mile 219) in percent. (www.tenneswim. org)

Since the filter as carrier material (aluminum oxide) itself absorbs radiation in the IR range below 1600 cm^{-1}, there can be misinterpretations in this area, because in this so-called fingerprint area characteristic absorption peaks for the different polymers can be found. With a silicon oxide filter, the uncertainty range can be reduced (absorption < 1300 cm^{-1}).

Since there are no uniform standards for the measurement methodology regarding the filters, the recording modes and the stored spectrum databases, there can be discrepancies in the assignment of the spectra to the corresponding plastics. This applies in particular to plastics with a similar fingerprint area such as SAN and ABS or the polyesters PEST, PBT and PET as well as PU and PSU. This should be taken into account when comparing the results of different microplastic investigations in the same medium if the exact type distribution plays an important role.

The infrared spectroscopy in combination with a microscope is not the only method to investigate microplastic occurrence and distribution. Since each method has its limitations, it is important to adapt the investigation method to the investigation task in such a way that the most reliable result is obtained, or to know a complementary method that supplements or confirms a result. Therefore, in Sect. 2.7, all so far important and validated methods with their strengths and weaknesses are presented.

2.7 Alternative and Complementary Microplastic Analysis Methods in Comparison

2.7.1 Raman Microscopy

Not only with infrared and microwave spectroscopy, but also with the Raman spectroscopy can rotation and vibration spectra of molecules, macromolecules, and polymers be investigated. When Raman spectroscopy (Brandmüller and Moser 1962) is combined with a microscope, microplastic types can be analyzed complementarily to IR spectroscopy. In contrast to IR spectroscopy (absorption spectrum), an emission spectrum is obtained due to different physical principles and the excitation of the sample. Raman spectroscopy is based on the Raman effect, which the Indian physicist C. V. Raman experimentally confirmed in 1928 (Nobel Prize in Physics 1930), after A. Smektal had already theoretically predicted it five years earlier. The effect is based on the different interaction of matter with monochromatized light, for example, that of a laser. While the majority of the laser light penetrates the sample unhindered (transmission), a very small part of the light (0.01%) is scattered in all spatial directions. This scattered radiation, the so-called Rayleigh radiation, has the same frequency as the incident light, due to the elastic collisions with the molecules of the sample. An even smaller proportion (10–6%) of the monochromatic light leads to inelastic collisions with the molecules of the sample (Hesse et al. 2015). This results in the absorption and emission of a part of the radiation, depending on the vibration and rotation excitation or relaxation. The measurable absorbed and re-emitted radiation energy depends on the characteristic vibration and rotation levels of the excited molecules in the sample and thus makes their analysis with a photoelectronic detector possible. In Fig. 2.164, the structure of a Raman spectrometer is schematically illustrated.

The schematic energy level representation in Fig. 2.165 explains the formation of the typical Raman emission spectrum depicted below. If the energy of the irradiated electromagnetic radiation is not large enough to achieve an electron excitation to a higher energy level, the excitation radiation is elastically scattered and emitted unchanged as v_0 or the excited molecules do not fall back into the ground state E_0, but into a first excited vibrational state E_1 (inelastic scattering), so that the energy difference corresponds to the vibrational energy, which can also be found in the IR spectrum as an absorption band. The emitted radiation is therefore less energetic and can be seen as a Stokes line in the longer wavelength range, to the left of the Rayleigh band in Fig. 2.152. The Anti-Stokes band, equidistant from the Rayleigh band on the right side, arises because some molecules, according to the Boltzmann distribution, are already in a thermally excited state E_1 and after the interaction with the excitation frequency v_0, they transition to the ground state, thereby emitting a higher energetic shorter wavelength radiation (Anti-Stokes radiation). Since at room temperature most molecules are in the vibrational ground state (Boltzmann distribution), the Stokes scattering is more intense than the Anti-Stokes scattering.

The frequency difference between Stokes and Anti-Stokes radiation is often given as a relative value to the Stokes frequency, the so-called Raman shift, in which the excitation frequency v_0 is set to zero and the Stokes frequency differs from the Anti-Stokes frequency or energy ($E = hc/\lambda$) only in sign.

The Raman spectroscopy is therefore to be considered complementary to IR spectroscopy, as the Raman effect only exists when the polarizability of the atoms involved in the bond changes within a vibration, while molecules are

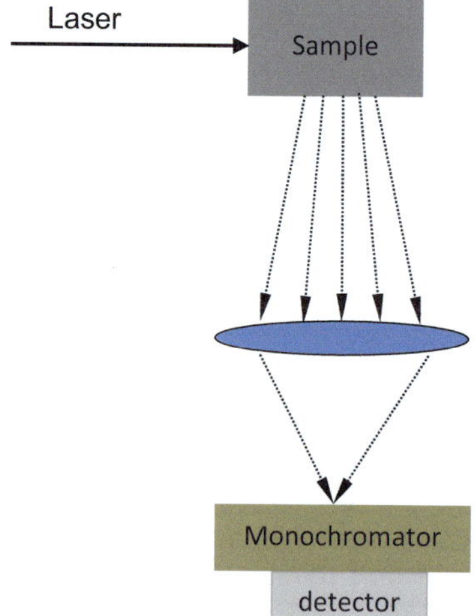

Fig. 2.164 Structure of a Raman spectrometer

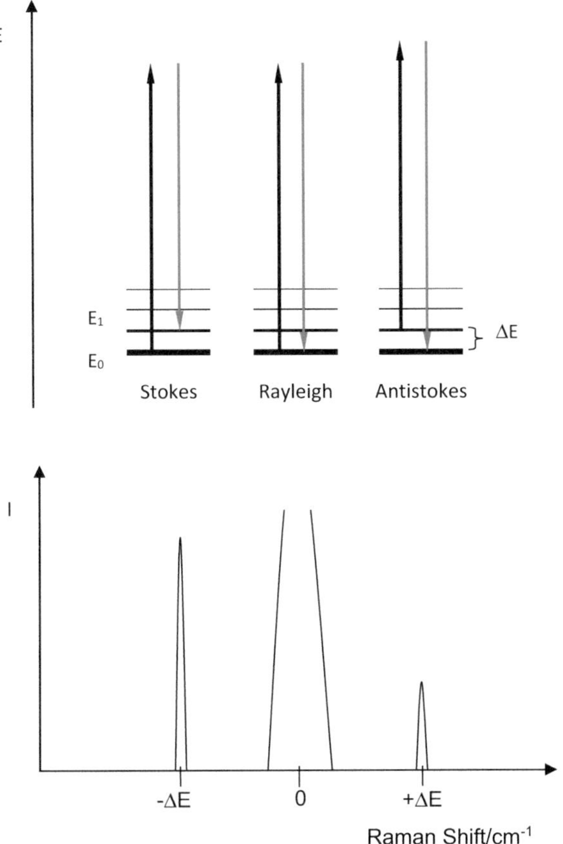

Fig. 2.165 Schematic level representation of the Stokes, Rayleigh, and Anti-Stokes lines (above, black arrow = incident light; gray arrow = emitted light) with the associated Raman spectrum (below)

only IR-active when a change in dipole moment occurs during the vibration. The more easily deformable the electron cloud around an atom is, the greater its polarizability.

Both selection rules are complementary to each other. For example, if we consider vibrations that occur symmetrically to the center of symmetry of a molecule, such as the symmetric stretching vibration in the linear CO_2 molecule, this is IR-inactive, but Raman-active. In general, no vibration can be simultaneously Raman and IR-active.

Within the spectroscopic analysis methods of organic chemistry, Raman spectroscopy has emerged from its shadowy existence to become a powerful complement to established IR spectroscopy due to recent developments in laser technology, coupled with the use of semiconductor detectors and computer-aided evaluation programs.

Both types of spectroscopy are the most widely used, non-destructive, optical examination methods for the identification of microplastics, which in combination with a powerful microscope can visualize particle sizes around 10 µm (Löder and Gerdts 2015; Löder et al. 2015b) (FTIR) up to 0.5 µm (Enders et al. 2015; Imhof et al. 2012 2016; Fischer et al. 2015; Ivleva and Nießner 2015) (Raman) in diameter and enable their analysis. This also makes the particle size distribution on a filter or in a soil sample (Fath et al. 2019) feasible.

The ATR technique is usually analyzed in transmission mode, as the fixation of the ATR crystal on smaller particles is difficult to position without moving the sample. Usable signals in transmission mode are only obtained in areas where the filter material is also IR-permeable. With an aluminum oxide filter with a pore size of 25 µm (Anodisk 25), the radiation permeability rapidly decreases at 1300 wave numbers. With an etched silicon dioxide filter, the IR range can be extended so that signals below 1300 cm^{-1} can also be detected (Wiesheu et al. 2016).

The biggest disadvantage of optical methods lies in the very high time expenditure, even when working with a multi-diode detector (FPA detector). For a filtered water volume of 1000 liters, the isolation of the microplastic particles from the biomass and from the sand is very time-consuming until a dried filter lies under the IR microscope sample table. With a filter area of 284 mm^2, the sample table in Fig. 2.166 only shows a fraction of 4 mm^2. In this section, each particle is manually approached, focused, then a spectrum is taken from it and compared with the stored plastic spectra library. If there is a match, the particle is marked as the corresponding plastic. Not every particle is a "hit", as the sample preparation of the individual filters can have a very different result. With good samples, the accompanying materials on the filter are low. With less good isolation, sand, shell lime or other indefinable substances are found next to the microplastics despite density separation. From this description, it becomes understandable that it can become difficult to completely evaluate a sample window within the measurement time, as long as there is still liquid nitrogen available, which must cool the heat detector. If the software crashes every now and then and you have to approach and focus the just evaluated measuring point again, this method can be described as very tedious.

The main measurement error of a sample is mainly based on the evaluation of only a few filter sections (3–5), the average of which is then extrapolated to the entire filter area. The standard deviation in this process is between 5 and 20%. With such a high particle density in a measuring field, it cannot be ruled out that the smallest plastic particles overlap and are not evaluated as individual particles.

With the FPA detector technique, you are much faster, but you lose accuracy due to the automated scanning of the surface, especially when the filter carries a lot of microplastic overlapping material. In addition, an unmanageable amount of data must be processed and software must be available that can handle this without problems.

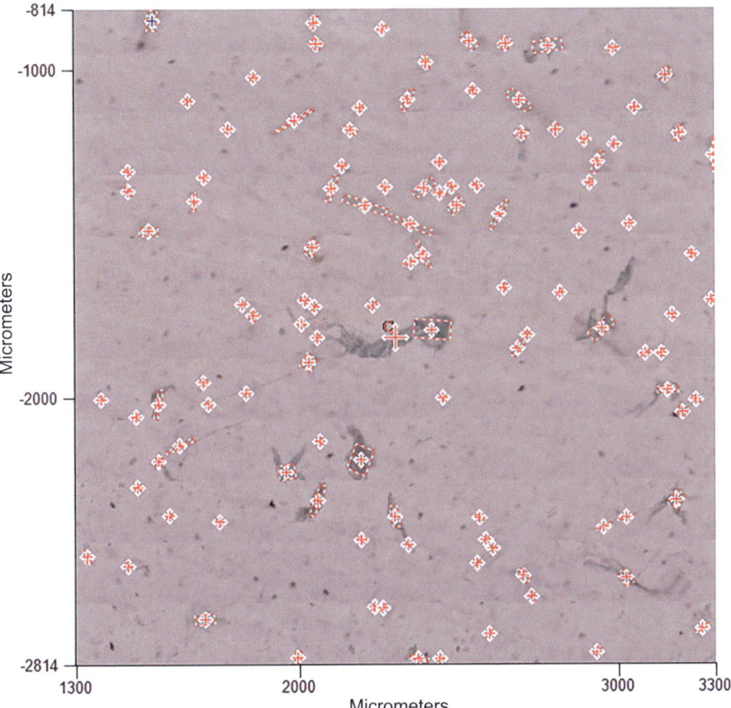

Fig. 2.166 4-mm^2 section of an aluminum oxide filter under the IR microscope. The red crosses indicate already detected microplastic particles. The filter area shows 112 measuring points and is 75% read out

2.7.2   Liquid Extraction

Another, significantly faster method for microplastic analysis, which can also process larger sample volumes, is liquid extraction. This method is based on conventional laboratory technologies in organic synthesis and uses chemical extraction and subsequent separation of the extracts by chromatographic separation using reversed-phase liquid chromatography (RP-LC). By using gel permeation chromatography (GPC) or size exclusion chromatography (SEC = Science-Exclusion-Chromatography) for the different extracts, additional information about the degree of decomposition of the polymer particles can be obtained and statements about the origin can be made by comparing with a database.

Although the extraction method loses information about the particle size and the particle size distribution because the plastics are dissolved in a solvent, it allows for exact concentration specifications. After the retention values of the polymer standards have been determined under fixed column parameters and their peak areas in the chromatogram have been evaluated and incorporated into the creation of a calibration line, the plastic concentration in the extract can be analyzed chromatographically.

With the help of GPC, the molar mass distribution of the individual polymers can be determined. This is a significant aspect for the investigation of microplastics from environmental samples when it comes to detecting aging processes through oxidation products or decomposition products. While the oxidative decomposition process of polypropylene, for example, can be detected by the formation of hydroxy, carbonyl, and double bonds in the polymer using IR and Raman spectroscopy (Käppler et al. 2015), bond breaks within the long-chain molecules can be detected by a change in the molecular weight distribution (Elert et al. 2017).

In the case of water samples, it is of course also possible, despite the dissolution of the filtered out microplastic particles by a previously labor-intensive fractionated filtration, to obtain information about MP particle sizes of different polymer types, even without making them visible under a microscope.

Suitable extraction agents, which are often used to extract microplastics from soil samples or filter residues, are DMF (dimethylformamide), THF (tetrahydrofuran), DMAc (dimethylacetamide), and HFIP (1,1,1,3,3,3-hexafluoro-2-propanol).

A disadvantage of this method is that the most common plastics in the environment, PP and PE, do not dissolve in any of the mentioned solvents. However, THF dissolves polystyrene (PS) and PVC very well, and HFIP is a suitable solvent for PET and other polyesters (Elert et al. 2017; Fuller and Gautam 2016).

2.7.3 Thermal Extraction and Desorption (TED-GC-MS)

A polymer type characteristic dry extraction, in which thermally specific decomposition products are released, collected on a sorbent and thermally desorbed again, represents a valuable addition to the previously presented investigation methods for MP, especially since with this method all plastics, including the above-mentioned solvent-insoluble PP and PE, can be captured and analyzed and quantified by GC-MS. Even with this non-destructive method, no particle sizes and particle size distributions can be determined, unless one pyrolyzes fractionated filter residues as already mentioned.

Just like the ABS presented in Sect. 2.3.5, every other plastic within a thermogravimetric analysis, in which a weighed sample is pyrolyzed (thermal extraction) or heated above the glass transition temperature in the absence of oxygen, releases polymer-specific gaseous decomposition products. The weight loss is measurable and the gaseous products of the thermally induced reaction can then be quantified in a gas chromatograph. Since polymers only decompose at temperatures above 300 °C, the soil matrix or residual biomass of the filter residues can be thermally separated. To concentrate the process gases of plastic pyrolysis, they are bound to polydimethylsiloxane (PDMS)-solid phase adsorbers, to then be heated a second time in the thermal desorption unit. In this process, the previously adsorbed decomposition products are desorbed and transported into a cold injection system by an inert gas stream (helium), where they condense and concentrate at -100 °C.

After this cryofixation, the decomposition products are once again evaporated under controlled conditions and separated via a chromatographic column (GC-MS). The separation depends on the boiling points of the analytes and the interaction between the mobile and stationary phase of the chromatographic column. For the separation of polymer-specific decomposition products, reversed-phase separation phases are used due to their nonpolar character.

A comparison of the obtained mass spectra and the fragmentation pattern of the substances separated in the GC column with the stored spectral database identifies the plastic, and the respective plastic mass fraction in the weighed sample can be determined via the peak areas in the chromatogram.

In Fig. 2.167, the TED-GC-MS chromatogram of the non-destructively measured Tennessee filter sample T8 (Mile 219) shown in Fig. 2.163 can be seen. Statements about the mass fraction of the different polymers in the filter cake can be made based on the intensities (TIC = *Total Ion Counts*) of the peaks at different retention times and the associated fragmentation in the mass spectrum. For complex samples with a high non-polymer content, such as soil samples or untreated filter samples, it is important to know the polymer-specific signals.

In the chromatogram, the typical peaks in the reference chromatogram are appropriately marked, and in Figs. 2.155, 2.156, and 2.157, the corresponding structures of the decomposition product are depicted. If all these signals are also visible in the sample chromatogram, then the presence of the corresponding polymer in the sample is confirmed without a doubt.

Polyethylene (PE)

It is known that during a pyrolysis of polyethylene, the long-chain linear aliphatic plastic decomposes into shorter saturated, mono- and di-unsaturated aliphatic compounds (Serrano et al. 2005; Sojak et al. 2017). This is also applied in the course of thermal recycling. In the chromatogram of the decomposition products of pure polyethylene, the most commonly used plastic, a multitude of multiplets can be seen along the retention time scale, depending on the chain length. Depending on the height of the chromatographic resolution, triplets to pentets can be found at a certain retention time. The signals of the multiplets belong to the corresponding alkanes, alkenes, and dienes and the associated *cis*and *trans* isomers.

However, during the thermal treatment of an environmental sample, not all decomposition products are polymer-specific, as lipids and fatty acids can still be present in both the soil matrix and the filter matrix, despite enzymatic treatment, which also release saturated and mono-unsaturated aliphatic compounds during pyrolysis (Kebelmann et al. 2013). Only with a long aliphatic polymer chain can

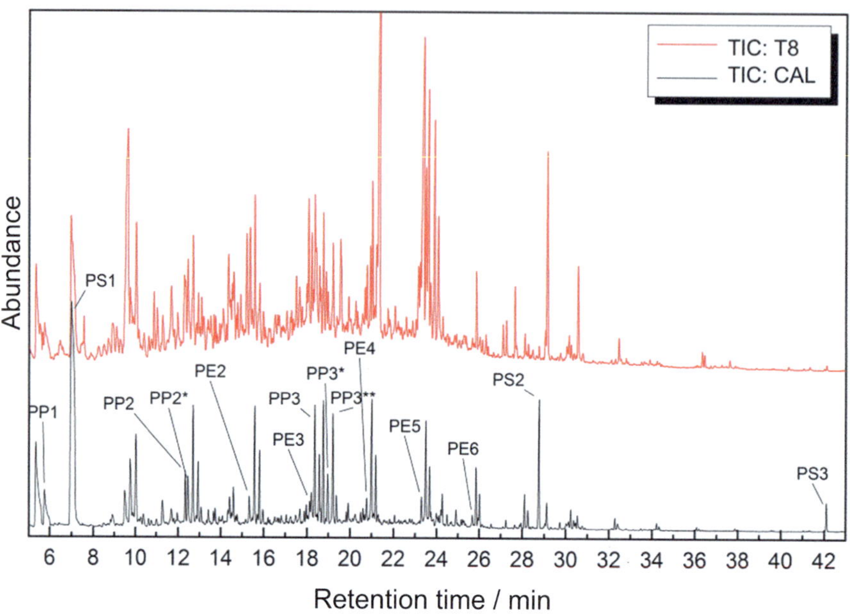

Fig. 2.167 TED-GC-MS chromatogram of a filtered (500 liters) water sample from the Tennessee River (Filter T8 above) compared to a reference sample with the polymers PE, PP, and PS (below)

the four α,ω-dienes representative of PE shown in Fig. 2.168 be produced from the macromolecule in a concerted double cleavage. They are therefore used for identification (Dümichen et al. 2015).

PP

For the second most commonly used polymer, polypropylene, the structures of its thermal degradation products are more complex due to the branching or the methyl substituent on the polymer chain. The tacticity leads to different diastereomeric products during cleavage. For this reason, six different fragments marked with PP1-PP6 are shown in the chromatogram (Figs. 2.167 and 2.169). As with PE, branched alkanes, alkenes, and dienes are formed, with the alkenes showing the highest intensity in the chromatogram (Sojak et al. 2017). The depicted PP degradation products are used for the identification of PP and are independent of the sample preparation, even if the sample was exposed to strong oxidative conditions (Dümichen et al. 2017).

Polystyrene (PS)

The aromatic polystyrene thermally decomposes successively at the chain ends, producing mesomerically stabilized allyl-benzyl radicals, which stabilize

Fig. 2.168 Double unsaturated hydrocarbons, characteristic pyrolysis products of polyethylene; PE1= 1,12-Tridecadiene, PE2 = 1,13-Tetradecadiene, PE3 = 1,14-Pentadecadiene, PE4 = Hexadecadiene

Fig. 2.169 Characteristic thermal degradation products of polypropylene; PP1 = 2,4-Dimethylhept-1-ene, PP2 and PP3 = Diastereomers of 2,4,6-Trimethylnon-1-ene, PP4, PP5 and PP6 = Diastereomers of 2,4,6,8-Tetramethylundec-1-ene

themselves via a cyclic transition state from the main chain with a proton and reform the monomer styrene (Recycling) or intermolecularly recombine with further allyl-benzyl radicals to form di- and trimers.

The main products of polystyrene pyrolysis are therefore styrene (PS1) and the oligomers of styrene, mainly the dimer 2,4-diphenyl-1-butene (PS2) and the trimer 2,4,6-triphenyl-1-hexene (PS3) (Fig. 2.170).

Polyethylene terephthalate from PET bottles can be identified by the thermal degradation products vinyl benzoate and benzoic acid as well as ethyl benzoate and 1,1-biphenyl.

In the polyester, there is a thermal α-cleavage, in which ethyl benzoate and benzoic acid are formed in the first step. This decarboxylates to a phenyl radical, which dimerizes with another phenyl radical to form biphenyl (Dimitrov et al. 2013). In the main degradation mechanism of the ester via a six-membered cyclic transition state, a proton of the β-C-atom is transferred to the carbonyl group. The end product of this thermally induced cleavage is, in addition to the vinyl benzoate, the benzoic acid, the starting product of the polymerization (Buxbaum 1968; DePuy and King 1960; Holland and Hay 2002). This also allows for a thermal recycling or partial return of the polymer to its educts.

Polyamide 6 from textile fibers mainly thermally decomposes homolytically into the volatile cyclic starting product of nylon production, ε-caprolactam (Braun et al. 2010). By-products of the pyrolysisare aldehydes and nitriles (Düssel et al. 1976; Ohtani et al. 1982; Dümichen et al. 2017).

Based on the retention times of the polypropylene-specific and polystyrene-specific peaks in the TED chromatogram, the Tennessee filter sample (500 liters of river water filtered) on the filter yields 20 µg (25–500 µm particle diameter) of polypropylene and 3 µg of polystyrene microplastic particles, with a standard deviation of 20–30%; (Dümichen et al. 2017; Imhof et al. 2012). The proportion of about 10% PS compared to PP also results from the optical analysis using the IR microscope. With an average particle size of the MP particles of 60 µm × 60 µm × 10 µm (no spherical particles, mainly flakes), the 20 µg would correspond to about 1500 PP particles. 1487 were counted on the filter with a standard deviation of 25%. The result of both investigations shows that both methods complement each other very well. With the TED-GC-MS method, the mass of the plastic particles can be determined and with the IR microscopic method the number of particles as well as their shape and size.

For the Tennessee River, a microplastic load can thus be better estimated than with optical methods. The filter results of water samples are normalized to 1000 liters, so that 40 µg PP and 6.3 µg PS, multiplied by the discharge volume of 2000 m³/s (Table 2.38), extrapolated to a year, result in a load of 2.5 tons of PP microplastic and 0.4 tons of PS that are added to the Ohio River each year. However, the proportion of PP and PS of the total microplastic amount is only just under 19% (Fig. 2.163) so that the total microplastic load in the Tennessee River is around 15 tons.

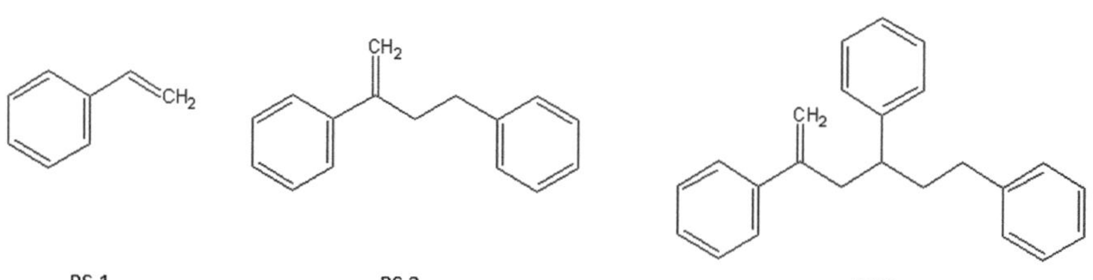

Fig. 2.170 Characteristic thermal degradation products of polystyrene; PS1 = Styrene, PS2 = 2,4-Diphenyl-1-butene, PS3 = 2,4,6-Triphenyl-1-hexene

2.8 Outlook on Microplastic Development

If one assumes that the number of microplastic particles increases inversely proportional to their particle diameter, as is the case with the water samples from the Rhine (Fath 2016), and as the curve in Fig. 3.34 describes, which is based on a cascading particle division, then there is an innumerable number of nanoplastic particles in our environment. Because the processes of plastic particle decomposition naturally do not stop at the transition from the micro to the nanometer dimension. A filtration to detect nanoplastic particles in waters involves a significantly higher technical and temporal effort than is already the case with microplastics. This is akin to the often-cited search for a needle in a haystack, as the smallest plastic particles must be isolated and analyzed from a high proportion of biomass. Cleaning waters contaminated with microplastics presents an as yet unsolvable task, so the removal of nanoplastics from the air, soil, or waters is not even being considered.

On the one hand, there is no scientific knowledge about the influence of the smallest artificial polymer particles on plant, animal, and human life, and on the other hand, there are many open questions, such as whether the aforementioned life forms are to be considered as a filter system in which the synthetic micro and nanoparticles get stuck and if so, in which organs and what damage do they cause there? Are foreign nanoplastic particles, for example, able to penetrate the human intestinal wall to enter the bloodstream, where they could promote arteriosclerosis? Or, due to their size and nonpolar character, are they even able to overcome the blood-brain barrier and be a trigger for a stroke by blocking the brain's blood vessels, as they could also influence blood clotting? Are nanoplastic particles capable of passing through cell walls and disrupting cell division, so that it can proceed less controlled and thus increase the risk of cancer? These are all questions that urgently need an answer, especially since we inhale micro and probably also nanoplastic particles with the air (Rillig 2012) and ingest them with food (Karami et al. 2017; Jander 2017; Fath et al. submitted) be it through maritime table salt, seafood, food made from dough prepared in a plastic bowl or whipped cream. Do we then completely excrete the micro and nanoplastics without them being able to cause damage, or will the "miracle material" plastic, after all studies have been completed, suffer the same fate as the once miracle material of the 1960s and 1970s, asbestos, the use of which is now banned due to its confirmed carcinogenic effect?

I do not want to be misunderstood here. The above critical remarks are not a plea against the use of plastic. Plastic is a smart, versatile material, with an almost infinite range of applications and properties from maximum elasticity to metallic strength properties and that at a low weight. Without the lightweight plastic components, for example, the fuel consumption and thus the CO_2 emissions of vehicles could not have reached the current low values. Plastic is an excellent material for many applications, that is undisputed, but due to the massive pollution of our world's oceans, coasts, rivers, lakes, and all other habitats, criticism is growing, and primarily environmental associations, but also politics and society will question whether plastic should be used for every purpose. At least until we have the material cycle of plastics under control, the use of plastics must be restricted. If the avoidance of plastic goods, where possible, does not meet with understanding, it can also be enforced by bans. In all African states, there is already a ban on plastic bags. Kenya was the last African country to follow its neighbors in 2018. For six years now, Rwanda has had the strictest law against plastic bags, where a specially appointed plastic police punish passers-by with plastic bags. Also in Europe, something is moving in terms of plastic bans. Sweden and recently also the UK ban microplastics in cosmetics, toothpaste, shower gel, and generally in drugstore products. France bans plastic dishes and plastic cutlery, and the USA has banned microplastics from cosmetic products since July 2017 with the "Microban".

Banning substances that are hazardous to health or the environment in general does not make sense. Our technological progress would not be possible without the use of hazardous substances, but we have learned to control these substances and their decomposition products to a large extent, so that they continue to be the "fuel" of our production plants. The diesel exhaust scandal shows that control can also be lost again. Nevertheless, no one would think of banning the extraction of petroleum or the production of fuel, especially since petroleum is also an important resource for synthesizing vital medicines. The prohibition of disposing of these liquids in the environment makes sense to (almost) everyone. With plastic, we still have a longer path of enlightenment ahead of us, to which this book hopefully contributes. It has already begun and will pick up even more speed in the future, when new insights about the negative effects of micro and nanoparticles could potentially even reverse the balance between the benefits and harms of plastics. As long as we do not yet have complete global control over our plastic waste in all stages from macro to meso to micro and nanoplastic, the avoidance strategy, where it is possible and makes sense, is the best course of action.

2.9   Avoidance of Microplastics

The environmental problem of "microplastics" encompasses a scale that will concern us far into the future. Further studies are needed to investigate the effects of microplastics, especially on humans. The consequences for human health are not yet known and can only be guessed at, but countless studies on animals show alarming results.

After listing the sources of plastic entry into the Rhine, which also applies to other inland waters, there are at least three ways to reduce the microplastic load of the Rhine.

1. Switching all cosmetic products to microplastic-free by using biodegradable natural products.
2. Not providing the Rhine with grinding material (macroplastics). Here, consumer awareness needs to be addressed and education about the effects of improper disposal is necessary.

3. Avoid entry via surface waters, for example through plastic-contaminated fertilizer.

In summary, there is a risk to humans from contamination with microplastics for two reasons. Due to their small size, but in some cases very large surface area (Fig. 3.6) and low polarity, microstructured plastics are very capable of adsorbing organic substances. We have also taken advantage of this property in the form of the passive sampler to detect organic pollutants by accumulating them on a plastic membrane. The author always wore this sampler on his right lower leg during the swimming phases. A fish that consumes microplastic particles may thus potentially ingest a higher toxic potential than if it were to drink a liter of water. Microplastic particles can be thought of as a magnet for organic substances. Surface-active pollutants such as perfluorinated surfactants (PFT) seek large surfaces to attach themselves to (Fath et al. 2016).

The fact that fish cannot distinguish microplastic particles from their food is shown by the infrared image of the gastrointestinal tract of an asp (Figs. 2.63 and 2.64), which was caught in the Rhine near Karlsruhe in April of the year 2015.

Visible are the green polyamide fibers, possibly from our fleece textiles. During the washing process, fibers always flow towards the sewage treatment plant with the wastewater from the washing machine. These are not completely retained there. A swab on the dryer's sieve shows us how much abrasion occurs per wash cycle. Also visible are microparticles of polypropylene (red) and polyethylene (blue).

We do not eat the fish's stomach-intestinal content, but the pollutants that stick to the microplastic particles are absorbed into its tissue during food breakdown. The second danger is that the listed plastic ingredients, such as plasticizers, are dissolved out of the plastic matrix by stomach secretions and also stored in the tissue. At the end of the food chain is the human who consumes contaminated fish.

Microplastic particles are a vehicle (Trojan horse) that can introduce pollutants into our food chain. Thus, every chemical accident or scandal, such as the recent PFT scandal, which reports how poisoned Düsseldorf groundwater flows into the Rhine, is of great concern, as the "poison" not only flows downstream, but also remains in the local flora and fauna with the help of microplastics.

Microplastic particles and microfibers are not only found in the gastrointestinal tract, but also under the scales and between the gills of fish (Fig. 2.158). The danger to humans from grilling a fish, where the plastic can also burn over the embers and potentially release toxic gases, is currently being investigated at HFU.

References

Abts, G. (2014). *Kunststoff-Wissen für Einsteiger* (4. ed.). München: Hanser.
Alonso, M., & Finn, E. J. (2000). *Physik* (3. ed.). München: Hanser.
Andrady, A. L. (2011). Microplastics in the marine environment. *Marine Pollution Bulletin, 62*, 1596–1605. https://doi.org/10.1016/j.marpolbul.2011.05.030.

ARD. (2015). Kontraste: Öko-Irrweg Biotonne: Plastikverseuchter Kompost macht Äcker zu Müllhalden. https://www.rbb-online.de/kontraste/ueber_den_tag_hinaus/wirtschaft/oekoirrweg-biotonne.html.

Atkins, P. W. (2013). *Physikalische Chemie* (5. ed.). Weinheim: Wiley-VCH.

BAG (Bundesamt für Gesundheit). (2017). Gesundheitsgefährdung durch Kunstrasen? Faktenblatt. https://www.bag.admin.ch/bag/de/home/gesund-leben/umwelt-und-gesundheit/chemikalien/chemikalien-a-z/kunstrasen.html.

Bakir, A., Rowland, S. J., & Thompson, R. C. (2014). „Enhanced desorption of persistent organic pollutants from microplastic under simulated physiological conditions". *Environmental Pollution, 185*, 16.

Barnes, D. K. A. (2005). Remote islands reveal rapid rise of southern hemisphere, sea debris. *ScientificWorldJournal, 5*, 915–921.

Barnes, D. K. A., Galgani, F., Thompson, R. C., & Barlaz, M. (2009). Accumulation and fragmentation of plastic debris in global environments. *Philosophical Transactions of the Royal Society of London Series B, 364*, 1985–1998. https://doi.org/10.1098/rstb.2008.0205.

Bergmann, M., Sandhop, N., Schewe, I., et al. (2016). Observations of floating antropogenic litter in the barents sea and fram strait, arctic. *Polar Biology, 39*, 553. https://doi.org/10.1007/s00300-015-1795-8.

BfR (Bundesinstitut für Risikobewertung). (2003a). Quellen für Acrylamid in Kosmetika. Stellungnahme vom 24. März 2003. http://www.bfr.bund.de/cm/343/quellen_fuer_acrylamid_in_kosmetika.pdf.

BfR (Bundesinstitut für Risikobewertung). (2003b). „Weichmacher DEHP: Tägliche Aufnahme höher als angenommen?" Stellungnahme vom 23. Juli 2003. https://mobil.bfr.bund.de/cm/343/taegliche_aufnahme_von_diethylhexylphthalat.pdf.

BfR (Bundesinstitut für Risikobewertung). (2005). Übergang von Weichmachern aus Schraubdeckel-Dichtmassen in Lebensmittel. Stellungnahme Nr. 010/2005. http://www.bfr.bund.de/cm/343/uebergang_von_weichmachern_aus_schraubdeckel_dichtmassen_in_lebensmittel.pdf.

BfR (Bundesinstitut für Risikobewertung). (2010). Endokrine Disruptoren: Substanzen mit schädlichen Wirkungen auf das Hormonsystem. A/2010, 19.04.2010. https://www.bfr.bund.de/de/presseinformation/2010/A/endokrine_disruptoren__substanzen_mit_schaedlichen_wirkungen_auf_das_hormonsystem-50488.html.

Bibra Toxicology Advice & Consulting. (2005). Toxicity profile for ethylene bis stearamide. https://www.bibra-information.co.uk/downloads/toxicity-profile-for-ethylene-bis-stearamide-2005.

Birnbaum, L. S., & Staskal, D. F. (2004). Brominated flame retardants: Cause for concern? *Environmental Health Perspectives, 112*, 9–17. https://doi.org/10.1289/ehp.6559.

BMG (Bundesministerium für Gesundheit). (Hrsg.) (2005). Stoffmonographie Di(2-ethylhexyl)phthalate (DEHP)-Referenzwerte für 5oxo-MEHP und 5OH-MEHP im Urin. *Bundesgesundheitsblatt – Gesundheitsforschung – Gesundheitsschutz, 48*(6), 706–722.

Bohren, C. F., & Huffman, D. R. (2008). *Absorption and scattering of light by small particles*. New York: Wiley-VCH.

Bombelli, P., Howe, C. J., & Bertocchini, F. (2017). Polyethylene bio-degradation by caterpillars of the wax moth Galleria mellonella. *Current Biology, 27*(8), 292–293. https://doi.org/10.1016/j.cub.2017.02.060.

Boucher, J., & Friot, D. (2017). *Primary microplastics in the oceans: A global evaluation of sources* (p. 43). Gland: IUCN.

Brandmüller, J., & Moser, H. (1962). Einführung in die Ramanspektroskopie. *Wissenschaftliche Forschungsberichte. Naturwissenschaftliche Reihe* 70. Steinkopff: Darmstadt.

Braun, U., Bahr, H., & Schartel, B. (2010). Fire retardancy effect of aluminium phosphinate and melamine polyphosphate in glass fibre reinforced polyamide 6. *EPolymers 41*, 1–14.

Braun, G., Brüll, U., Alberti, J., & Furtmann, K. (2001). „*Vorkommen von Phthalaten in Oberflächenwasser und Abwasser*". Essen: Landesumweltamt NRW

Bravo, R., et al. (2012). Plastic ingestion by harbour seals in the Netherlands. *Marine Pollution Bulletin, 67,* 200–2002.

Briehl, H. (2008). *Chemie der Werkstoffe* (2. ed.). Wiesbaden: Teubner.

Browne, M. A., Galloway, T., & Thompson, R. (2007). Microplastic – An emerging contaminant of potential concern? *Integrated Environmental Assessment and Management, 3,* 559–566.

Browne, M. A., Galloway, T., & Thompson, R. (2010). Spatial patterns of plastic debris along estuarine shorelines. *Environmental Science & Technology, 44*(9), 3404–3409. https://doi.org/10.1021/es903784e.

Browne, M. A., Crump, P., Niven, S. J., Teuten, E., Tonkin, A., Galloway, T., et al. (2011). Accumulation of microplastic on shorelines woldwide: Sources and sinks. *Environmental ScienceTechnology, 45,* 9175–9179. https://doi.org/10.1021/es201811s.

Bruker Optik. (2008). Einführung in die FT-IR-Spektroskopie, Version 2.0; Tutorial Bruker.

BUND (Bund für Umwelt und Naturschutz Deutschland). (2014). Stoppt Mikroplastik in Alltagsprodukten – Umweltbewusst einkaufen! https://klimaschutzfonds-wedel.de/pdf/140527-bund-mikroplastik_produktliste.pdf.

BUND (Bund für Umwelt und Naturschutz Deutschland). (2018). Mikroplastik und andere Kunststoffe in Kosmetika. Der BUND-Einkaufsratgeber. https://www.bund.net/fileadmin/user_upload_bund/publikationen/meere/meere_mikroplastik_einkaufsfuehrer.pdf.

Buxbaum, L. H. (1968). The degradation of Poly(ethylene terephthalate). *Angewandte Chemie International Edition in English, 7,* 182–190.

Carrington, D. (6. September 2017). Plastic fibres found in tap water around the world, study reveals. *The Guardian.* https://www.theguardian.com/environment/2017/sep/06/plastic-fibres-found-tap-water-around-world-study-reveals.

Catarino, A. I., Macchia, V., Sanderson, W. G., Thompson, R. C., & Henry, T. B. (2018). Low levels of microplastics (MP) in wild mussels indicate that MP ingestion by humans is minimal compared to exposure via household fibres fallout during a meal. *Environmental Pollution, 237,* 675–684.

Cheng, Z., Nie, X. P., Wang, H. S., & Wong, M. H. (2013). "Risk assessments of human exposure to bioaccessible phthalate esters through market fish consumption". *Environment International, 57–58,* 75–80. https://doi.org/10.1016/j.envint.2013.04.005.

Chi, Z., Wang, D., & You, H. (2016). „Study on the mechanism of action between dimethyl phthalate and herring sperm DNA at molecular level". *Journal of Environmental Science and Health, Part B, 51*(8), 553–557.

Choi, K., et al. (2012). In Vitro metabolism of di(2-ethylhexyl)phthalate (DEHP) by various tissues and Cytochrome P 450s of human and rat. *Toxicology in Vitro: An International Journal Published in Association with BIBRA, 26*(8), 315–322.

CIRS (Chemical Inspection and Regulation Service). (2008). Reach SVHC candidate list. http://www.cirs-group.com/uploads/soft/140227/3-14022F92616.pdf.

Claessens, M., de Meester, S., van Landuyt, L., de Clerck, K., & Janssen, C. R. (2013). Occurrence and distribution of microplastics in marine sediments along the Belgian coast. *Marine Pollution Bulletin, 10,* 2199–2204. https://doi.org/10.1016/j.marpolbul.2011.06.030.

Codina-García, M., Militão, Teresa, Moreno, Javier, & González-Solís, Jacob. (2013). Plastic debris in Mediterranean seabirds. *Marine Pollution Bulletin, 1–2,* 220–226. https://doi.org/10.1016/j.marpolbul.2013.10.002.

Cole, M., Lindeque, P., Halsband, C., & Galloway, T. S. (2011). Microplastics as contaminants in the marine environment: A review. *Marine Pollution Bulletin, 62*(12), 2588–2597. https://doi.org/10.1016/j.marpolbul.2011.09.025.

Cole, M., Lindeque, P., Fileman, E., Halsband, C., Goodhead, R., Moger, J., et al. (2013). Microplastic ingestion by zooplankton. *Environmental Science & Technology, 12,* 6646–6655. https://doi.org/10.1021/es400663f.

Collard, F., Gilbert, B., Compère, P., Eppe, G., Das, K., Jauniaux, T., et al. (2017). Microplastics in livers of European anchovies (*Engraulis encrasicolus,* L.). *Environmental Pollution, 229,* 1000–1005.

Cozar, A., et al. (2014). "Plastic debris in the open ocean". *Proceedings of the National Academy of Sciences USA, 111,* 10239–10244.

Dekant, W., & Vamvakas, S. (1994). *„Toxikologie für Chemiker und Biologen".* Heidelberg: Spektrum Akademischer.

DePuy, C. H., King, R. W., & Cozar, A. (2014). Pyrolytic Cis eliminations. *Chemical Reviews, 60,* 431–457.

Derraik, J. G. B. (2002). The pollution of the marine environmental by plastic debris: A review. *Marine Pollution Bulletin, 44,* 842–852.

Dickmann, R. (1933). Studies on the waxmoth Galleria mellonella with particular reference to the digestion of wax by the larvae. *Journal of Cellular and Comparative Physiology, 3,* 223–246.

Dimitrov, N., Kratofil Krehula, L., Ptiček Siročić, A., & Hrnjak-Murgić, Z. (2013). Analysis of recycled PET bottles products by pyrolysis-gas chromatography. *Polymer Degradation and Stability, 98,* 972–979.

Dris, R., et al. (2015). Beyond the ocean: Contamination of freshwater ecosystems with (micro-) plastic particles. *Environmental Chemistry, 12,* 539–550.

DRZE (Deutsches Referenzzentrum für Ethik in den Biowissenschaften). (2018). Planet Plastik. http://www.drze.de/bibliothek/presseschau/artikel?aid=43810&set_language=de.

Dümichen, E., Barthel, A.-K., Braun, U., Bannick, C. G., Brand, K., Jekel, M., et al. (2015). Analysis of polyethylene microplastics in environmental samples. *Water Research, 85,* 451–457.

Dümichen, E., Eisentraut, P., Bannick, C. G., Barthel, A.-K., Senz, R., & Braun, U. (2017). Fast identification of microplastics in complex environmental samples by a thermal degradation method. *Chemosphere, 174,* 572–584.

Düssel, H. J., Rosen, H., & Hummel, D. O. (1976). Feldionen- und Elektronenstoß-Massenspektrometrie von Polymeren und Copolymeren, 5. Aliphatische und aromatische Polyamide und Polyimide. *Macromolecular Chemistry and Physics, 177,* 2343–2368.

Dutescu, R. M. (2011). "Expressionsanalyse der nukleären Rezeptoren PPAR-α/γ-1/γ-2 und der Transkriptionsfaktoren T-bet und GATA-3 nach Stimulation von dermalen Endothelzellen mit den Weichmacher, Di(2-ethylhexyl)phthalat-Metaboliten 2-Ethylhexanol und 4-Heptanon"; Inauguraldissertation, Institut für Klinische Chemie und Molekulare Diagnostik des Fachbereichs Medizin der Philipps-Universität Marburg.

Eerkes-Medrano, D., Thompson, R. C., & Aldridge, D. C. (2015). Microplastics in freshwater systems: A review of the emerging threats, identification of knowledge gaps and prioritisation of research needs. *Water Research, 75,* 63–82.

EFSA (European Food Safety Authority). (2016). Presence of microplastics and nanoplastics in food, with particular focus on seafood. EFSA Panel on Contaminants in the Food Chain (CONTAM). *EFSA Journal, 14*(6), 4501.

Elert, A. M., Becker, R., Duemichen, E., Eisentraut, P., Falkenhagen, J., Sturm, H., & Braun, U. (2017). Comparison of different methods for MP detection: What can we learn from them, and why asking the right question before measurements? *Environmental Pollution,* Dec 231(Pt 2), 1256–1264. https://doi.org/10.1016/j.envpol.2017.08.074. Epub 2017 Sep 21.

Enders, K., Lenz, R., Stedmon, C. A., & Nielsen, T. G. (2015). Abundance, size and polymer composition of marine microplastics ≥10 mm in the Atlantic Ocean and their modelled vertical distribution. *Marine Pollution Bulletin, 100,* 70–81.

Engler, R. E. (2012). The complex interaction between marine debris and toxic chemicals in the ocean. *Environmental Science & Technology, 46,* 12302–12315.

Farrell, P., & Nelson, K. (2013). Trophic level transfer of microplastic: *Mytilus edulis* (L.) to *Carcinus maenas* (L.). *Environmental Pollution, 117,* 1–3.

Fath, A. (2010). Hansgrohe – Wassersymposium.

Fath, A. (2016). *Rheines Wasser – 1231 Kilometer mit dem Strom.* München: Hanser.

Fath, A. et al. (2016). „Electrochemical decomposition of fluorinated wetting agents in plating industry waste water". *Water Science & Technology, 73*(7), 1659–1666.

Fath, A., Juri Jander, A., Birte Beyer, B., Jonas Loritz, A., Erik Dümichen, C., Martin Knoll, D., & Gunnar Gerdts, B. (2019). *Quantification and identification of microplastics in the surface waters of the river Rhine and the river Tennessee.* Anthropocene: Elsevier.

Fath, A. et al. (eingereicht). Scientific Reports. „*Microplastic entry in homemade food*".

Feldhahn, T. (2008). Thesisarbeit, Hochschule Offenburg.

Fischer, D., Kaeppler, A., & Eichhorn, K.-J. (2015). Identification of microplastics in the marine environment by raman microspectroscopy and imaging. *American Laboratory, 47,* 32–34.

Fraunhofer UMSICHT. (2014). *Biowachspartikel Heals Alternative zu Mikroplastik.* http://www.umsicht.fraunhofer.de/de/presse-medien/2014/140612-mikroplastik.html. Stand: 11.09.2014.

Fromme, H., Becher, G., Hilger, B., & Völkel, W. (2016). Brominated flame retardants – Exposure and risk assessment for the general population. *International Journal of Hygiene and Environmental Health, 219*(1), 1–23. https://doi.org/10.1016/j.ijheh.2015.08.004, PMID 26412400.

Fuller, S., & Gautam, A. (2016). A procedure for measuring microplastics using pressurized fluid extraction. *Environmental Science & Technology, 50*(11), 5774–5780.

Gächter, R., & Müller, H. (1993). *Plastics additives handbook.* München: Hanser.

Gaihre, B., & Jayasuriya, A. C. (2016). Fabrication and characterization of carboxymethyl cellulose novel microparticles for bone tissue engineering. *Materials Science and Engineering, 69,* 733–743. https://doi.org/10.1016/j.msec.2016.07.060. Epub 2016 Jul 22.

Galgani, F., Hanke, G., Werner, S., & De Vrees, L. (2013). Marine litter within the European Marine Strategy Framework Directive. *Ices Journal of Marine Science, 70,* 1055–1064.

González-Castro, M. I., Olea-Serrano, M. F., Rivas-Velasco, A. M., Medina-Rivero, E., Ordon̄ez-Acevedo, L. G., & De León-Rodríguez, A. (2011). "Phthalates and Bisphenols migration in Mexican food cans and plastic food containers". *Bulletin of Environmental Contamination and Toxicology, 86*(6), 627–631. https://doi.org/10.1007/s00128-011-0266-3.

Günzler, H., & Heise, H. M. (2003). *IR-Spektroskopie – Eine Einführung* (4. ed.). Weinheim: Wiley-VCH.

Halang, V. (o. J.). Ist Mikroplastik wirklich gefährlich? *enorm,* http://enorm-magazin.de/ist-mikroplastik-wirklich-gefaehrlich.

Hanser Kundencenter. (2017). Mikrokunststoff in Binnengewässern – Untersuchungen am Beispiel des Rheins. https://www.kunststoffe.de/fachinformationen/online-beitraege/artikel/mikrokunststoff-in-binnengewaessern-3988323.html?article.page=5.

Harrison, J. P. et al. (2012). The applicability of reflectance micro-Fourier-transform infrared spectroscopy for the detection of synthetic microplastics in marine sediments. *Science of the Total Environment, 416,* 455–463.

Harsch, A., & Kirschner, N. (2014). Entwicklung eines Schnelltests zur Bestimmung des Weichmachergehalts in Kunststoffen. Hochschule Furtwangen. Villingen-Schwenningen: s.n., p. 12, Projektarbeit.

Hart, H., Craine, L. E., & Hart, D. J. (2002). *Organische Chemie, 2. vollständig überarbeitete und aktualisierte Auflage.* Weinheim: Wiley-VCH.

Hartline, N. L., Bruce, N. J., Karba, S. N., Ruff, E. O., Sonar, S. U., & Holden, P. A. (2016). Microfiber masses recovered from conventional machine washing of new or aged garments. *Environmental Science & Technology, 50*(21), 11532–11538. https://doi.org/10.1021/acs.est.6b03045.

HELCOM BASE Project. (2014). Preliminary study on synthetic microfibers and particles at a municipal waste water treatment plant. http://helcom.fi/Lists/Publications/Microplastics%20at%20a%20municipal%20waste%20water%20treatment%20plant.pdf. Stand: 11. Sept. 2014.

Hesse, M., Meier, H., & Zeeh, B. (1987). *Spektroskopische Methoden in der organischen Chemie* (3. ed.). Stuttgart: Thieme.

Hesse, M., Meier, H., Zeeh, B., Tagg, A. S., Sapp, M., Harrison, J. P., & Ojeda, J. J. (2015). Identification and quantification of microplastics in wastewater using focal plane array-based reflectance micro-FT-IR imaging. *Analytical chemistry, 87,* 6032–6040.

Hesse, M., Meier, H., & Zeeh, B. (2016). *Spektroskopische Methoden in der organischen Chemie* (9, überarbeitete ed.). New York: Georg Thieme.

Hinterbuchner, T. (2006). Das Verhalten von Benzotriazolen in Abwasserreingungsanlagen Als DIPLOMARBEIT eingereicht an der Fachhochschule Wels zur Erlangung des akademischen Grades Diplom-Ingenieur (FH) von Weber, W. H. & Müller, A & Weiss, Stefan & Seitz, W & Schulz, Wolfgang. (2009). 1H-benzotriazole and tolyltriazoles in the aquatic environment. Occurrence in ground, surface and wastewater. *Vom Wasser, 107,* 16–24.

Holland, B. J., & Hay, J. N. (2002). The thermal degradation of PET and analogous polyestersmeasured by thermal analysise – Fourier transform infrared spectroscopy. *Polymer, 43,* 1835–1847.

Hüffer, T., & Hofmann, T. (2016). Sorption of non-polar organic compounds by micro-sized plastic particles in aqueous solution. *Environmental Pollution, 214,* 194–201.

Hummel, D. (2017). *Untersuchung der Sorption wässrig gelöster organischer Substanzen an Polymerpartikel.* Furtwangen: Studiengang NBT.

Imhof, H. K., Schmid, J., Niessner, R., Ivleva, N., & Laforsch, C. (2012). A novel, highlyefficient method for the separation and quantification of plastic particles in sediments of aquatic environments. *Limnology and Oceanography: Methods, 10,* 524–537.

Imhof, H. K., et al. (2013). Contamination of beach sediments of a subalpine lake with microplastic particles. *Current Biology, 23,* 867–868.

Imhof, H. K., Laforsch, C., Wiesheu, A. C., Schmid, J., Anger, P. M., Niessner, R., et al. (2016). Pigments and plastic in limnetic ecosystems: a qualitative and quantitative study on microparticles of different size classes. *Water Research, 98,* 64–74.

IPASUM (Institut und Poliklinik für Arbeits-, Sozial- und Umweltmedizin der Universität Erlangen-Nürnberg). (o. J.). Phthalate – Weichmacher – DEHP. https://www.arbeitsmedizin. uni-erlangen.de/forschung/studien/phthalate.shtml.

Ivar do Sul, J. A., & Costa, M. F. (2014). The present and future of microplastic pollution in the marine environment. *Environmental Pollution, 185,* 352–364.

Ivashechkin, P. (2005). „Kurzfassung des Berichts zum Vorhaben: Literaturauswertung zum Vorkommen gefährlicher Stoffe im Abwasser und in Gewässern". AZ IV 9 – 042 059, für das Ministerium für Umwelt und Naturschutz Landwirtschaft und Verbraucherschutz des Landes Nordrhein-Westfalen.

Ivleva, N. P., & Nießner, R. (2015). Kunststoffpartikel im Süßwasser. *Nachrichten Aus der Chemie, 63,* 46–50.

Jambeck, J. R., et al. (2015). Plastic waste inputs from land into the ocean. *Science, 347,* 768–770. https://doi.org/10.1126/science.1260352.

Jander, J. (2017). *Mikroplastik in Flüssen und Lebensmitteln.* HFU: Masterthesis.

Käppler, A., Windrich, F., Löder, M. G. J., Malanin, M., Fischer, D., Labrenz, M., et al. (2015). Identification of microplastics by FTIR and Raman microscopy: A novel silicon filter substrate opens the important spectral range below 1300 cm_1 for FTIR transmission measurements. *Analytical and Bioanalytical Chemistry, 407,* 6791–6801.

Karami, A., Golieskardi, A., Choo, C. K., Larat, V., Galloway, T. S., & Salamtinia, B. (2017). The presence of microplastics in commercial salts from different countries. *Scientific Reports.* https://doi.org/10.1038/srep46173.

Kebelmann, K., Hornung, A., Karsten, U., & Griffiths, G. (2013). Intermediate pyrolysis and product identification by TGA and Py-GC/MS of green microalgae and their extracted protein and lipid components. *Biomass Bioenergy, 49,* 38–48.

Kemmlein, S., Hahn, O., & Jann, O. (2003). Emissionen von Flammschutzmitteln aus Bauprodukten und Konsumgütern. project no. (UFOPLAN reference no.) 299 65 321, Environmental Research Programme of the Federal Ministry for Environment, Nature Conservation and Nuclear Safety, commissioned by the Federal Environmental Agency (UBA), UBA-FB 000475, Berlin.

Kershaw, P. J. (2014). Sources, fate and effects of microplastics in the marine environment: A global assessment. *Report & Studies GESAMP, 90*(96), 2015.

Klein, S., Worch, E., & Knepper, P. (2015). Occurrence and spatial distribution of microplastics in river shore sediments of the rhine-main area in Germany. *Environmental science & technology, 49,* 6070–6076.

Klöpffer, W. (2012). *Verhalten und Abbau von Umweltchemikalien* (2. ed.). Weinheim: Wiley-VCH.

Koch, H. M. (2006). Institut und Poliklinik für Arbeits-, Sozial- und Umweltmedizin der Universität Erlangen; „Phthalate". https://www.arbeitsmedizin.uni-erlangen.de/forschung/studien/phthalate.shtml.

Kole, P. J., Löhr, A. J., Van Belleghem, F.-G., & Ragas, A. M. J. (2017). "Wear and tear of tyres: A stealthy source of microplastics in the environment". *International Journal of Environmental Research and Public Health, 14*(10), 1265. https://doi.org/10.3390/ijerph14101265.

Kurzenberger, I. (2010). Benzotriazole in der aquatischen Umwelt – Entwicklung einer spurenanalytischen Bestimmungsmethode und Verhalten bei der Trinkwasseraufbereitung, Diplomarbeit, Universität Hohenheim, Institut für Lebensmittelchemie.

LAGA (Bund/Länder-Arbeitsgemeinschaft Abfall). (2013). Abschlussbericht LFP-Vorhaben L1.11. „Erarbeitung eines PAK-Schnellerkennungsverfahrens zur Abfalluntersuchung". Zugegriffen: 28. Mai 2013.

Lagerberg, J. W., et al. (2015). In vitro evaluation of the quality of blood products collected and stored in systems completly free of di(2-ethylhexyl)-phthalates plasticized materials. *Transfusion, 55*(3), 322–531.

Lart, W. (2018). Sources, fate, effects and consequences for the seafood industry of micro and nanoplastics in the marine environment. Seafish Information Sheet No FS 92.04.19. Grimsby, Seafish.

Leser, C. (2015). *Zukunftsfähige Verwertungswege des Gärrests von Nawaros und Abfällen nach Kreislaufwirtschaftsprinzip.* Hochschule Furtwangen: Thesisarbeit.

Lewin-Kretzschmar, U. (o. J.). Prävention, Kompetenz-Center Gefahrstoffe und biologische Arbeitsstoffe. Berufsgenossenschaft Rohstoffe und chemische Industrie, Leuna.

LfU (Bayerisches Landesamt für Umwelt). (2016). Publikationen des Bayerischen Landesamts für Umwelt. https://www.lfu.bayern.de/publikationen/doc/publikationskatalog_des_lfu.pdf.

Liebezeit, G., & Dubaish, F. (2012). Mikroplastik – Quellen, Umweltaspekte und Daten zum Vorkommen im Niedersächsischen Wattenmeer. *Zeitschrift der Naturschutz- und Forschungsgemeinschaft Mellumrat, 11*(1), 21–31.

Liebezeit, G., & Liebezeit, E. (2014). Synthetic particles as contaminants in German beers. *Food Additives & Contaminants. Part A, Chemistry, Analysis, Control, Exposure & Risk Assessment, 9,* 1574–1578. https://doi.org/10.1080/19440049.2014.945099.

Lobelle, D., & Cunliffe, M. (2011). Early microbial biofilm formation on marine plastic debris. *Marine Pollution Bulletin, 62,* 197–200.

Löder, M. G. J., & Gerdts, G. (2015). Methodology used for the detection and identification of microplasticsda critical appraisal. In M. Bergmann, L. Gutow, & M. Klages (Hrsg.), *Marine Anthropogenic Litter* (S. 201–227). Cham: Springer.

Löder, M. G. J., Kuczera, M., Mintenig, S., Lorenz, C., & Gerdts, G. (2015a). FPA-based micro-FTIR imaging for the analyses of microplastics in environmental samples. *Environmental Chemistry, 12,* 563–581. https://doi.org/10.1071/EN14205.

Löder, M. G. J., Kuczera, M., Mintenig, S., Lorenz, C., & Gerdts, G. (2015b). Focal plane array detector-based micro-Fourier-transform infrared imaging for the analysis of microplastics in environmental samples. *Environmental Chemistry, 12,* 563–581.

Lopez L, R., & Mouat, J. (2009). *Marine litter in the Northeast Atlantic Region.* London: OSPAR Commission.

Loritz, J., (2014). "Mikroplastikbelastung im Rhein", Bachelor-Thesisarbeit, HFU.

LUBW (Landesanstalt für Umwelt Baden-Württemberg). (2018). Mikroplastik in Binnengewässern Süd- und Westdeutschlands. http://www4.lubw.baden-wuerttemberg.de/servlet/is/274206/.

Lunder, S., Sharp, R., Ling, A., & Colesworthy, C. (2008). *Study finds record high levels of toxic fire retardants in breast milk from American mothers.*

Lusher, A. L., McHugh, M., & Thompson, R. C. (2013). Occurrence of microplastics in the gastrointestinal tract of pelagic and demersal fish friom the English Channel. *Marine Pollution Bulletin, 67*(1–2), 94–99. https://doi.org/10.1016/j.marpolbul.2012.11.028.

Lusher A, Hollman P, & Medonza-Hill. (2017). Microplastics in fisheries and aquaculture. Status of knowledge on their occurrence and implications for aquatic organisms and food safety. *FAO Fisheries and Aquaculture Technical Paper No 615.* Rome: FAO.

Maier, R.-D., & Schiller, M. (2016). *Handbuch Kunststoff-Additive* (4. ed.). München: Hanser.

Mani, T., et al. (2015). Micoplastic Profile along the River Rhine. *Scientific Reports, 5,* 17988. https://doi.org/10.1038/srep17988.

Manickum, T., & John, W. (2014). Occurrence, fate and environmental risk assessment of endocrine disrupting compounds at the wastewater treatment works in Pietermaritzburg (South Africa). *The Science of the Total Environment, 468–469,* 584–597.

Mato, Y., Isobe, Tomohiko, Takada, Hideshige, Kanehiro, Haruyuki, Ohtake, Chiyoko, & Kaminuma, Tsuguchika. (2001). Plastic resin pellets as a transport medium for toxic chemicals in the marine environment. *Environmental Science & Technology, 2,* 318–324. https://doi.org/10.1021/es0010498.

Meeker, J. D., Sathyanarayana, S., & Swan, S. H. (2009). Phthalates and other additives in plastics: Human exposure and associated health outcomes. *Philosophical Transactions of the Royal Society of London Series B, Biological Sciences, 1526,* 2097–2113. https://doi.org/10.1098/rstb.2008.0268.

Metrio, G. de, Corriero, A., Desantis, S., Zubani, D., Cirillo, F., Deflorio, M., Bridges, C. R., Eicker, J., de la Serna, J. M., Megalofonou, P., & Kime, D. E. (2003). Evidence of a high percentage of intersex in the Mediterranean swordfish (Xiphias gladius L.). *Marine Pollution Bulletin, 3,* 358–361, https://doi.org/10.1016/s0025-326x(02)00233-3.

Moore, C. J. (2008). Synthetic polymers in the marine environment: A rapidly increasing, long-term threat. *Environmental Research, 2,* 131–139. https://doi.org/10.1016/j.envres.2008.07.025.

Moore, C. J., Lattin, G. L., & Zellers, A. F. (2011). Quantity and type of plastic debris flowing from two urban rivers to coastal waters and beaches of Southern California. *Revista de Gestão Costeira Integrada, 1,* 65–73. https://doi.org/10.5894/rgci194.

Morritt, D., Stefanoudis, P. V., Pearce, D., Crimmen, A., & Clark, P. F. (2014). Plastic in the Thames: A river runs through it. *Marine Pollution Bulletin, 78,* 196–200.

NABU. (o. J.). Recycling und der gelbe Sack—It's complicated. https://www.nabu.de/umwelt-und-ressourcen/abfall-und-recycling/recycling/21113.html.

Naumer, H., & Heller, W. (1997). *Untersuchungsmethoden in der Chemie* (2. ed.). Stuttgart: Thieme.

NDR. (2010). „45 Min – Gefahr Weichmacher": Warum sind immer mehr Männer nur noch eingeschränkt fruchtbar? https://www.ndr.de/der_ndr/presse/mitteilungen/pressemeldungndr5930.html.

Neek, P., Weinschrott, H., & Fath, A. (2017). *Galvanotechnik – „Polycarbonatgehaltsbestimmung mittels Infrarotspektroskopie",* Bd. 1 (pp. 30–33). Leuze.

Ntv. (2017). Schwer wie 822.000 Eiffeltürme – Acht Milliarden Tonnen Plastik gibt es bereits. https://www.n-tv.de/wissen/Acht-Milliarden-Tonnen-Plastik-gibt-es-bereits-article19944523.html.

Oehlmann, J., Schulte-Oehlmann, U., Kloas, W., Jagnytsch, O., Lutz, I., Kusk, K. O., Wollenberger, L., Santos, E. M., Paull, Gr. C., Van Look, K. J. W., & Tyler, C. R. (2009). A critical analysis of the biological impacts of plasticizers on wildlife. *Philosophical Transactions of the Royal Society of London. Series B, Biological Sciences, 1526,* 2047–2062. https://doi.org/10.1098/rstb.2008.0242.

Ohtani, H., Nagaya, T., Sugimura, Y., & Tsuge, S. (1982). Studies on thermal degradation of aliphatic polyamides by pyrolysis-glass capillary chromatography. *Journal of Analytical and Applied Pyrolysis, 4,* 117–131.

Ortner, J., & Hensler, G. (1995). Beurteilung von Kunststoffbränden. PDF, 54 S.

Patel, M. M., Goyal, B. R., Bhadada, S. V., Bhatt, J. S., & Amin, A. F. (2009). Getting into the brain: Approaches to enhance brain drug delivery. *CNS drugs, 1,* 35–58.

Pillard, D. A., Cornell, J. S., Dufresne, D. L., & Hernandez, M. T. (2001). Toxicity of benzotriazole and benzotriazole derivatives to three aquatic species. *Water Research, 35,* 557–560.

PlasticsEurope. (2013). Plastics – The facts 2013. An analysis of European latest plastics production, demand and waste data. http://www.plasticseurope.org/documents/document/20131018104201-plastics_the_facts_2013.pdf. Zugegriffen: 3. Okt. 2014.

Rajesh Kumar, S., Asseref, P. M., Dhanasekaran, J., & Krishna Mohan, S. (2014). A new approach with prepregs for reinforcing nitrile rubber with phenolic and benzoxazine resins. *RCS Advances, 24,* 12526–12533.

Reemtsma, T., Miehe, U., Dünnbier, U., & Jekel, M. (2010). Polar pollutants in municipal wastewater and the water cycle: Occurrence and removal of benzotriazoles. *Water Research, 44,* 596–604.

Rillig, M.-C. (2012). Microplastic in terrestrial ecosystems and the soil? *Environmental Science & Technology, 46*(12), 6453–6454. https://doi.org/10.1021/es302011r.

Rios, L. M., Moore, C., & Jones, P. R. (2007). Persistent organic pollutants carried by synthetic polymers in the ocean environment. *Marine Pollution Bulletin, 8,* 1230–1237. https://doi.org/10.1016/j.marpolbul.2007.03.022.

Rippen, G. (1993). „Datensammlung über Umweltchemikalien – Naphthalin". *Handbuch Umweltchemikalien,* Ecomed, Loseblattsammlung, 20. Erg. Lieferung.

Rosado-Berrios, C. A., et al. (2011). Mitochondrial permeability and toxicity of diethylhexyl and monoethylhexyl phthalates on TK6 human lymphoblasts cells. *Toxicology in Vitro: An International Journal Published in Association with BIBRA, 25*(8), 2010–2016.

Ruff, M., & Singer, H. (2013). "20 Jahre Rheinüberwachung". Aqua & Gas Nr. 5, pp. 16–25.

Ryan, P. G., Moore, Charles J., van Franeker, Jan A., & Moloney, Coleen L. (2009). Monitoring the abundance of plastic debris in the marine environment. *Philosophical Transactions of the Royal Society of London Series B, Biological Sciences, 1526,* 1999–2012. https://doi.org/10.1098/rstb.2008.0207.

Saechtling, H., & Baur, E. (2007). *Saechtling-Kunststoff-Taschenbuch* (30. ed.). München: Hanser.

Selke, S. E. M., & Culter, J. D. (2016). *Plastics packaging – Properties, processing applications and regulations* (3. ed.). München: Hanser.

Serrano, D. P., Aguado, J., Escola, J. M., Rodríguez, J. M., & San Miguel, G. (2005). An investigation into the catalytic cracking of LDPE using PyeGC/MS. *Journal of Analytical and Applied Pyrolysis, 74,* 370–378.

Sjödin, A., Hagmar, L., Klasson-Wehler, E., Kronholm-Diab, K., Jakobsson, E., & Bergman, Å. (1999). Flame retardant exposure: Polybrominated diphenyl ethers in blood from Swedish workers. *Environmental Health Perspectives, 107*(8), 643–648.

Skrzypek, K. P. (2003). „*Austauschprozesse von organischen Umweltchemikalien mit biogenen Tensiden in quellfähigen Tonmineralien*". Dissertation, vorgelegt beim Fachbereich Geowissenschaften der Wolfgang-Goethe-Universität Frankfurt am Main.

Sojak, L., Kubinec, R., Jurdakova, H., Hajekova, E., & Bajus, M. (2007). High resolution gas chromatographic-mass spectrometric analysis of polyethylene and polypropylene thermal cracking products. *Journal of Analytical and Applied Pyrolysis, 78,* 387–399.

Spangenberg, B. (o. J.). Laborvorschrift zum Versuch IR-Spektroskopie, HS Offenburg.

Spiegel online. (20. Januar 2016). Umweltschutz: Schwimmbarrieren sollen Müll aus Meer fischen. http://www.spiegel.de/wissenschaft/technik/umweltschutz-schwimmbarrieren-sollen-muell-aus-dem-meer-fischen-a-1070868.html.

Streitwieser, A., Heathcock, C. H., & Kosower, E. M. (1994). *Organische Chemie.* Weinheim: Wiley-VCH.

Stryer, L. (1990). *Biochemie.* Heidelberg: Spektrum der Wissenschaft.

Suchentrunk, R. et al. (2007). „*Kunststoffmetallisierung*" (3. ed.). Leuze.

Süddeutsche Zeitung. (2011). 25 Jahre Sandoz-Katastrophe. Als im roten Rhein die Fische starben. http://www.sueddeutsche.de/wissen/jahre-sandoz-katastrophe-als-im-roten-rhein-die-fische-starben-1.1177611.

SWR2. (2016). Dick durch Weichmacher – Plastikverpackungen untersucht. https://www.swr.de/swr2/wissen/plastikverpackungen-untersucht-dick-durch-weichmacher/-/id=661224/did=16849468/nid=661224/8j0s6a/index.html.

Taylor, M. L., et al. (2016). Plastic microfibre ingestion by deep-sea organisms. *Scientific Reports, 6,* 33997. https://doi.org/10.1038srep33997.

Teuten, E. L., Saquing, J. M., Knappe, D. R. U., Barlaz, M. A., Jonsson, S., Björn, A., Rowland, S. J., Thompson, R. C., Galloway, T. S., Yamashita, R., Ochi, D., Watanuki, Y., Moore, C., Viet, P. H., Tana, T. S., Prudente, M., Boonyatumanond, R., Zakaria, M. P., Akkhavong, K., Ogata, Y., Hirai, H., Iwasa, S., Mizukawa, K., Hagino, Y., Imamura, A., Saha, M., & Takada, H. (2009). Transport and release of chemicals from plastics to the environment and to wildlife. *Philosophical Transactions of the Royal Society of London. Series B, Biological Sciences, 1526,* 2027–2045. https://doi.org/10.1098/rstb.2008.0284.

Thalheim, M. (2016). „Phthalate: Innovation mit Nebenwirkung". *Dtsch Arztebl, 113*(45), A-2036/B-1704/C-1688.

Thompson, R. C., et al. (2004). Lost at sea: Where is all the plastic? *Science, 304,* 838.

Thompson, R. C., et al. (2009). Our plastic age. *Philosophical Transactions of The Royal Society B Biological Sciences, 364,* 1973–1976.

Torre, M., Digka, N., Anastasopoulou, A., Tsangaris, C., & Mytilineou, C. (2016). Anthropogenic microfibres pollution in marine biota. A new and simple methodology to minimize airborne contamination. *Marine Pollution Bulletin, 113*(1–2), 55–61.

Troitzsch, J. (2012). Flammschutzmittel. Anforderungen und Innovationen. *Kunststoffe, 11,*84.

Umweltbundesamt. (2007). „Phthalate – Die nützlichen Weichmacher mit den unerwünschten Eigenschaften". https://www.umweltbundesamt.de/sites/default/files/medien/publikation/long/3540.pdf.

Umweltbundesamt. (2008). Bromierte Flammschutzmittel – Schutzengel mit schlechten Eigenschaften? https://www.umweltbundesamt.de/sites/default/files/medien/publikation/long/3521.pdf.

Umweltbundesamt. (2011). Telegramm: Umwelt und Gesundheit, Ausgabe 01/2011; „Neue Weichmacherin Kunststoffen". https://www.umweltbundesamt.de/sites/default/files/medien/.../Ausgabe01-2011.pdf.

Umweltbundesamt. (2013). „Häufige Fragen zu Phthalaten bzw. Weichmachern. https://www.umweltbundesamt.de/themen/gesundheit/umwelteinfluesse-auf-den-menschen/chemische-stoffe/weichmacher/haeufige-fragen-zu-phthalaten-bzw-weichmachern#textpart-1.

Umweltbundesamt. (2016). Polyzyklische Aromatische Wasserstoffe – Umweltschädlich! Giftig! Unvermeidbar? https://www.umweltbundesamt.de/sites/default/files/medien/376/publikationen/polyzyklische_aomatische_kohlenwasserstoffe.pdf.

Van Cauwenberghe, L., & Janssen, C. R. (2014). Microplastics in bivalves cultured for human consumption. *Environmental Pollution, 193,* 65–70.

Van Cauwenberghe, L., Vanreusel, A., Mees, J., & Janssen, C. R. (2013). Microplastic pollution in deep-seea sediments. *Environmental Pollution, 182,* 495–499. https://doi.org/10.1016/j.envpol.2013.08.013.

van der Meer, P. F., & Devine, D. V. (2017). Alternatives in blood operations when choosing non-DEHP bags. *View Issue TOC, 112*(2), 183.

Vianello, A., et al. (2013). Microplastic particles in sediments of Lagoon of Venice, Italy: First observations on occurrence, spatial patterns and identification. *Estuarine, Coastal and Shelf Science, 130,* 54–61.

von Moos, N. (2010). *Histopathological and cytochemical analysis of ingested polyethylene powder in the digestive gland of the blue mussel.* Switzerland: Basel.

von Moos, N., Burkhardt-Holm, P., & Köhler, A. (2012). Uptake and effects of microplastics on cells and tissue of the blue mussel Mytilusedulis L. after an experimental exposure. *Environmental Science & Technology, 46,* 11327–11335.

Wagner, M., & Oehlmann, J. (2009). Endocrine disruptors in bottled mineral water: Total estrogenic burden and migration from plastic bottles. *Environmental Science and Pollution Research, 16*, 278–286.

Wang, W., Ndungu, A. W., Li, Z., & Wang, J. (2017). Microplastics pollution in inland freshwaters of China: A case study in urban surface waters of Wuhan, China. *The Science of the Total Environment, 575*, 1369–1374.

Weber, C. (2002). „Plastikmüll mit Infrarotspektroskopie sortieren". *Physik Journal, 1*(7–8), 116–119.

Welle, F., Wolz, G., & Franz, R. (2004). „Study on the migration behaviour of DEHP versus an alternative plasticiser, Hexamoll® DINCH, from PVC tubes into enteral feeding solutions", Poster presentation at the 3rd international Symposium on Food Packaging, 17–19 November 2004, Barcelona. https://www.ivv.fraunhofer.de/content/dam/ivv/en/documents/Forschungsfelder/Produktsicherheit-und-analytik/Study_on_the_migration_behaviour_of_DEHP.pdf.

Welle, F., Wolz, G., & Franz, R. (2005). Migration von Weichmachern aus PVC-Schläuchen in enterale Nährlösungen. *Pharma International, 3*, 17–21.

Wick, A., Jacobs, B., Kunkel, U., Peter Heininger, P., & Ternes, T. (2016). Benzotriazole UV stabilizers in sediments, suspended particulate matter and fish of German rivers: New insights into occurrence, time trends and persistency. *Environmental Pollution, 212*, 401–412.

Wiesheu, A. C., Anger, P. M., Baumann, T., Niessner, R., & Ivleva, N. P. (2016). Ramanmicrospectroscopic analysis of fibers in beverages. *Analytical Methods, 8*, 5722–5725.

Wiig, O., Derocher, A. E., Cronin, M. M., & Skaare, J. U. (1998). Female pseudohermaphrodite polar bears at Svalbard. *Journal of Wildlife Diseases, 4*, 792–796. https://doi.org/10.7589/0090-3558-34.4.792.

Wilhelm, S. (2008). „*Wasseraufbereitung*" (7. ed.). Heidelberg: Springer.

World Economic Forum. (2016). The new plastic economy: Rethinking the future of plastics. http://www3.weforum.org/docs/WEF_The_New_Plastics_Economy.pdf.

WSV (Wasser- und Schifffahrtsverwaltung des Bundes). (o. J.). Bereitgestellt durch die Bundesanstalt für Gewässerkunde (BfG).

Yamada-Onodera, et al. (2001). Degradation of polyethylene by fungus. *Polymer Degradation and Stability, 72*, 323–327.

Yoshida, S., et al. (2016). A Bakterium that degrades and assimilates poly(ethylene terephthalate). *Science, 351*, 1196–1199.

Zarfl, C., & Matthies, M. (2010). Are marine plastic particles transport vectors for organic pollutants to the Arctic? *Marine Pollution Bulletin, 60*, 1810–1814.

Zhang, Z. et al. (2017). *Nature Communications, 8*, 14585.

Ziccardi, L. M., Edgington, A., Hentz, K., Kulacki, K. J., & Kane Driscoll, S. (2016). Microplastics as vectors for bioaccumulation of hydrophobic organic chemicals in the marine environment: A state-of-the-science review. *Environmental Toxicology and Chemistry, 35*(7), 1667–1676.

Zubris, K. A. V., & Richards, B. K. (2005). *Synthetic fibers as an indicator of land application of sludge*. Environmental pollution (Barking, Essex: 1987): 2, pp. 201–211, https://doi.org/10.1016/j.envpol.2005.04.013.

Microplastics as an Opportunity

3

Contents

© The Author(s), under exclusive license to Springer-Verlag GmbH, DE, part of
Springer Nature 2024
A. Fath, *Microplastic*, https://doi.org/10.1007/978-3-662-69844-0_3

Special plastics are used as adsorber materials for a large number of trace substances in bodies of water. In contrast to a water sample taken directly on site for analysis, a plastic membrane is used that adsorbs trace substances over a longer period of time. This so-called passive sampler represents the sample from which the adsorbates can finally be analyzed after extraction.

3.1   Passive Samplers as Water Filters

Passive samplers have been used for the investigation of bodies of water since 1990. They have a simple structure, require no power supply, and can be used in remote areas. The results are also reproducible (Rüdel et al. 2007). Passive samplers are referred to as "artificial mussels" or "artificial fish skin" and reveal the substances with which water organisms come into contact. Passive samplers are low-maintenance and more cost-effective than other sampling systems. The schematic structure of a passive sampler is simplified in Fig. 3.1. Due to diffusion processes, trace substances from the environment accumulate on the passive sampler. The trace substance diffuses from the water phase into the collection phase due to the difference in chemical potentials (Kraus et al. 2015; Górecki and Namieśnik 2002).

Different plastics are used as passive samplers, but the basic structure is almost identical. The dissolved substances are in the water, which flows around the passive sampler. The collection phase can consist of different materials, e.g. silicone,

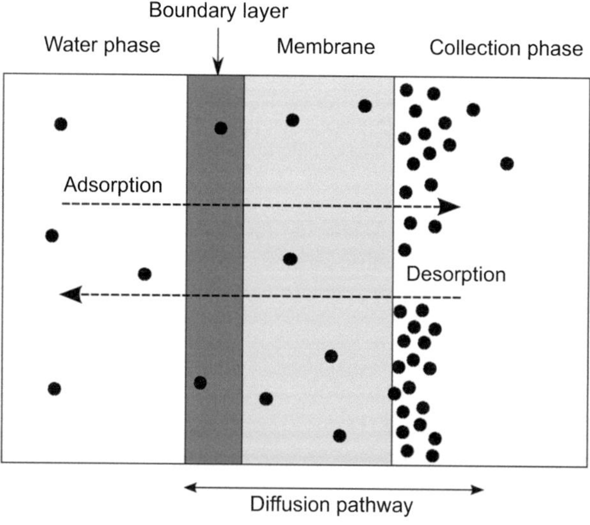

Fig. 3.1 Schematic structure of a passive sampler. The chemicals (black dots) diffuse from the water phase across the boundary layer and the membrane into the collection phase. (Modified after Kraus et al. 2015; Mills et al. 2007)

polyethylene or adsorber resins. The surface of the collection phase can be modified in such a way that the accumulation of the substances being examined is improved. In some passive samplers, a membrane is additionally applied to the collection phase. This membrane can, for example, consist of polyethylene, polyethersulfone or cellulose acetate (Kraus et al. 2015). A semipermeable membrane ensures that only the desired chemicals reach the collection phase. This excludes solids and other unwanted substances (Alvarez et al. 2005).

The substances are bound in the collection phase by adsorption (Kraus et al. 2015). Adsorption is the process in which dissolved substances accumulate on a surface. The adsorptive or adsorbate are the free particles in the solution (trace substance) and the adsorbent is the surface of the passive sampler in the described application (Job and Rüffler 2011).

Passive samplers enable an accumulation of pollutants, so that even small concentrations in bodies of water can be detected (Vrana et al. 2005). The amount of the trace substance that accumulates in the collection phase depends on the concentration of the trace substance in the body of water. It is also crucial how long the passive sampler is used in the body of water. The temperature and the flow conditions also play a role in the accumulation (Rüdel et al. 2007).

Passive samplers capture trace substances that are polar and can accumulate in organisms. With the help of the passive sampler, continuous sampling can be carried out over several days to weeks. Various types of passive samplers are used for the investigation of hydrophilic substances in bodies of water, e.g. the POCIS (Polar Organic Chemical Integrative Sampler) or the Chemcatcher (Ricking 2009). The collection phase of the Chemcatcher usually consists of styrene-divinylbenzene (SDB). The membrane is made of polyethersulfone (PES), but the membrane is only used optionally. In POCIS, the collection phase is located between two membranes, which are also made of PES (Moschet et al. 2015).

Vrana et al. (2005) provide a comprehensive overview of various passive samplers (data on the construction method, the substances that can be analyzed, etc.). With the help of different passive samplers, various substances can be detected. The substances that adhere to a specific passive sampler are in a similar range of hydrophobicity. Chemicals adhere to POCIS (Polar Organic Chemical Integrative Sampler) that have a log-K_{OW} of 0–3, whereas on the Chemcatcher trace substances with a log-K_{OW} of 1-6 adsorb (Vrana et al. 2005).

Passive samplers have the advantage that they can remain permanently in the water and thus provide an overview of all substances that occur in the water during this period. Spot samples only capture the substances at a certain point in time, which means that one-off events can remain undetected, such as accidents in which chemicals enter the water (Alvarez et al. 2005). Guidelines for the use of passive samplers in water analysis (preparations, execution of the experiment, and analysis of the passive sampler) were already established in 2010 (Alvarez 2010).

The passive sampler SDB-RPS (Styrene-Divinylbenzene-Reversed-Phase-Sulfonate) "Chemcatcher" (Fig. 3.2) consists of a divinylbenzene polymer, which was modified with sulfonic acid groups to achieve a certain hydrophilicity.

Fig. 3.2 Left: SDB-RPS "Chemcatcher". Right: SDB-RPS "Chemcatcher", protected with aluminum wire mesh and tacked onto a neoprene sock so that it can be carried by a swimmer

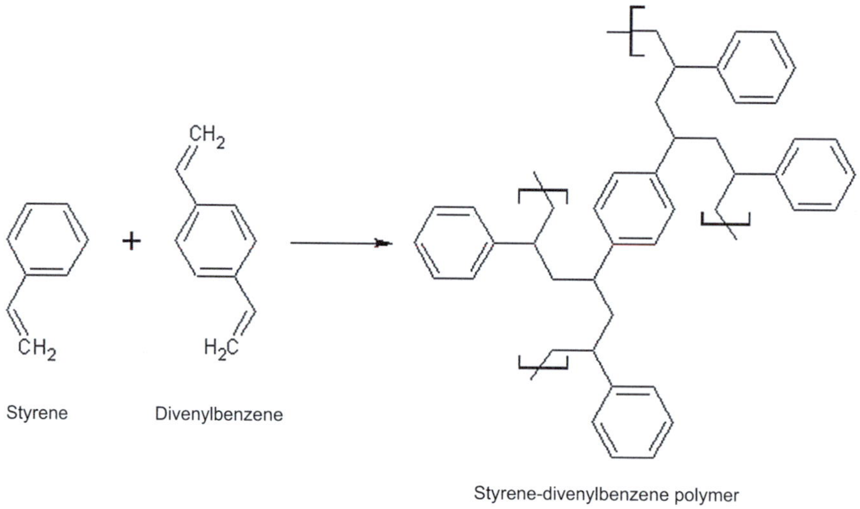

Fig. 3.3 Production of the styrene-divinylbenzene polymer

In Fig. 3.3, the production of the SDB-RPS "Chemcatcher" from the two monomers styrene and divinylbenzene is shown, and Table 3.1 displays the IR spectrum of the polymer.

Passive collectors with a microstructured plastic surface are mainly used in water monitoring. Figure 3.4 shows the use of passive collectors, which were

Table 3.1 Characteristic IR vibration frequencies ($\tilde{v}$) of polystyrene divinylbenzene (SDB)

Wavenumber	Vibration
3443	
3019	C–H valence vibrations of aromatic substances
2960	Asymmetric valence vibration of the C–H bond from the functional group CH_3
2923	Asymmetric valence vibration of the C–H bond from the functional group CH_2
1603	Benzene finger between 1650 and 2000
1486	Asymmetric deformation vibration of the C–H bond from the functional group CH_2
1448	Asymmetric deformation vibration of the C–H bond from the functional group CH_3
1243	
1212	
1155	Valence vibration of the sulfonic acid (?)
1095	Fingerprint area:
1041	Out-of-plane vibrations Deformation vibration in aromatics
1017	
828	
794	
707	

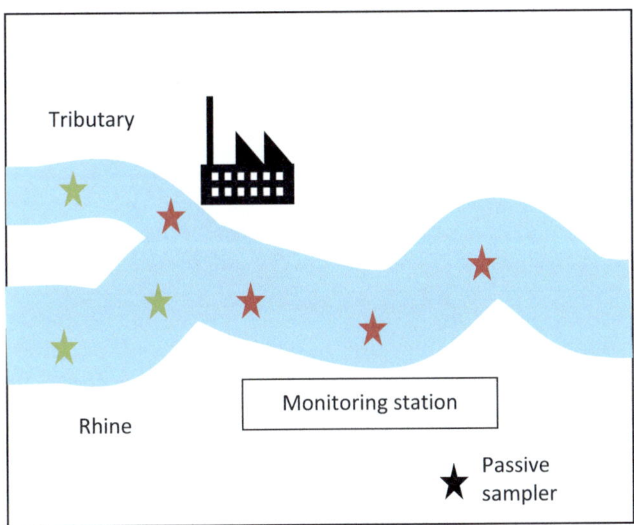

Fig. 3.4 Use of passive collectors for water monitoring

placed at fixed distances from each other along a river and a tributary. If there are abnormalities at a river monitoring station, the origin of a contamination can be narrowed down by evaluating the collectors. In the hypothetical case presented, the tributary is to be held responsible for the contamination, on whose banks an industrial plant discharges its wastewater.

In the project "Rheines Wasser" (Fath 2016), the author of the non-fiction book swam with the passive collector attached through the Rhine for four weeks to find out which organic substances water organisms, such as the salmon, which has been resettled in recent years, come into contact with in the Rhine. In Fig. 3.5, the author is seen during the project "Rheines Wasser", with the passive collector fixed to a neoprene sock on the right lower leg. This passive type of sampling reflects the natural living conditions of water organisms. The detected pollutants can potentially accumulate in the organs or adipose tissue of aquatic organisms (Fath 2017).

With the help of the passive collector, 128 substances were detected. These 128 substances are those that adsorbed to the membrane in a detectable amount. Of course, these are by far not all trace substances that are in the Rhine.

Most of the adsorbates are pharmaceuticals and pesticides and their metabolites (Fath 2017). The project was able to show that the concentration of trace substances increases the further the Rhine flows. This increase is due to anthropogenic

Fig. 3.5 Andreas Fath during the project "Rheines Wasser" in the summer of 2014 with a fixed passive collector on his leg. (Braxart/Furtwangen University, Rheines Wasser project)

causes, e.g. private households, industry, agriculture etc. (Furtwangen University 2014). The increase in anthropogenic influence can be seen, for example, in the population density. Near Lake Constance, there are an average of 120 inhabitants/km^2 and on the Lower Rhine (from Bonn to Düsseldorf) 680 inhabitants/km^2 (IKSR 2009).

3.2 Microplastics as Water Filters

If a structured plastic membrane, like that of the passive collectors, can adsorb harmful trace substances from waters, the question inevitably arises whether it would not be possible to eliminate larger quantities of unwanted substances from contaminated waters by adsorption by increasing the sorbent surface.

If microplastic particles act like "magnets" on pollutants, one could use specifically specified microplastics to clean waters. Whether such a goal is achievable depends on the adsorption properties of microplastic particles of different sizes and structures. It would also be important to determine whether there are selective accumulations on certain types of plastic with different sizes and surface structures. If this were the case, one could grind up technical plastics that have served their purpose and use them for the selective cleaning of waters and, after a desorption, recover the organic compound, which is often a valuable active ingredient, and reuse the microplastic filter material.

Microplastics are not technically used in the form we find them in our waters. Two microscope images from the study of the Rhine are exemplarily depicted (Fig. 3.6).

This effect of a large surface area for the accumulation of pollutants is already known. It is used in the purification of wastewater with activated carbon. The electron microscopic image of the activated carbon in Fig. 3.7 shows the highly structured surface.

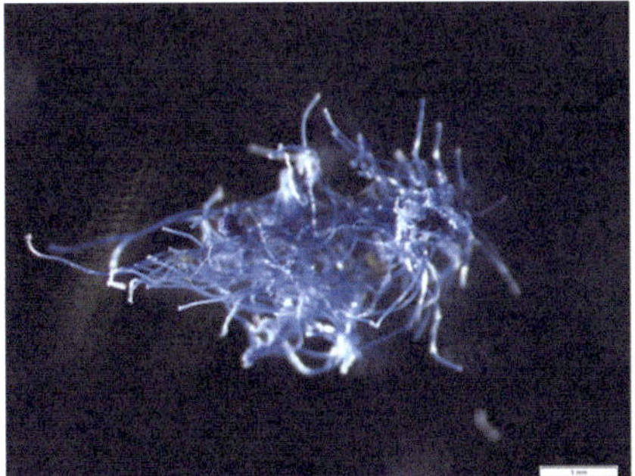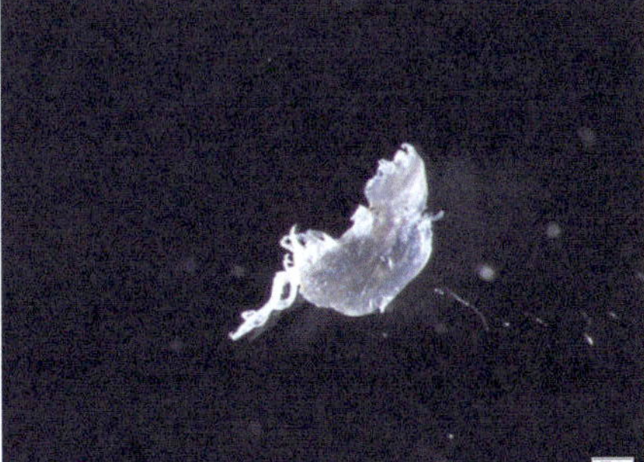

Fig. 3.6 Left: Fiber structure of polyamide. Right: Highly fissured polyethylene. Both particles are smaller than half a millimeter and have a large surface area, which is important for the accumulation of pollutants

Fig. 3.7 SEM image of activated carbon

Table 3.2 Frequently used adsorber materials

Adsorbent	Surface area in m^2/g
Activated carbon	300–2500
Silica gel	300–350
Aluminum oxide	200–500
Molecular sieves (zeolites)	500–1100

This microparticulate "mountain and valley landscape" results in one gram of activated carbon having the surface area the size of a football field. As a fourth treatment stage in some sewage treatment plants, activated carbon filtration is therefore added to eliminate trace substances from the inflow into our rivers.

In addition to different activated carbons (coal, coconut, etc.), other adsorber materials are also used for physisorption or chemisorption of molecules to eliminate them from the fluid phase (desiccants) or to separate substance mixtures (column chromatography). These are always adsorber materials with a large surface area. In Table 3.2, a selection of various frequently used adsorbents with their active surface area is listed.

Microplastic particles also accumulate organic pollutants (Wedler 1970). If it is possible to produce filter materials for water or wastewater treatment from plastic waste, the plastic waste would gain a function and thus a value that would reduce thoughtless "disposal" in aquatic habitats and thus protect the water resource. As a basic material, enormous amounts of plastic waste are available, which are currently only used for thermal recovery. This type of recovery would still be possible with the loaded plastic filter material, especially since the pollutants are also burned at the same time. On the other hand, with MP there is the opportunity to combine the very good adsorption properties of activated carbon with the recovery of trace substances and the reusability of the filter material.

That the sorption of pollutants on polymers can potentially be used for water purification has already been shown (Muhandiki 2008; Matsuzawa 2010).

Further investment in basic research is still required until a filter material made from plastic waste, which is effective and competitive with activated carbon, is fully developed. This involves topics such as sorption studies of different substances on different plastic materials, optimization of the particle size distribution, increasing the surface area and thus the adsorption capacity, selection of plastic types or combinations of plastics for universal filter material, pre-treatment of plastics, selectivity studies, packing density for corresponding flow water volumes, regeneration, eluate processing, and much more. These basic investigations have only just begun.

Basic investigations to determine whether microplastic particles strongly, less strongly, or not at all accumulate certain substances on their surface include measurements of adsorption, which are carried out in the same way as the measurement of adsorption on soils (Klöpffer 2012).

3.2.1   Measurement of the Adsorption of Substances on Microplastics

For the measurement of adsorption, an adsorption isotherm is determined. It is important that the temperature is kept constant, as both adsorption and desorption rates are temperature-dependent and a state of equilibrium can only be established at a constant temperature after a certain time. This equilibrium state is the basis for determining the adsorption isotherm. The adsorption isotherm is a characteristic curve, the evaluation of which provides information about the interactions between adsorbate and adsorbent and about the surface of the adsorbent. For the measurement of the isotherm, the water solubility of the substance to be adsorbed should be known. In further preliminary tests, the optimal test conditions are determined. The optimal ratio of the amount of microplastic to the solution can be predicted using the K_{ow} value (Hummel 2017). Determining the waiting time until equilibrium is established and also checking the stability of the substance over the entire equilibration time are also important preliminary tests.

The adsorption isotherm provides information about the distribution equilibrium of a substance in solution in relation to the adsorbed concentration on an adsorbent at a constant temperature. The concentration of the substance in the solution in g/l is plotted on the x-axis and the sorbed concentration on a solid, e.g. microplastic, in g/kg on the Y-axis.

To record the adsorption isotherms, different concentrations of the test substance are prepared in a 0.01 M $CaCl_2$ solution and stirred or shaken vigorously with plastic powder, the dry weight of which is known, until the distribution equilibrium has been established. The necessary shaking time must be determined analytically beforehand. It is reached when the concentration of the test substance in the solution no longer changes significantly after periodic measurement intervals. After the separation of the two phases by centrifugation, the concentration of the test substance in the aqueous and in the solid phase is determined (direct method). A concentration determination of the test substance on microplastic particles requires a very high effort, especially since the test substance can both adsorb and absorb and is not necessarily fluorescence-active. A quantitative desorption with a suitable solvent would be necessary. To quantify the sorption on microplastic particles, the indirect method is more suitable, even if it becomes inaccurate with very low and very strong adsorption. In the indirect method, the difference in concentration in the aqueous phase before and after the addition of microplastic particles is determined. Since some plastic particles still float on the surface of the liquid after centrifugation, the sample is taken from the middle of the centrifuge tube. To prevent any carried-over plastic particles from entering the HPLC separation column, a pre-column is installed which can intercept them. Despite the separation effort of centrifugation compared to filtration, this method has established itself, as influences due to sorption of the test substance on different filter materials are eliminated. For the same reason, it is recommended to use glassware for further sample processing, as the least sorption is to be expected with glass (Walker and Watson 2010). The concentration range over which adsorption isotherms extend depends on two factors. While the upper concentration limit is determined by the solubility of the test substance, the lower limit is determined by the measurement accuracy of the analytical method. The interpretation of the recorded adsorption isotherms is done after a fit with the adsorption models presented in section 3.2.2.

3.2.2 Basics of Adsorption

Solids have the ability to bind atoms and molecules from their gaseous environment or their fluid environment to their surface. This is referred to as adsorption. If, in the further course of this interaction, molecules penetrate into the solid phase, this is called absorption. Both processes are combined under the term "sorption". Gases can not only absorb into liquids, but also into solids, provided their structure allows it. To understand the terminology of adsorption on solid particles, such as microplastics, the different terms of the individual states are shown in Fig. 3.8.

The solid, is referred to as "adsorbent" and the adsorbed substances as "adsorptive" or as "adsorpt" or "adsorbate" in the adsorbed state. The process of detaching the adsorbate back into the liquid phase is called "desorption".

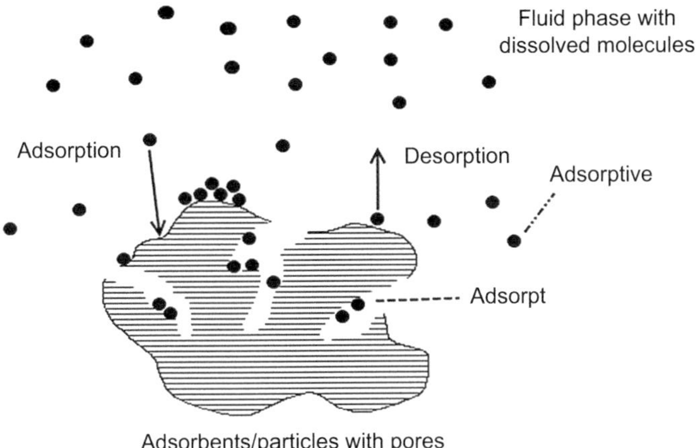

Fig. 3.8 Basic terms of the adsorption of molecules or atoms from the liquid or gas phase onto a particle with a structured (porous) surface

Between adsorption and desorption, after a finite exposure, an equilibrium state is established in which both speeds are equal and thus the concentrations of the adsorbed substances and the substances dissolved in the fluid phase remain constant.

The fact that an attractive interaction occurs when an adsorptive approaches a solid adsorbent is exploited in many technical applications such as coating technology, heterogeneous catalysis and wastewater treatment.

Depending on the strength of the interaction between adsorbent and adsorptive, a distinction is made between physisorption and chemisorption. The difference between a physically bound adsoptive and a chemically bound adsorptive is determined solely by the reaction enthalpy of the exothermic adsorption reaction that is released.

A physical interaction arises from dipole-dipole interactions between the approaching molecules, due to the displacement of their electron shells. The dipole induction is either caused by the influence of the sorbent molecules or the adsoptive already has a dipole character and can interact with a nonpolar sorbent. The Van der Waals forces involved only reach low energy amounts up to the range of hydrogen bondings of 40–50 kJ/mol. In Fig. 3.10, the physisorption of a fatty acid on an activated glass surface by the formation of two hydrogen bonds is shown. In the case of a strong affinity of the adsorbent to the adsorptive, the physical binding of the adsorbate can be considered a precursor for a chemical binding. As soon as the activation energy is overcome, a chemical surface binding of the adsorbate can occur. This releases significantly larger energy amounts >100 kJ/mol. This process is shown in the potential diagram in Fig. 3.9. The chemical bonding of gases to solids, for example, is the basis for important heterogeneous catalyses such as the oxidation of nitrogen catalyzed by a gold surface. In this

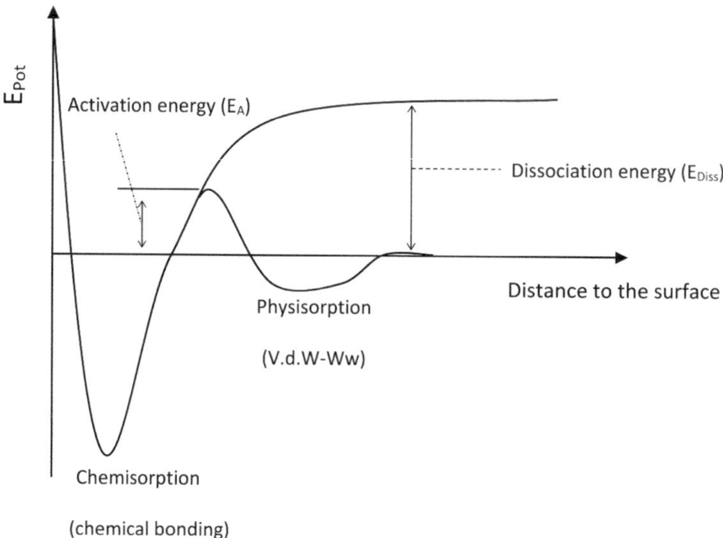

Fig. 3.9 Energy potential diagram of a molecule approaching the solid surface

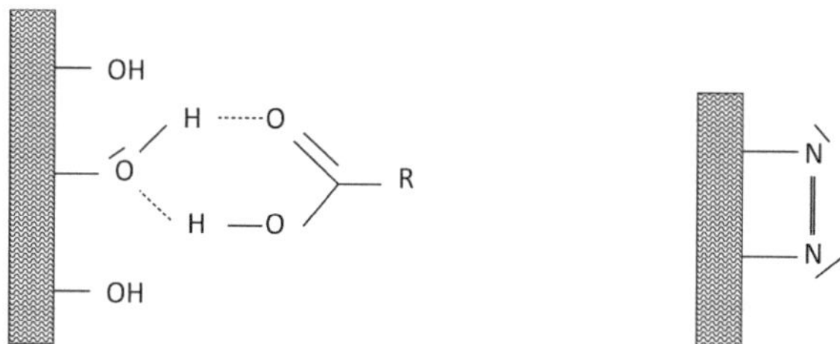

Fig. 3.10 Physisorption of an undissociated fatty acid on an activated glass surface (left) and Chemisorption of nitrogen on a gold surface

process, the triple bond in nitrogen is broken in favor of the formation of a surface bond, making the nitrogen accessible for a bond with oxygen (Fig. 3.10 right).

Not only in catalysis are the chemical surface bonds important, but also in coating technology, where for example silver surfaces form an Ag-S bond with thiols a SAM*(self assembled monolayer)* layer as a so-called primer, which can be further functionalized with a polysiloxane layer (Fath 2009). The layer sequence can be seen in Fig. 3.11.

On the silver surface, an exothermic redox reaction takes place according to the following Eq. 3.1:

$$Ag(s) + RSH(aq) \rightarrow AgSR(s) + H_2(g) \tag{3.1}$$

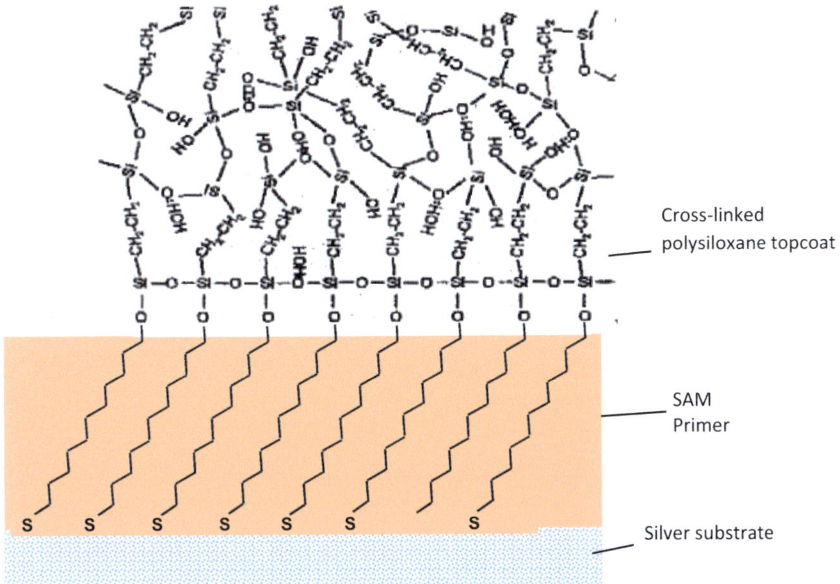

Fig. 3.11 Chemical attachment of a *self assembled monolayer* (SAM) to a silver surface

The chemical bond has both ionic and covalent character due to the polarizability of the soft sulfur (HSAB concept).

The chemical fixation of biomolecules e.g. of proteins to glass carrier materials (Hartmann 1993) is of great importance in biochemical analysis. Through protein immobilization by chemical bonding to a solid substrate surface, antibodies can be fixed and used as biosensors or immunosensors (Schmidt and Bilitewski 1992).

While chemical adsorption plays a decisive role in catalysis and for coating technology, physical separations such as chromatography are based on the weaker physisorption, where the physical interaction is reversible and the carrier materials can thus be regenerated.

3.2.3 Adsorption Isotherms

The adsorption behavior of substances from the gas phase or dissolved substances in the liquid phase can be described using adsorption isotherms or sorption isotherms. The experimentally determined isotherms provide information about the distribution equilibrium of a substance between the liquid or gaseous phase and the solid phase at a constant temperature. Since there is no universal equation due to the complex processes for adsorption and desorption equilibrium and the different influencing factors from the adsorbent and adsorptive, various empirical models are used that best mathematically describe the experimentally determined isotherms. Statements about the interaction between adsorbate and adsorbent and the nature of the surface of the adsorbent can be made via the adsorption model with the best fit. The distribution of a substance in the solid phase and the liquid phase can be calculated with the appropriate model. The interaction of the adsorbates with each other also has an influence on the shape and thus the equation of the adsorption isotherm. It makes a big difference whether a monolayer or multilayer adsorption takes place. The most important models are presented below under the assumption of their specific simplifications.

Linear (Henry's) Adsorption Isotherm

For the adsorption of substances on solid surfaces, the occupancy density or the coverage or loading degree θ is used instead of the interfacial concentration.

The loading degree θ is a function of the concentration c_i of the adsorptive i in the solution or the partial pressure p_i of the adsorptive in the gas phase. Here, θ is the quotient of surface concentration c_{ad} and the surface saturation concentration c_{max}.

$$\theta = f(c_i)_T \text{ oder } \theta = f(p_i)_T \left(\text{für gasförmige Adsoptive}\right) \tag{3.2}$$

For gas adsorption, the following applies accordingly $\theta = V_\sigma / V_{mono}$. Here, V_σ is the specific surface volume in cm^3/g solid and V_{mono} is the specific surface volume at maximum surface coverage.

In surface waters, the adsorption of dissolved substances to suspended matter plays a major role in analytics. New synthetic suspended matter, including microplastic contaminants, are increasingly joining biological and inorganic suspended matter. At low surface occupancy c_{ad} due to a low dissolved concentration c_w, the adsorption isotherms show a linear course far from surface saturation, derived from Henry's law.

Henry's law from 1803 is the oldest distribution law ever (Henry 1803). It states that the vapor pressure of a dissolved substance i is directly proportional to its concentration in solution. The proportionality constant is the so-called Henry coefficient H_i:

$$p_i = H_i \cdot x_i \text{ (Henry'sches Gesetz)} \tag{3.3}$$

Here, x_i is the mole fraction of the component x_i with $x_i = n_i / (n_i + n_A)$.

Through the linear relationship $H_i = c_{i,g} / c_{i,w}$, which describes the concentrations of a substance in the gas phase and the aqueous phase at equilibrium, and the knowledge of the Henry coefficient H_i, it is possible to determine the concentration of a substance i, for example DEHP $\left(H = 5{,}98 \times 10^{-5}\right)$ (Klöpffer 2012), in aqueous solution using the head-space technique (Kolb 1999) by gas chromatography. The Henry coefficient of a poorly soluble substance can also be calculated from the saturation vapor pressure of the substance and its water solubility (Klöpffer 2012). The initial linear course of the distribution equilibrium between adsorbed concentration c_{ad} and dissolved concentration at equilibrium c_w can thus be very well modeled (described) with a linear equation, with a distribution coefficient K_H analogous to Henry's law.

The surface concentration on a microplastic particle can be calculated in this linear range with the equation $c_{ad} = K_H \cdot c_w$. Instead of K_H, the literature also uses K_d (d$=$*distribution*) for the linear distribution coefficient.

Thus, the degree of coverage also has a linear relationship with the equilibrium concentration of the adsorbate in aqueous solution. This Henry isotherm, shown in Fig. 3.12, applies:

$$\theta = c_{ad}/c_{max} = K \cdot c_w \tag{3.4}$$

Freundlich's Adsorption Isotherm

In the adsorption isotherms derived by Freundlich (1906), the adsorption of substances on heterogeneous surfaces is described. This means that the interactions of the active centers of the surface with the adsorbate in the monolayer are not the same everywhere. As a result, the adsorption energy decreases exponentially with increasing coverage. Equation 3.5, which describes Freundlich's adsorption isotherm, takes into account for $n_F > 1$ (the more common case) the fact that as the concentration in solution increases, the surface concentration approaches saturation:

$$c_{ad} = K_f \cdot c_w^{1/n_F} \tag{3.5}$$

However, due to the exponential growth, a complete loading of the surface is not achieved by the equation and cannot be represented or described with the Freundlich model. The case $n_F = 1$ (linear; Henry behavior) represents a limit case in the Freundlich's view, in which the Freundlich coefficient K_f or the Freundlich's adsorption constant corresponds to the distribution coefficient K_d.

By plotting the equilibrium concentration in aqueous solution against the loading of the sorbent on a double logarithmic scale, as shown in Fig. 3.13, a straight line is obtained, the slope of which can be used to calculate the Freundlich exponent n_F. The Freundlich coefficient K_F can be determined from the intercept of the line (Fig. 3.13).

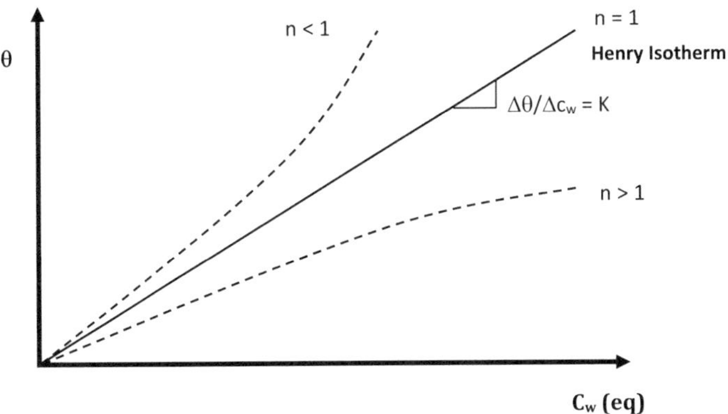

Fig. 3.12 Henry's adsorption isotherm (solid line) and its deviations according to Freundlich (dashed curves)

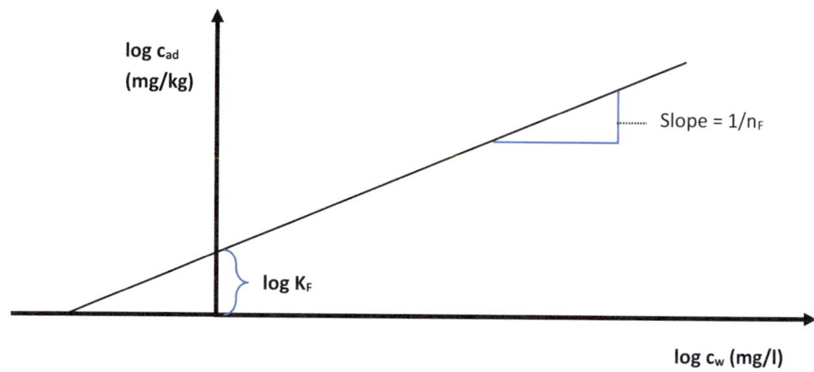

Fig. 3.13 Double logarithmic plot of Freundlich's adsorption isotherm

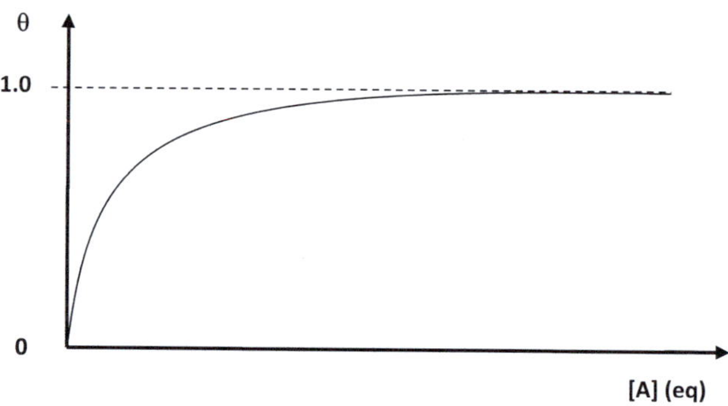

Fig. 3.14 Langmuir's adsorption isotherm

Langmuir's Adsorption Isotherm

The Langmuir model describes classic isotherms that approach saturation. Figure 3.14 shows such an isotherm. The course of the isotherms is based on the assumption that only a monolayer of adsorbates accumulates on a surface, with the interactions between adsorbate and adsorbent being the same everywhere on the homogeneous surface. The Langmuir model is therefore a kind of parking lot model, where parking is only in the first row and only a limited number of identical parking spaces, depending on the surface size, are available. The occupancy of a "parking space" is independent of whether spaces in the neighborhood are already occupied or not. This means that both the interactions within the adsorbate molecules and the interactions of the adsorbates with the solvent molecules are negligible. With a complete occupancy of the equivalent binding sites on the surface of an adsorbent, the surface of the sorbent, for example that of microplastic particles, can be calculated after determining the saturation occupancy at $\theta = 1$.

The equation of the Langmuir adsorption isotherms can be derived from the reaction kinetics for the reversible reaction equation (Eq. 3.6):

$$S + A \underset{k_{De}}{\overset{k_{Ad}}{\rightleftharpoons}} SA \tag{3.6}$$

The speed of adsorption of a gaseous or dissolved adsorbate A onto a solid substrate S in the forward reaction is described by the rate constant k_{Ad}, while the speed of the desorption as a reverse reaction is given by the rate constant k_{De}. Once the equilibrium state is reached, both reactions are equally fast, and in this dynamic equilibrium state, neither the surface concentration of the adsorbate nor the dissolved concentration [A] changes, although both processes continue at the molecular level, but no change can be detected macroscopically, for example by analytics.

In the equilibrium state, the degree of coverage θ no longer changes. Under these equilibrium conditions, the process of adsorption proceeds as quickly as desorption, so that:

$$V_{Ad} = V_{De} \tag{3.7}$$

The reaction rate of adsorption is proportional to the partial pressure of A or the concentration of A. and to the remaining number of binding sites. The free number of binding sites can be described by the expression $N(1-\theta)$, where N indicates the total number of binding sites.

The reaction rate of desorption is a first order reaction and only depends on the number of adsorbed particles $N\theta$.

This results in the following rate laws for the reaction rate of both reactions:

$$V_{Ad} = d\theta/dt = K_{Ad}\,[A]\,N(1-\theta) \tag{3.8}$$

$$v_{De} = -d\theta/dt = k_{De}\,[A]\,N\theta \tag{3.9}$$

If both reaction rates are set equal, one obtains with the Langmuir adsorption coefficient $K_L = k_{Ad}/k_{De}$ Eq. 3.10 of the Langmuir isotherms shown in Fig. 3.14

$$\theta = \frac{K_L[A]}{1 + K_L[A]} \tag{3.10}$$

Equation 3.10 can be rearranged to

$$K_L[A]\theta + \theta = K_L[A]. \tag{3.11}$$

If the occupancy degree θ is replaced by the quotient of specific surface concentration c_{ad} and maximum surface concentration c_{max}, the Langmuir equation can be represented in a linearized form:

$$\theta = \frac{c_{ad}}{c_{max}} = \frac{K_L[A]}{1 + K_L[A]} \tag{3.12}$$

$$C_{\mathrm{ad}} = (1 + K_{\mathrm{L}}[A]) = c_{\max} K_{\mathrm{L}}[A]$$

$$\frac{[A]}{c_{\mathrm{ad}}} = \frac{1 + K_{\mathrm{L}}[A]}{c_{\max} K_{\mathrm{L}}}$$

$$\frac{[A]}{c_{\mathrm{ad}}} = \frac{1}{c_{\max}}[A] + \frac{1}{K_{\mathrm{L}} c_{\max}}$$

Linearized Form of the Langmuir Equation

If one plots the quotient of the particle concentration in the solution at equilibrium $[A]$ and the surface concentration c_{ad} on the y-axis against $[A]$ on the x-axis, a straight line results with a slope that corresponds to the reciprocal of the saturation occupancy. The Langmuir adsorption coefficient K_{L} can then be determined from the saturation concentration $c_{\max}$ via the intercept of the Langmuir.

The specific surface area A_{spez} of particles can be calculated using the saturation occupancy $c_{\max}$ $[\mu g/g]$ by the following equation:

$$A_{\mathrm{spez}} = n_{\max} N_{\mathrm{A}} A_{\mathrm{ad}} \tag{3.13}$$

$$n_{\max} = c_{\max} / M_{\mathrm{ad}} \tag{3.14}$$

The specific surface determination of solids is mainly determined by a gas adsorption, which is recorded volumetrically or gravimetrically. If the gas molecules only attach in a monolayer on the surface, then after using $\theta = V_\sigma / V_{\mathrm{Mono}}$ in the Langmuir equation (Eq. 3.10), the following linearized form results:

$$\frac{p}{V_\sigma} = \frac{p}{V_{\mathrm{Mono}}} + \frac{1}{K_{\mathrm{L}} V_{\mathrm{Mono}}}. \tag{3.15}$$

Here, p is the partial pressure of the used gaseous adsorbate.

For example, for an adsorption of carbon monoxide on activated carbon at 0 °C, the following specific surface coverage volumes V_σ, depending on the set partial pressure p_{co}, are obtained (Table 3.3):

By plotting the quotient p_{co}/V_σ against p_{co}, a straight line with a slope of 0.0091 is obtained. The reciprocal of the slope gives the volume of the CO monolayer and in this case is 110 cm³. Thus, according to Eq. 3.16, the surface area A_{C} of one gram of the activated carbon used is:

$$A_{\mathrm{c}} = n\, N_{\mathrm{A}}\, A_{\mathrm{M}} \left(A_{\mathrm{M}} = \text{Platzbedarf des adsorbierten Gasmoleküls}\right) \tag{3.16}$$

Table 3.3 Surface coverage volume of activated carbon depending on the carbon monoxide partial pressure at 0°C

p_{co} [kPa]	13.4	26.6	39.9	53.3	66.8	79.9	93.4
V_σ [cm³]	10.3	18.7	25.4	31.5	37.0	41.5	46.2

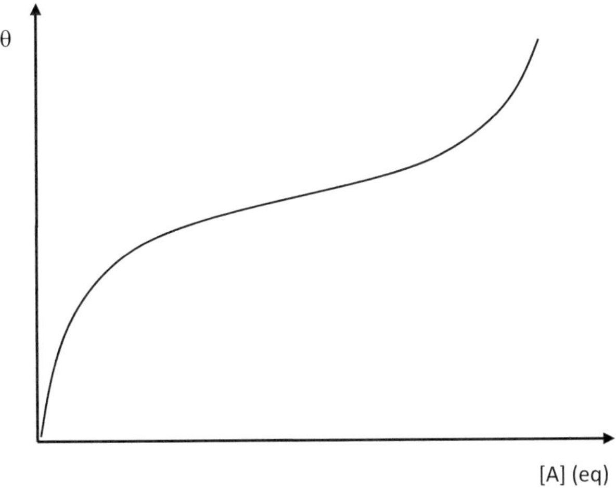

Fig. 3.15 Example of a BET adsorption isotherm

Assuming that due to the comparable bond length and the averaged atomic diameter of the nitrogen molecule, which is isoelectronic to carbon monoxide, a similar space requirement of $16{,}2 \cdot 10^{-20}\,\mathrm{m}^2$ is needed, this would result in a specific activated carbon surface area of:

$$A_\mathrm{c} = \left(V_\mathrm{Mono}/22400\right) 6{,}022\ 10^{23}\ 16{,}2 \cdot 10^{-20}\,\mathrm{m}^2 = 479\,\mathrm{m}^2/\mathrm{g} \qquad (3.17)$$

BET Adsorption Isotherm

Brunauer et al. (1938) extended the Langmuir model to the solubility limit or saturation concentration of the adsorbates. The fundamental extension of the BET isotherms named after them is based on multilayer adsorption. After the chemisorption of the first monolayer, the BET model describes the physisorption of further, up to infinitely many adsorptive layers in a multilayer system. In the higher concentration range of the adsorption isotherm (Fig. 3.15), the occupancy level θ increases again after reaching surface saturation. Not all places in the first row need to be occupied to continue layering. The slope in the "plateau area" depends on the difference between chemisorption and physisorption energy. The BET isotherm thus shows a typical S-shaped course, as shown in Fig. 3.15.

The derivation of Eq. 3.18 for the BET adsorption isotherms is also based on a kinetic consideration, where adsorption and desorption are in dynamic equilibrium:

$$\theta = \frac{K[A]}{([A]_\mathrm{s} - [A])\left[1 + (K - 1)[A]/[A]_\mathrm{s}\right]} \qquad (3.18)$$

Here, the constant K is the BET adsorption coefficient and $[A]_\mathrm{s}$ indicates the solubility or saturation concentration of the adsorbate in aqueous solution.

The assumptions that the surface behaves energetically homogeneous and thus the adsorption energy for all positions is the same and independent of the degree of coverage, corresponds to the Langmuir's approach. There are no interactions between the adsorbed molecules of a layer, while multilayer adsorption is initiated by van der Waals interactions between the molecular layers. In this process, the condensation energy of adsorbed gas molecules is released.

The BET model finds its practical application in determining the active surface area of sorbents, including the number of pores and pore distribution. The model thus provides the basis for the development of effective heterogeneous catalysts and adsorber resins, whose performance is determined by the mentioned surface characteristics.

Determination of the Specific Surface Area of Microplastic Particles by Gas Adsorption

Using the BET isotherm, the specific surface area of microplastic particles and other materials with a large surface-to-volume ratio can be calculated through the adsorption of gas molecules, mainly nitrogen. To do this, one must first determine the amounts of adsorbed substances at different adsorbate concentrations. These measurements are mainly carried out volumetrically for gases in special apparatus (Best and Spingler 1972). However, they can also be done gravimetrically. In a storage vessel, a preliminary pressure p_0 of the gaseous adsorbate is first set. Then a second valve to a second vessel, in which a weighing of the adsorbent is located, is opened. After a certain time, an equilibrium pressure p has been established. The volumes of both vessels are known. If no adsorption of nitrogen were to occur, a higher equilibrium pressure would result than with adsorption. This means that the amount of gas in both vessels has changed by Δn due to the adsorption. With the equation of state for ideal gases, the specific surface volume V_σ of nitrogen can be calculated. The BET equation for the described gas adsorption can be linearized as follows:

$$\frac{p}{V_\sigma(p_0 - p)} = \frac{p(K-1)}{p_0 K V_{\text{Mono}}} + \frac{1}{K V_{\text{Mono}}} \tag{3.19}$$

The BET constant K is a dimensionless measure of the ratio of the chemisorption energy q_1 of the monolayer to the physisorption energy of the multilayers q_2:

$$K = e^{\frac{q_2 - q_1}{RT}} \tag{3.20}$$

If one now plots $p/V_\sigma(p_0 - p)$ against p/p_0 in a coordinate system, one obtains a straight line, from whose slope m and the y-intercept y_0 both the specific surface area A_s and the BET coefficient K can be calculated:

$$m = \frac{K-1}{K V_{\text{Mono}}} \text{ und } y_0 = \frac{1}{K V_{\text{Mono}}} \tag{3.21}$$

$$m = \frac{K}{K V_{\text{Mono}}} - \frac{1}{K V_{\text{Mono}}}$$

$$m = \frac{1}{V_{\text{Mono}}} - y_0$$

$$\frac{1}{V_{\text{Mono}}} = m + y_0$$

$$V_{\text{Mono}} = \frac{1}{m + y_0}$$

$A_s = n\,N_A A_M$ (A_M = space requirement of the adsorbed gas molecule)
$n = V_{\text{Mono}}/22.400\,\text{cm}^3$ (molar volume V_m of an ideal gas at 273 K)
K is calculated from the axis intercept y_0

$$y_0 = \frac{1}{K V_{\text{Mono}}} \tag{3.22}$$

$$y_0 = \frac{m + y_0}{K}$$

$$K = \frac{m}{y_0} + 1$$

Determination of Pore Size and Number of Pores on the Particle Surface by Gas Desorption

For the characterization of porous solids with respect to pore distribution and pore size number of pores are also performed gas adsorption and desorption measurements. With the method of nitrogen low-temperature adsorption, macro- ($\varnothing > 50$ nm), micro- ($\varnothing < 2$ nm) and mesopores ($\varnothing = 2$–50 nm) can be detected and measured. With the already described volumetric measurement, the amount of an adsorptive A at constant temperature (77 K for nitrogen as adsorptive) is determined in relation to the relative pressure p_0/p. From the results of the measurements, adsorption or desorption isotherms can be recorded, from whose course the specific surface area, the pore sizes, the pore size distribution and the average pore diameters can be calculated (Reichert 1987). In the evaluation of the isotherms, the so-called probe molecule also plays an important role, especially when it comes to the determination of micropores. Smaller probe molecules like CO_2 or H_2 "see" a larger adsorbent surface than N_2, as the adsorbate molecules can penetrate into smaller pores and not larger ones. Already the shape of the isotherms gives information about the type of pores in a porous adsorbent. A microporous surface provides a Langmuir isotherm, while macro pores often show a BET isotherm. In the case of mesopores, a hysteresis is shown within the BET isotherms when a desorption is carried out after the adsorption, that is, the relative pressure is reduced again. The hysteresis arises because the adsorption on a surface with mesopores proceeds differently than the desorption. If one considers a mesopore as a cylinder, then the chronological filling can be illustrated in relation to the corresponding isotherms, as shown in Fig. 3.16. First, the interior of the cylinder is occupied with

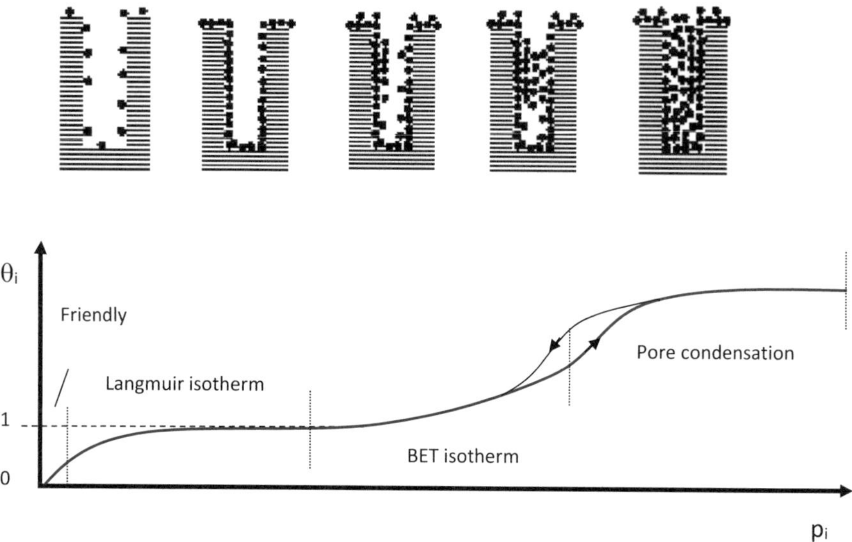

Fig. 3.16 Formation of mono- and polymolecular layers on porous surfaces and development of a hysteresis

adsorbate. After the complete surface occupation with a monolayer, the cylinder becomes narrower until it is completely closed and filled with adsorber molecules. The desorption, on the other hand, does not proceed on the same path back. It proceeds hemispherically deeper into the pore. Since a different pressure prevails over a hemisphere than over a cylinder, a measurable difference between ad- and desorption results. The difference can be mathematically captured by the Kelvin equation (Eq. 3.23), and the shape of the hysteresis provides information about the size of the mesopores. Assuming a cylindrical pore geometry, the radius r_K of a cylinder can be calculated with

$$r_K = \frac{-2\sigma V_m}{RTln\left(\frac{p}{p_0}\right)}$$

(3.23)

(Barrett et al. 1951). Here, σ is the surface tension of the probe molecule at boiling temperature and V_m is its molar volume in the liquid state.

3.2.4   Sorption of Aqueously Dissolved Organic Substances on Microplastic Particles

In principle, organic substances dissolved in water, depending on their polarity, adsorb more or less strongly to suspended solids. The less water-soluble the organic substances are, the higher the sorption to solid phases is expected to be.

This is particularly evident in the case of PAHs. To analyze their concentration in bodies of water, a solid-phase extraction of the suspended solids in the water sample is required. Microplastic particles belong to a new and still little-studied category of suspended solids in bodies of water, whose sorption capacity is an important prerequisite for assessing whether bodies of water could be cleaned using the sorption properties of microplastic filters.

To investigate this question, the sorption of hormones to different polymer particles was studied (Hummel et al. submitted).

Due to their steroid basic structure, hormones are, depending on the substitution on the basic structure, slightly water-soluble and non-polar, thus a predestined adsorptive for non-polar plastics. Another reason for choosing hormones as adsorptive is their strong influence on the reproductive behavior of aquatic organisms such as amphibians (Meeker et al. 2009). The damaging hormone effect leads to the feminization of amphibian populations and thus to the long-term extinction of amphibian species in the event of sustained influence. The WWF estimates a biodiversity loss of over 70% in aquatic animal species since the 1970s (WWF 2014). Hormones may possibly share some of the blame for this. Both natural and artificial estrogens are found in the effluent of sewage treatment plants, where they are only partially degraded or even reactivated and discharged into rivers (Tamschick et al. 2016). 17α-Ethinylestradiol (EE2) is a synthetic estrogen used in contraceptive pills. Although it does not occur in the environment, it is excreted in part through the urine due to its low bioavailability and is only incompletely degraded in sewage treatment plants. As a result, the synthetic hormone enters the aquatic environment in biologically active concentrations and disrupts the hormone balance of the aquatic habitats there.

The concentrations of 17α-Ethinylestradiol on the reproduction and development of fish are at much lower concentrations than those for acute toxicity. Ecotoxicological tests show that the threshold at which endocrine effects are detected is in the range of 0.3–1 ng/l. The 17α-ethinylestradiol concentrations in European surface waters are below the detection limit of 0.1–0.3 ng/l, but values around 1 ng/l have occasionally been detected. The presence of 17α-ethinylestradiol in bodies of water, even at very low concentrations close to the detection limit, poses a potential risk to fish (and other vertebrates). It should be noted that the determination of this safety margin is only partially possible due to analytical limitations, as the detected concentrations of 17α-ethinylestradiol in surface waters are below or around the detection limit of the analytical method (IKSR n.d.).

To investigate the sorption of hormones to microplastics, a selection of representatives found in our aquatic environment was made. These include 17α-ethinylestradiol (EE2), norethisterone (Nor), and estrone (E1).

The sorption of 17α-Ethinylestradiol (EE2), norethisterone (Nor), and estrone (E1) to PE, PVC, PVC-DINCH, PA-6, PA-12, and ABS is determined by measuring the sorption isotherms. The concentrations of the hormones before sorption in the liquid phase and after sorption in the liquid phase (indirect method) are recorded after equilibrium has been established using the HPLC analysis

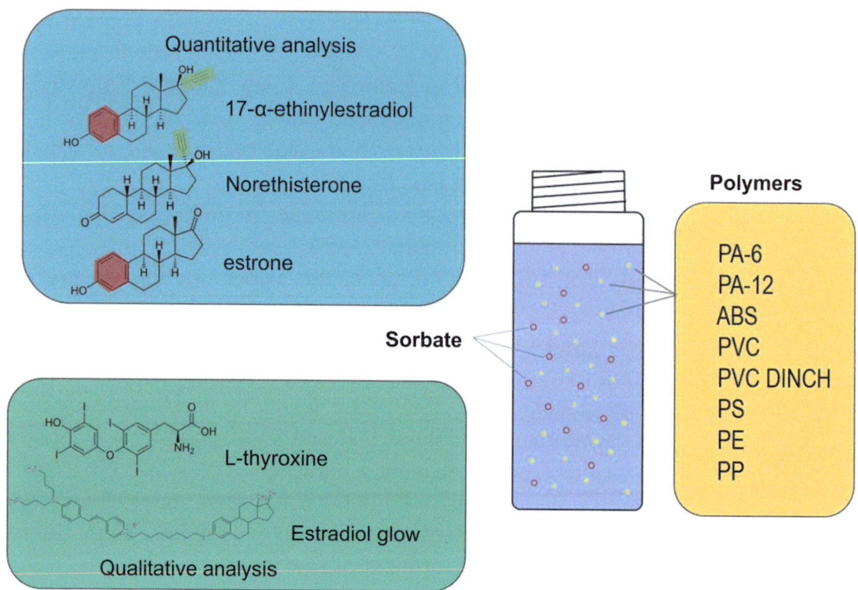

Fig. 3.17 Overview of the adsorption of hormones onto polymer particles. (Hummel 2017)

technique. For EE2 and E1, a fluorescence detector (emission: 275 nm; detection: 310 nm) is used, and for norethisterone, a (240 nm) UV detector is used (Walker and Watson 2010).

The diagram in Fig. 3.17 provides an overview of the combinations between the types of microplastics as sorbents and the different hormones as sorbates.

Grinding of Polymers to Produce Microplastics

For the grinding of polymers into microplastic particles <50 µm in diameter, a cryogenic swing mill, as shown in Fig. 3.18, is used. For this purpose, the grinding cup, which contains a stainless steel impact ball, is filled with cut-up polystyrene spoons. The grinding cup is then screwed into the cooling jacket of the cryogenic mill. During the grinding process, the grinding container is cooled down to −196 °C using liquid nitrogen. The grinding container is vibrated, up to a frequency of 30 Hz. The deeply cooled, brittle plastic is continuously shattered by the repeated impact of the stainless steel ball and thus ground into a fine powder (Fig. 3.18). In addition, the cooling removes the frictional heat generated inside the grinding container, thus preventing a chemical change in the plastics due to their temperature increase.

The resulting polymer powder can be produced from various plastic granules, reject parts, or plastic waste.

Larger microplastic particles up to 5 mm in diameter can also be obtained by a conventional shredder, which is used to recover thermoplastic recyclate from sprue or reject parts. The result of this cutting process can be seen in Fig. 3.19.

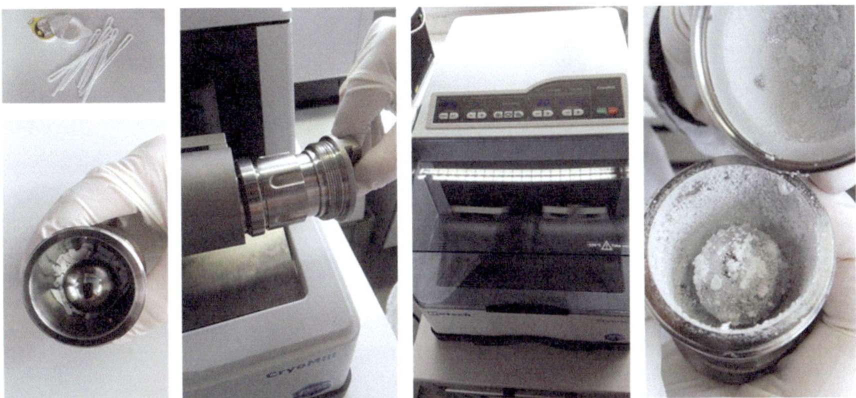

Fig. 3.18 Production of PS powder from plastic waste

Fig. 3.19 Production of recyclate (microplastics) from sprue parts in a plastic injection molding plant

Characterization of the Produced Microplastics

The particle size distribution of the polymers in aqueous calcium chloride solution was determined using laser diffraction spectrometry. The values given in Table 3.4 are the mean values of a triple determination.

Table 3.4 Results of the particle size measurement of all polymers using a laser diffraction spectrometer, indicating the measurement range. (Weighting: Number)

Polymer	d_{10}	d_{50}	d_{90}	Measurement range start [µm]	Measurement range end [µm]
PA-6	0.89	2.38	8.92	1	300
	12.19	24.08	55.30	10	2000
PA-12	19.33	39.25	63.63	1	300
	1.01	3.69	28.96	10	2000
ABS	0.89	2.38	11.60	1	300
	13.08	24.98	57.09	10	1000
PVC	13.68	29.44	89.20	10	2000
PVC-DINCH	21.11	52.03	103.18	10	2000
PS	11.30	19.33	43.71	10	2000

The specific surface area and pore volume (Table 3.5) of the polymers were determined using N_2-BET. The specific surface area within a relative pressure range $\left(\frac{p}{p_0}\right)$ from 0.074 to 0.297 as well as the pore volume with $\frac{p}{p_0} = 0{,}989$.

The low measurement values for the pore volume show that no pores are formed during the production of technical microplastics, for example for the operation of 3-D printers or for rapid prototyping in selective laser sintering (SLS). No pores are formed or exposed during the crushing process in the cryogenic mill either. This is different from the microplastics that have been in the environment (Rhine) for some time (Fig. 3.6).

Quantitative Results of Adsorption

After a balance has been established in the solution between the adsorption and desorption of hormones to the different added types of microplastics over several days, the remaining concentration of hormones in the liquid phase is determined chromatographically by HPLC after the liquid phase has been separated from the solid phase by centrifugation and the amount sorbed to the polymers is calculated based on the difference to the hormone input. The adsorption isotherms determined from this are shown in Fig. 3.20. The dissolved concentration is plotted against the solid phase concentration.

Table 3.5 Specific surface area (BET-SSA, *specific surface area*) and pore volume (TPV, *total pore volume*) of the used polymers

	PS	PVC	PVC DINCH	PA-6	PA-12	ABS	PE
BET-SSA [m²/g]	2.317	0.689	0.517	0.241	5.315	0.681	0.857
TPV [cc/g]	3.48E-03	1.12E-03	7.06E-04	5.41E-04	1.00E-02	1.06E-03	1.79E-03

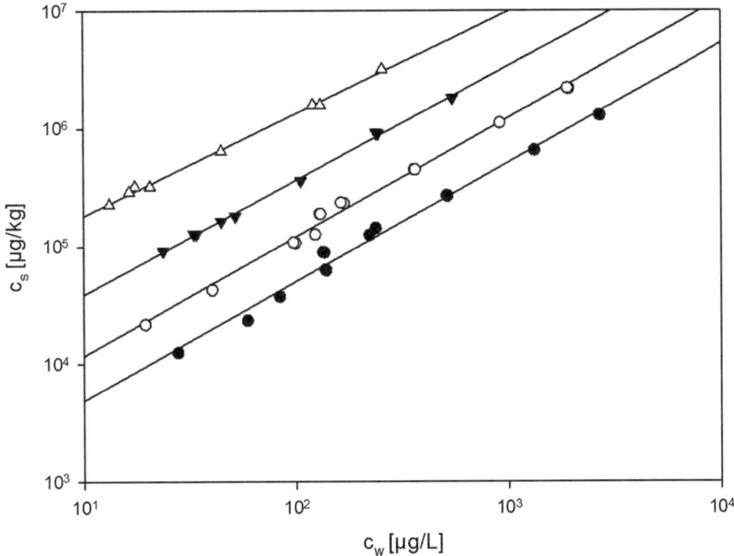

Fig. 3.20 Adsorption isotherms for the sorption of 17α-ethinylestradiol to PVC (●), PVC with the plasticizer DINCH (○), PA-6 (▼) and PA-12(△)

For all combinations of EE2 with the different polymers (for clarity, only PA and PVC are shown here), the isotherms show a linear course and are best interpreted with the linear or Freundlich's adsorption model. The steady increase suggests that a saturation of the hormone concentration on the plastic particle surface has not yet been reached and that further adsorption sites are still available.

The parameters of the adsorption models with the corresponding correlation factors are listed in Table 3.6. The higher the distribution coefficient K_d or the Freundlich coefficient K_F, the more EE2 adsorbs to the polymer particles.

The strongest sorption can be observed for PA-12 and PA-6. A K_d of 12.887 means according to the definition of $K_d = C_s / C_w$, that the concentration of EE2 on one kilogram of solid particles (C_s) is greater by the mentioned factor than the concentration of the hormone in one liter of water. The total concentration of, for example, EE2 in a body of water can thus be calculated using the distribution coefficient, or the deviation from an analysis result can be estimated, in which the filtrate is actually examined. Before or in an HPLC column (pre-column), water samples are filtered to prevent the actual separation column from being clogged by solids. The pre-filters in the HPLC system have a pore size of 0.5 µm. The substances sorbed to microplastics, which are retained there, are thus not detected in the analysis, but can also be detected by the eluent, whereby sorbed pollutants may possibly be extracted and thus at least partially included in the analysis result. If the desorption by the used eluent does not take place, the analysis result can be significantly affected, depending on the number and size of microplastic particles. According to the distribution coefficient of almost 13,000 for the hormone

Table 3.6 Parameters of the linear and Freundlich models for the adsorption of EE2 on polymer particles

Polymer	Linear Model		Freundlich Model		
	K_d	r^2	K_F	n_F	r^2
ABS	174.12	0.8925	276.24	0.9333	0.9049
PA-6	3497.46	0.9945	4125	0.9714	0.9942
PA-12	12,887.54	0.796	24,441.56	0.8688	0.9965
PVC	515.09	0.9727	437.9	1.021	0.9696
PVC-DINCH	1220.91	0.9815	1101.2491	1.0144	0.9802

EE2 on polyamide 12 listed in Table 3.6, the analysis for EE2 in a water sample would be about 10% too low if no solid phase extraction takes place after injection in the pre-column. With 100 PA-12 particles, the error would already be almost 50%. Especially in hormone analysis, where physiological impairment of aquatic habitats can already be detected at hormone concentrations that are at the detection limit of 1 ng/l or even below, the effects mentioned must be taken into account and investigated for a water assessment in order to be able to evaluate the actual load.

The sorption to PVC, from which DEHP was extracted, is weaker than the sorption to PVC mixed with DINCH. Based on the results obtained from the adsorption isotherms, the following affinity sequences of the polymers towards ethinylestradiol (EE2) can be determined.

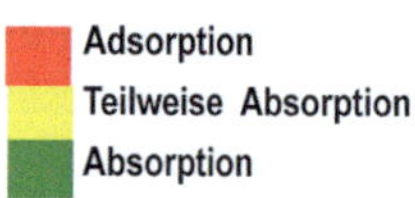

The fact that the sorption depends on the structure and polarity of the dissolved hormones is shown by the following result: When the sorption isotherms are normalized to the surface of the polymers, norethisterone shows the strongest sorption to PE, followed by PVC-DINCH. The course of the isotherms for PS, PA-6, PA-12 and PVC is similar (Hummel 2017).

The same tendency regarding the affinity of the hormones to the plastics can be observed for the sorption of EE2 and E1 on all polymers. The sequence of affinity also shows a certain dependence of the sorption on the water absorption capacity of the polymers. With increasing swelling capacity, the sorption capacity also increases, while the specific glass transition temperatures (Table 3.7) have no influence on the affinity.

By normalizing the isotherms to the surface, the loading of the surface can be represented in the case of pure adsorption. Normalization to the octanol-water distribution coefficient does not yield a significant correlation.

In order to better classify the sorption of the hormones, the distribution coefficients of other toxic or endocrine-acting substances, which can also be detected in waters, are listed in Table 3.8 from two different polymer articles.

The data in Table 3.8 show that the nonpolar polyaromatic hydrocarbon phenanthrene (Phe) and the plasticizer diethylhexyl phthalate have a higher affinity to PE and also to PVC than the hormones examined. If the polarity of the adsorbate increases compared to the nonpolar plastic, the distribution coefficient decreases and thus also the affinity. The strong influence of the type of plastic as a sorbent becomes clear in the comparison between PE and PVC in relation to Phe, DEHP and PFOA, whereas DDT (Fig. 3.21) does not seem to have a significant "preference" for either of the two plastics. However, this investigation does not examine the influence of the surface structure and does not make any statement about the sorption mechanism. In the case of PVC material, sorption, as can be seen with the hormones, plays a significant role.

The adsorption isotherms only show the distribution of the hormones in the liquid and solid phase, but do not provide any information about whether hormones or other substances are adsorbed by the microplastic particles, i.e., whether they only accumulate on the surface or whether they are absorbed and penetrate into the particle matrix. The sorption describes both mechanisms. To investigate the sorption of hormones on different types of microplastics, qualitative methods, such as

Table 3.7   Physical properties of different polymer particles. (Wypych 2016)

Polymer	$T_g{}^a$ [°C]	Water absorption [%]	SA^b [m^2/g]	d_{32} [µm]
HDPE	−118 to −133	0.005–0.01	0.857	k. A.
PP	−	0.02–0.04	k. A.	k. A.
PA-6	50–75	7.1–10	0.241	39.6
PA-12	55	1.3–1.5	5.315	50.7
PS	85–102	0.03–0.1	2.317	140.7
ABS	102–107^c; −58^d	0.7–1.03	0.681	45.0
PVC	82–87	0.04–0.4	0.689	114.4
PVC-DINCH	k. A.	k. A.	0.517	107.9

^{a}Glass transition temperature
^{b}Specific surface area, ^{c}Acrylonitrile-styrene mesophase, ^{d}Butadiene component d_{32} is the so-called Sauter diameter, a parameter for the particle size distribution; Def.: If the entire volume of all particles is transformed into spheres of the same size, whose volume to surface ratio is to be reproduced, then these spheres have the Sauter diameter

Table 3.8 Distribution coefficients of Phe, DDT, DEHP and PFOA on PE and PVC. (Bakir et al. 2014)

K_d	PE	PVC
Phe	52.000	2300
DDT	97.000	105.000
PFOA	500	7
DEHP	98.000	12.000

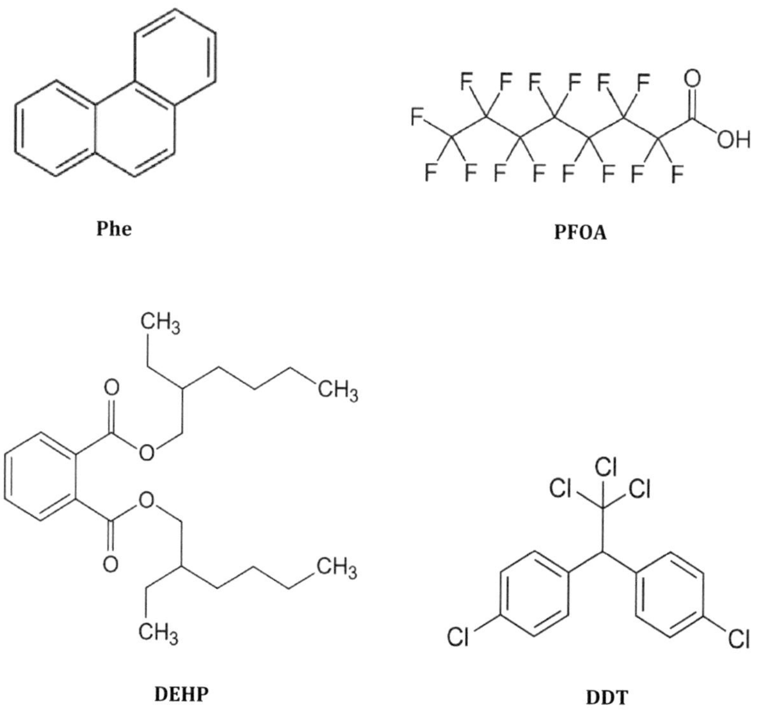

Fig. 3.21 Structural formulas of the POP *(persistent organic pollutants)*

confocal laser scanning microscopy, can be used, whereby the fluorescence of the adsorbate is used for imaging.

Adsorption or Absorption?

The hormone concentration at and in the polymer particle can also be determined qualitatively in a direct way. Either the adsorbate to be examined, such as the L-Thyroxine, shows fluorescence after appropriate stimulation, or a fluorescence-labeled hormone derivative such as Estradiol-Glow is used, whose concentration on the surface of the polymer particle can be determined by confocal laser scanning microscopy. Since laser scanning microscopy has a certain penetration depth into the polymer, this method can provide insights into whether a sorbate is adsorbed on the surface or whether it penetrates into the polymer matrix and is absorbed. The type of sorption and the penetration depth within the absorption are revealed by confocal fluorescence microscopy.

Confocal Laser Scanning Microscopy (qualitative method)

A confocal microscope is a special light microscope that only images areas that are in focus together. Unlike conventional light microscopy, not the entire object is illuminated, but at any time during scanning only a fraction of it, in many cases only a small spot of light that is in focus.

The method of confocal microscopy has the advantage over widefield microscopy that surface structures can be better represented and depending on the penetration depth of the laser beam, deeper layers can also be viewed. While in widefield microscopy the entire image of the sample is captured at once with the eyepiece, with confocal microscopy a sample is scanned point by point or in a line-like manner. The laser beam of a laser scanning microscope is focused on a specific point of the sample and the focus is moved within a plane of the sample to scan the entire plane. The focal plane can be freely chosen within the sample as long as it is within the range that can be focused by the objective and the sample condition allows the laser beam to penetrate. In reflection, the laser beam reflected by the sample is detected to capture the structure of an object. In fluorescence microscopy, a fluorophore is excited by its absorption wavelength and the light it emits is detected. The excitation wavelength is filtered out before the detection to avoid getting an image of the reflection of the examined object. The intensity of the detected radiation can be regulated by varying the laser power (Schwarzenbach et al. 2016).

3.2.5 Sorption of Hormones to Microplastics

For the investigation of the sorption of hormones using confocal fluorescence microscopy, three different methods are presented. On the one hand, the inherent fluorescence of selected hormones is used, whose fluorescence wavelength is in the range of the detection wavelength of the fluorescence microscope, to make them visible. A second approach is based on the use of already fluorescence-labeled hormones, which are commercially available, such as Estradiol-Glow. Since the functionalization of the hormone changes not only the structure but also the size and the polarity, and these changes also have an influence on the sorption properties, these compounds are only of limited significance. Nevertheless, the method is suitable for representing the type influence and the surface structural influence, i.e., a relative comparison of different microplastic particles, based on a model substance, and also for gaining insights into their mode of action as sorbents. The third method eliminates the influence of a fluorescence marker by coupling with a fluorophore only after sorption. However, restrictions must also be made with this method due to possibly different sorption properties of the hormones and their coupling reagents.

Thyroxine

The thyroid hormone thyroxine belongs to the group of non-proteinogenic α-amino acids (see structural formula in Fig. 3.22). In case of hypothyroidism, thyroxine-containing preparations are prescribed to substitute the missing own hormone thyroxine, which is needed for energy metabolism. Due to the increase in cardiovascular activity and the associated increased energy turnover, these hormone preparations are also misused as a "diet pill". Thyroxine is one of the five most frequently prescribed drugs worldwide (Löffler and Petridas 2014). Due to its frequent use and poor bioavailability, correspondingly high concentrations are excreted again and end up in wastewater and eventually in bodies of water.

At an excitation wavelength of 400 nm, thyroxine emits a fluorescence dye fluorescence wavelength at 454 nm (Hummel 2017).

This emitted radiation becomes visible in the fluorescence microscope and is shown in Fig. 3.23. The orange-red intrinsic fluorescence clearly shows that thyroxine is adsorbed on the surface of the polyamide-6 particle and even absorbed a few μm into the plastic.

The thyroxine crystals emit at a different wavelength (white) due to their morphology than the thyroxine sorbed to polyamide 6 (yellow-orange area).

Estradiol

With the already mentioned limitations, the sorption properties of 17β-estradiol can be studied with the fluorescence-labeled derivative Estradiol-Glow. The fluorescence labeling is necessary to shift the emitted fluorescence radiation into the measurable range of the fluorescence microscope. With a fluorescence excitation of $\lambda = 467$ nm, the Estradiol-Glow emits a fluorescence radiation of $\lambda = 618$ nm detectable by the microscope. It is easy to see from the structural formula in Fig. 3.24 that the attached long-chain chromophore, which is bound to the female steroid hormone via an ether bond, can have an influence on the sorption mechanism.

Fig. 3.25 shows a three-dimensional image of polyamide-6 particles, from which information about the shape and surface texture of microplastic particles can be obtained. The three-dimensional image was calculated and assembled from the results of the individual layer planes of the Z-axis. The distance of the individual images is 1 μm.

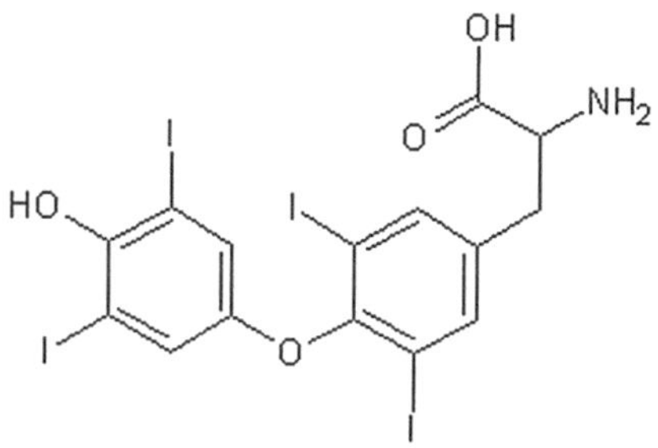

Fig. 3.22 Structural formula of thyroxine

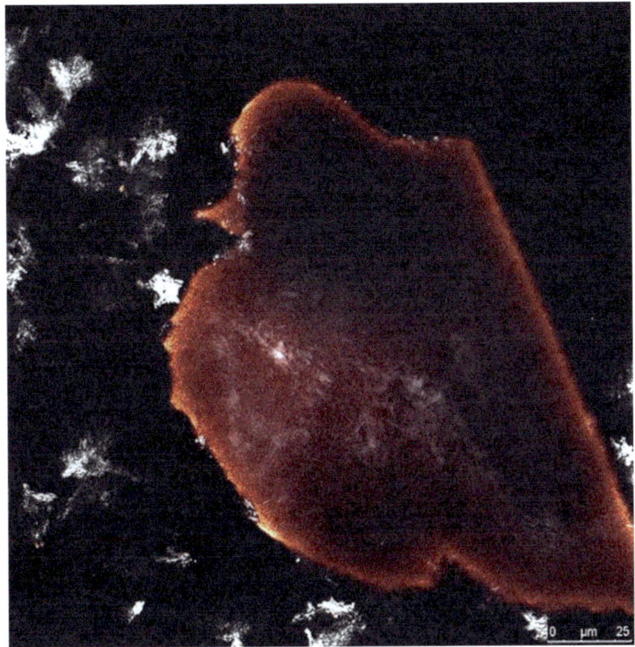

Fig. 3.23   Image of polyamide 6 with thyroxine with digital amplification of the fluorescence. (After Löffler and Petridas 2014)

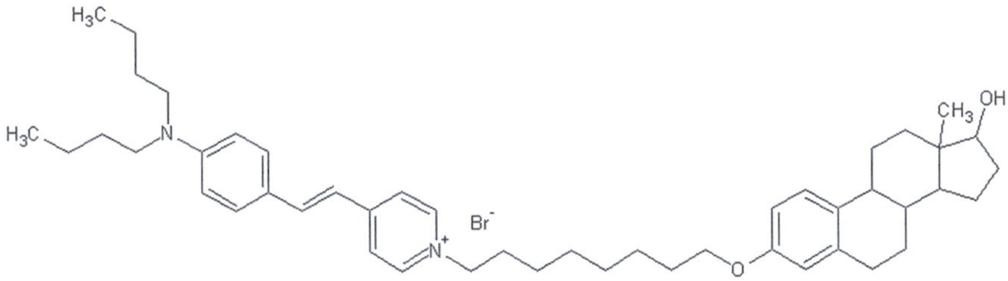

Fig. 3.24   Structural formula of Estradiol-Glow

The fluorescence on the stained particle surface indicates the distribution of Estradiol-Glow on it. The different concentrations of Estradiol-Glow on the surface correlate with the intensity of the fluorescence radiation, which is indicated by the color differences. This principle also applies to the two-dimensional images in Fig. 3.26.

In the left image of Fig. 3.26, the different intensities of the fluorescence on the surface of polyamide-6 microplastic particles can be seen, which indicate the presence of Estradiol-Glow. While in the right image of Fig. 3.26, the fluorescence signals at a penetration depth of 25 μm into the plastic particle are captured in the

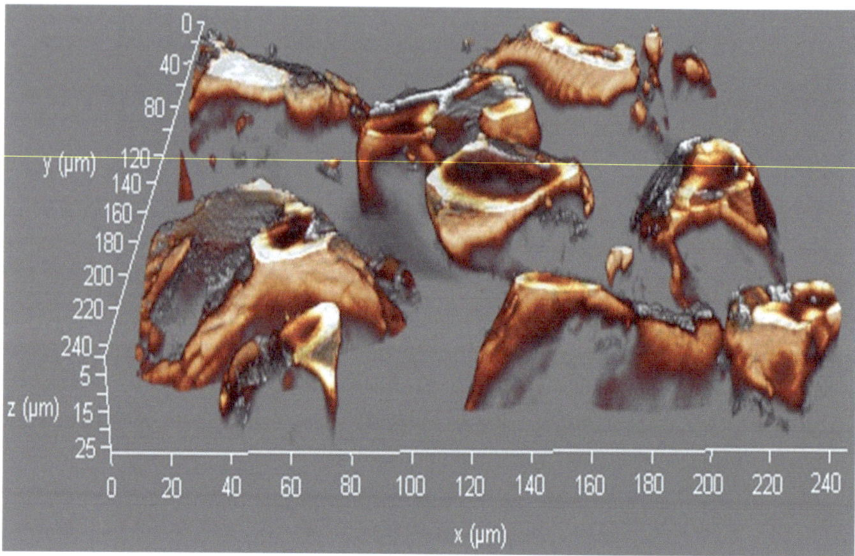

Fig. 3.25 3-D image of polyamide-6 particles with sorbed Estradiol-Glow, calculated from an image stack

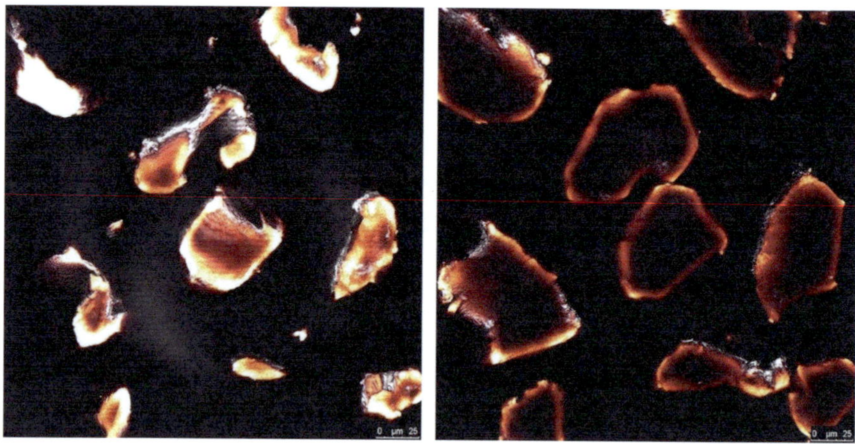

Fig. 3.26 Polyamide-6 particles with 0 µm sample penetration depth (left) and with 25 µm sample penetration depth (right)

image. It is clearly visible that there is no more Estradiol-Glow inside the polyamide-6 particle. Only from the edge areas of the particles is a narrow fluorescence band of about 5 µm thickness to be observed. Furthermore, the Estradiol-Glow has not penetrated after an exposure time of four days.

Based on the present images, in addition to adsorption, absorption of the Estradiol-Glow molecule can also be determined. However, absorption only occurs in an outer layer of the particles.

To better represent the polymer particles, in addition to the detection of the particles' fluorescence, their reflection was also recorded. For this purpose, the particles were irradiated with monochromatic light and the reflected light was detected.

The fact that the type of sorption and the penetration depth of substances depend on the type of plastic is shown by the result of the exposure of different microplastic granules in an Estradiol-Glow solution. In Fig. 3.27, the different results are visually represented.

While Estradiol-Glow only adsorbs to polypropylene, the labeled hormone penetrates a few μm into the PA, and in PVC with plasticizer, the test hormone is completely absorbed and penetrates the entire volume of the PVC particles, despite the increased molecule size due to the attached fluorophore.

Ethinylestradiol

Since the ethinylestradiol (EE2, structural formula see Fig. 3.28) has an absorption maximum at 275 nm and emits radiation with an emission maximum of 310 nm (Hummel 2017) and with the used fluorescence detectors only wavelengths >350 nm are detectable, the binding of the hormone to a plastic particle in the fluorescence microscope is not visible. By chemically binding the EE2 to another molecule, a so-called fluorescence marker, the EE2 can be detected in the confocal fluorescence microscope based on its marker.

Of course, one must assume that the sorption properties of ethinylestradiol are influenced by the reaction with the fluorescence dye. To exclude this influence, the coupling can also be carried out after the sorption.

Attachment of a Fluorescence Marker Through click chemistry

Since the click chemistry cannot directly detect the detection of 17α-ethinylestradiol on the surface of polymer particles for the reasons mentioned

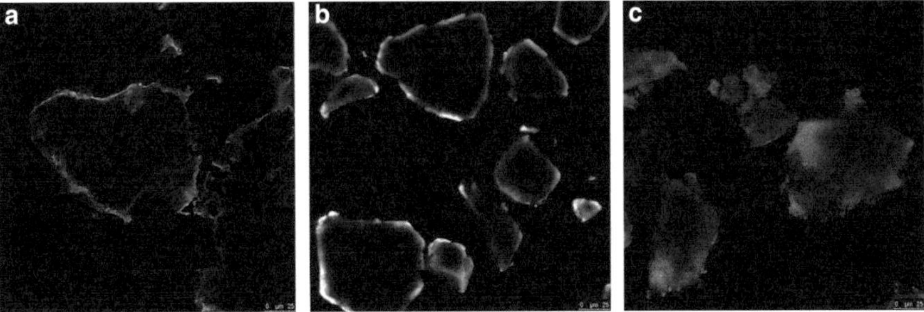

Fig. 3.27 Images of the fluorescence of Estradiol-Glow after the sorption on PP (**A**) (adsorption), PA-6 (**B**) (absorption with low penetration depth of the molecule) and PVC with the plasticizer DINCH (**C**) (absorption into the total volume of the particle), recorded using confocal laser scanning microscopy

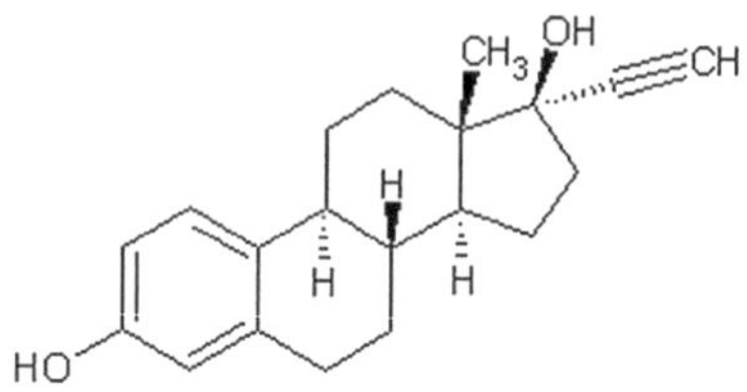

Fig. 3.28 Structural formula of 17α-ethinylestradiol

above, there is the possibility to chemically bind fluorescence dyes to the hormone using click chemistry and thus mark it. Click chemistry is a 4+2-Diels-Alder cycloaddition between the terminal alkyne group of ethinylestradiol and a terminal azide group of the fluorophore. The cyclization results in a thermodynamically stable aromatic triazole ring. The heterocycle connects the hormone EE2 with the chromophore.

Click reactions react almost completely and irreversibly (Kunz 2009). To accelerate the reaction, a catalyst is necessary. For this purpose, copper(I) ions are used. Since copper(I) salts are not stable in aqueous solution, they are formed by the reduction of $CuSO_4$ with sodium ascorbate and immediately stabilized by the polyhaptic chelate ligand THPTA (Tris(3-hydroxypropyltriazolylmethyl)amine) within the forming tetrahedral complex. Figure 3.29 shows the structure of the multidentate ligand. The copper(I)-THPTA complex must be able to penetrate into the polymer matrix to make absorbed EE2 detectable by the confocal fluorescence microscope. This also applies to the coupling reagent.

The coupling reaction of 17α-ethinylestradiol with the fluorophore 3-azido-7-hydroxycoumarin as a fluorescence indicator for EE2, shown in Fig. 3.30, takes place in a buffer system at pH 7. Since 3-azido-7-hydroxycoumarin shows fluorescence at an excitation wavelength of 405 nm and itself adsorbs and absorbs

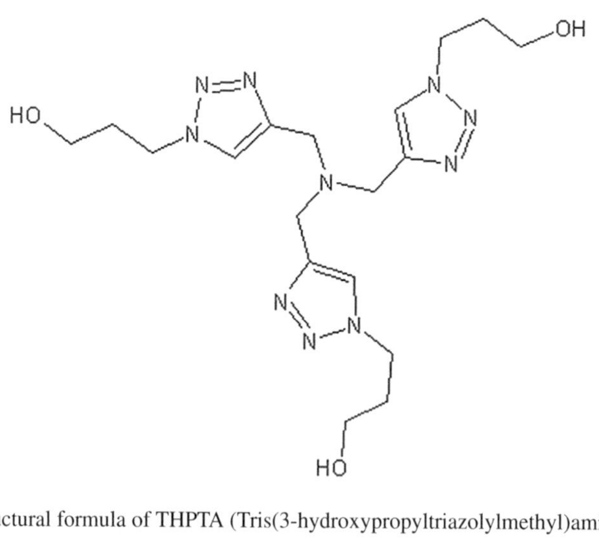

Fig. 3.29 Structural formula of THPTA (Tris(3-hydroxypropyltriazolylmethyl)amine)

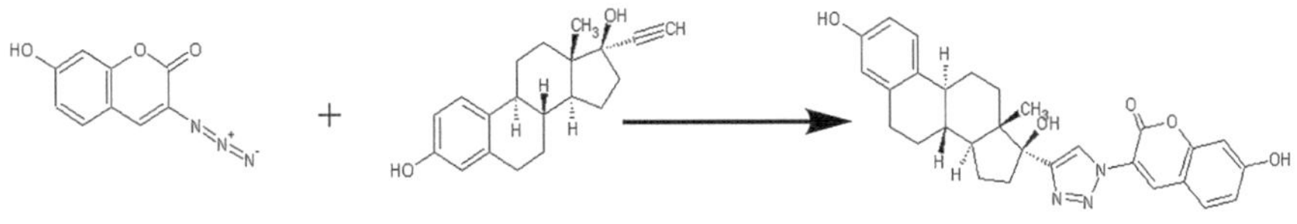

Fig. 3.30 Click reaction of 3-azido-7-hydroxycoumarin with 17α-ethinylestradiol

to polymer particles, the question arises how one can then detect the presence of hormones on the surface and in the polymer at all. For this purpose, the increase in fluorescence intensity by a multiple through the coupling of the azidocoumarin is used for imaging. The increase in fluorescence is the result of the expansion of the delocalized π-electron system by conjugation with the resulting triazole (Fig. 3.30).

For the click reaction with the acetylene group of ethinylestradiol, other azide-functionalized fluorescence dyes can naturally also be used, which differ greatly in their structure and polarity and thus determine their sorption properties on polymers and other sorbents. An example of this is the so-called Sufo-Cy5-Azide, the structure of which can be seen in Fig. 3.31.

The influence of the molecule size, the functional groups, and the polarity on the sorption behavior of a substance can be illustrated by comparing Sufo-Cy5-Azide with 3-Azido-7-hydroxycoumarin using confocal microscope images with fluorescence detection in Fig. 3.32. The microscope images show the fluorescence of polyamide-6 particles with a particle diameter of 50–100 μm, which were in one of the respective fluorescence dye solutions for three days.

In the images (Fig. 3.32), it can be seen from the different fluorescence intensities that the 3-Azido-7-hydroxycoumarin almost completely penetrates the PA-6

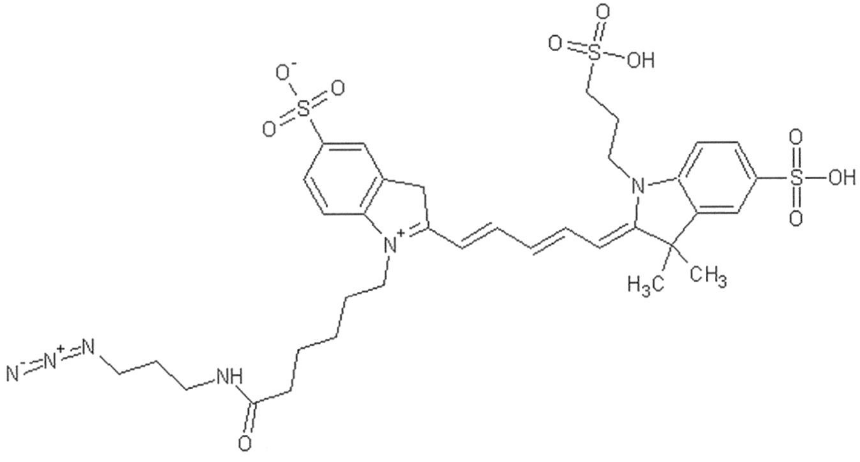

Fig. 3.31 Structural formula of Sufo-Cy5-Azide

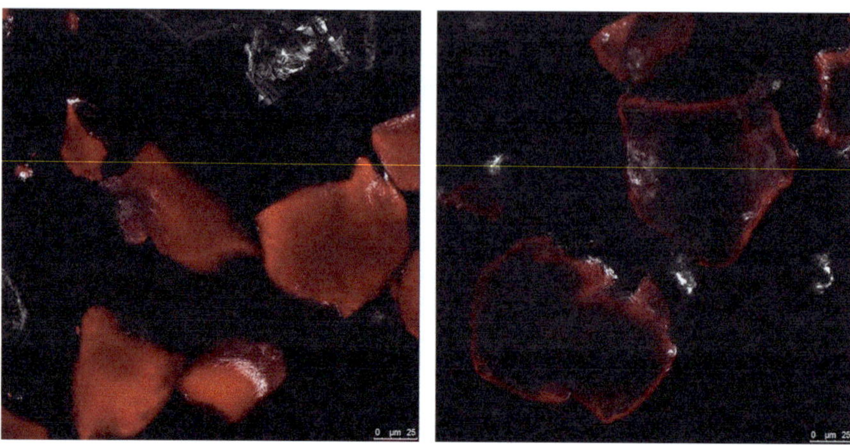

Fig. 3.32 PA-6 particles after three-day sorption of the dye 3-Azido-7-hydroxycoumarin (left) or after three-day sorption of the dye Sulfo-Cy5-Azide (right)

matrix (left), while the Sulfo-Cy5-Azide is only adsorbed on the surface and absorbed a few μm. One explanation for this is the smaller molecule size with a molar mass of 203.15 g/mol compared to the significantly bulkier Sulfo-Cy5-Azide with 833.01 g/mol and also the poorer water solubility of the coumarin derivative.

To investigate the sorption behavior of female steroid hormones, the labeling of EE2 was carried out using click chemistry after the sorption on polyamide-6. Since the sorption behavior is significantly influenced by the polarity of the sorbent, functional groups, and the molecule size, the labeling was carried out after the sorption of the hormone in order not to influence the sorption properties through chemical modification of the sorbent.

Coupling of Fluorescent Dyes with Ethinylestradiol to the Polymer Matrix

In order to investigate the sorption properties of Ethinylestradiol on microplastic particles, the plastic powders are stored for three days in the reaction solution with the respective fluorescent dye and together with the catalyst, and then scanned under the confocal microscope. In Fig. 3.33, the fluorescence signals of the chromophores without a binding of the hormone EE2 can be seen on the left half of the image (a and c). After the addition of the hormone, a clear increase in fluorescence can be seen on the right side of Fig. 3.33 (image b and d). The increase in fluorescence intensity (Fig. 3.33) after the addition of the hormone solution can indirectly demonstrate the coupling of the dye with the hormone EE2.

With Sulfo-Cy5-azide, the increase in intensity due to the coupling to EE2 is not associated with a change in the delocalization of the π-electron system, but with the polarity change in the immediate vicinity of the fluorophore that accompanies the coupling.

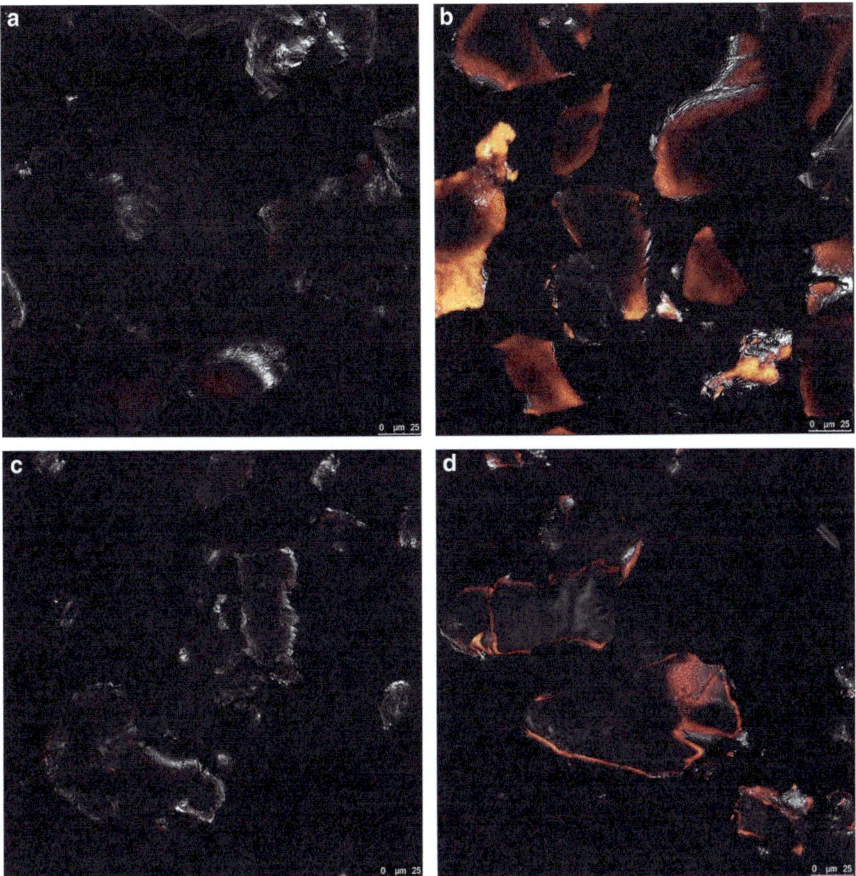

Fig. 3.33 PA-6 reference sample, mixed with 3-azido-7-hydroxycoumarin (**a**), PA-6 sample with sorbed EE2, mixed with 3-azido-7-hydroxycoumarin (**b**), PA-6 reference sample, mixed with Sulfo-Cy5 (**c**), PA-6 sample with sorbed EE2, mixed with Sulfo-Cy5 (**d**)

The emission spectrum of many fluorophores strongly depends on the polarity of their environment. A change of the solvent also leads in many cases to spectra that are altered in terms of their emission maximum and their intensity (Lakowicz 1983).

For a better optical distinction between sorbed fluorophore and EE2-coupled fluorophore, the radiation intensity of the microplastic samples with fluorophore without hormone is subtracted as a zero value from the samples mixed with hormone. In Fig. 3.33, these comparison images are juxtaposed. For the Sulfo-Cy5-azide coupled to EE2, mainly adsorption can be seen. For EE2-coupled 3-azido-7-hydroxycoumarin, absorption is observed, as increased fluorescence from the interior of the PA 6 microplastic particles can be seen compared to the reference sample.

The images shown of the samples mixed with Sulfo-Cy5 were taken with a sample penetration depth of 5 µm.

The fluorescence intensity for both used fluorescence markers is higher in the edge areas of the particles than in the interior of the particles. That the hormone actually not only adsorbs on the surface of the polymer particle, but penetrates deep into the interior, is clear in Fig. 3.33. The centers of the larger microplastic particles remain dark. An indication that no coupling with the fluorophore azido-7-hydroxycoumarin has taken place in these regions. That the coumarin derivative reaches these areas is shown in Fig. 3.32 (left). However, it cannot be ruled out that the hormone Ethinylestradiol is continuously absorbed in all depicted PA-6 polymer particles, because without absorption of the catalyst for the coupling in the corresponding areas, these remain dark due to the differential measurement.

The ligand THPTA ($M = 434.5$ g/mol), which stabilizes the catalyst, the copper(I) ion, has more than twice the molecular mass compared to 3-azido-7-hydroxycoumarin ($M = 203.15$ g/mol), so a lower penetration depth into the polymer particles may be the result.

Summary

For the sorbates examined, with the exception of norethisterone, polyamide-6 has the highest sorption capacity. The sorption capacity of polymers depends on the interplay of ad- and absorption. With strong additional absorption, plastic particles can bind larger amounts of substances than in the case of pure adsorption. For absorption, the water absorption capacity or the swelling ability of a plastic plays a crucial role (Hummel 2017). Plastics with a high plasticizer content are already swollen due to the liquid plasticizers and therefore show strong absorption compared to plasticizer-free polymers. It is not the case that the previous extraction of, for example, DEHD from the PVC matrix creates cavities that are filled by other substances from the solution, but rather the nonpolar hormones disperse into the liquid plasticizer phase. A correlation of the adsorption ability of a plastic, based on its glass transition temperature, is not apparent in the presented selection of sorbents and sorbates, as it is assumed that the lower the glass transition temperature of a plastic, the better its sorption properties (Guo et al. 2012). In the investigation of the sorption of phenanthrene, lindane, naphthalene or 1-naphthol on PE, the percentage of crystallinity in the polymer matrix also plays a role (Guo et al. 2012).

Confocal fluorescence microscopy has proven to be a meaningful tool for investigating sorption mechanisms. The presented results show that ad- and absorption depend on several factors and cannot be generalized. Both the molecule size, polarity and structure of the sorbates, as well as the different types of plastic as sorbents, have a strong influence on the type and amount of sorption, with a type of plastic itself again changing its sorption properties on different sorbates through its particle size, its swelling ability, its surface structure and additives. This is demonstrated by the example of PVC with and without DINCH (Hummel 2017).

The images of the confocal microscopy have shown that it would be sensible to relate the data of the adsorption isotherms to the particle surface and not to the polymer mass. In the representation of the adsorption isotherms, the plotting of the adsorbed substance amount per kg of adsorbent against the dissolved concentration has been established in the literature (Klöpffer 2012).

This type of plotting makes it difficult to draw a direct comparison without knowledge of the specific surface area of polymers, as a higher concentration at the particle surface is determined both by the affinity to the polymer and by the number of possible binding sites, i.e., the surface. A comparison through the particle size distribution is only limitedly possible, as laser diffraction spectrometry does not capture the surface structure, pores or channels in the particles. The particle size distribution, if no shape factor is included, is only valid for nearly spherical particles.

As can be seen from the results presented here, the size of the surface plays a major role in adsorption, especially with regard to the adsorption capacity to a certain amount of sorbent. This is one reason for the effectiveness of activated carbon, where 4 g have the same surface area as a football field. Since both the production of activated carbon under oxygen exclusion and the regeneration are very energy-intensive as temperatures up to 1000 °C must be reached, microplastic, which is generated from plastic waste, could represent a profitable alternative, provided that similarly large surfaces can be generated in the crushing and surface structuring process as with activated carbon.

The extent to which the surface of a solid increases through comminution is demonstrated by the following hypothetical example. If you divide a plastic ball with a volume of one liter, a diameter of 18 cm, and a surface area of 1037 cm^2 into two equal-sized balls, their total surface area increases by about 30% to 1035 cm^2 compared to the original ball. If you continue the comminution process of the plastic, with the respective predecessor fragments, to a spherical particle diameter below 5 mm, i.e., into the microplastic region, then the surface of the resulting 65,526 microplastic particles has increased to almost 42,000 cm^2 after 16 division steps—more than 40 times the original surface. The slope of the curve in Fig. 3.34 increases continuously. This means, the smaller the particles, the greater the surface increase after a division. The surface can be calculated for further comminution down to the μm scale using the given equation of the curve. The average particle diameters of the plastic powders used for the sorption of hormones were about 50 μm. This would correspond to a surface enlargement by a factor of 3620, starting from the original ball.

The above model does not do justice to the decay of macro to microplastics due to various mechanical, chemical, photochemical, and bacterial environmental influences, as the calculation of surface enlargement assumed an exact spherical shape and a smooth unstructured surface. Both assumptions do not reflect reality, as can be seen from the examination of microplastic findings in the Rhine (Fig. 2.158) and the decomposing plastic film in Fig. 3 (in the preface). Nevertheless, the model impressively shows the extreme growth of the surface with progressive comminution of plastic waste.

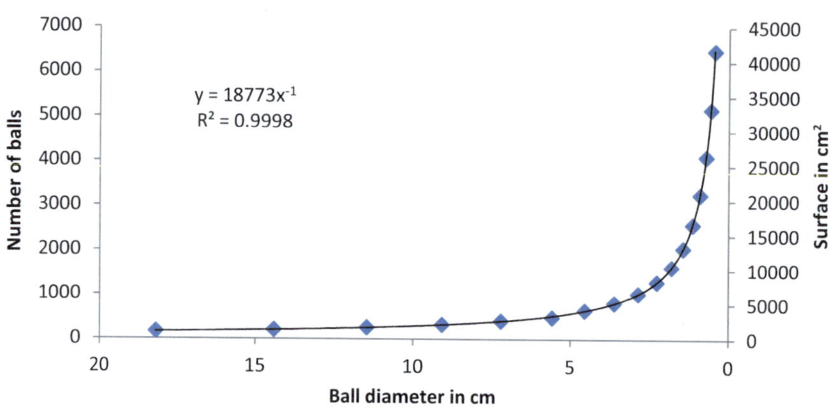

Fig. 3.34 Dependence of the sphere surface on the sphere diameter and the number of division steps

Regarding the utilization of microplastic particles as water filters, however, a critical limit in the particle size is reached when the packing density of the filter material becomes so large that unrealistic and impractical throughput times would be the result.

In the previous consideration, the factor "surface structuring" for surface enlargement and thus for increasing the "parking space offer" for hormones, for example, has not yet been taken into account. The potential for optimization that still exists here is demonstrated, for example, by the surface of a golf ball, which is covered with 300–450, dimples. The reason for this is not the 50% surface enlargement that is also achieved, but the lower air resistance compared to the smooth version. A more than 20-fold surface enlargement would be achieved by etching out the glass fibers in a glass fiber reinforced plastic. These are just a few examples that show the impact of structuring on the size of the surface.

Further basic research is still necessary until the development of an effective and competitive filter material made of plastic waste compared to activated carbon. These include, to name just a few, optimization of the particle size distribution, enlargement of the surface and thus the adsorption capacity, selection of plastic types or combinations of plastics for universal filter material, pre-treatment of plastics, selectivity studies, packing density for corresponding water volumes, regeneration, eluate processing, etc.

The first step, before we think about a meaningful recycling possibly even in the sense of upcycling, should be to reduce plastic waste. So the first priority is reduced use, the second is repeated use, and only the third is meaningful recycling.

References

Alvarez, D. A. (2010). *Guidelines for the Use of the Semipermeable Membrane Device (SPMD) and the Polar Organic Chemical Integrative Sampler (POCIS) in Environmental Monitoring Studies*. Chapter 4 of Section D, Water Quality, Book1, Collection of Water Data by Direct Measurement, Reston.

Alvarez, D. A., Stackelberg, P. E., Petty, J. D., Huckins, J. N., Furlong, E. T., Zaugg, S. D., et al. (2005). Comparison of a novel passive sampler to standard water-column sampling for organic contaminants associated with wastewater effluents entering a New Jersey stream. *Chemosphere, 61*(5), 610–622. https://doi.org/10.1016/j.chemosphere.2005.03.023.

Bakir, A., Rowland, S. J., & Thompson, R. C. (2014). Enhanced desorption of persistent organic pollutants from microplastic under simulated physiological conditions. *Environmental Pollution, 185,* 16 ff.

Barrett, E. P., Joyner, L. G., & Halenda, P. H. (1951). The determination of pore *volume and area distributions* in porous *substances*. I. Computations from nitrogen isotherms. *Journal of the American Chemical Society, 73*(1), 373–380.

Best, R., & Spingler, E. (1972). Messung von Adsorptions- und Desorptionsisothermen mit einer vollautomatischen Apparatur. *Chemie Ingenieur Technik, 44*(21), 1222. https://doi.org/10.1002/cite.330442108.

Brunauer, S., Emmett, P. H., & Teller, E. (1938). Adsorption of gases in multimolecular layers. *Journal of the American Chemical Society, 60,* 309.

Fath, A. (2009). Qualifizierung von Effektbeschichtungen. In R. Suchentrynk (Hrsg.), *Jahrbuch Oberflächentechnik* (vol. 65, p. 285). Bad Saulgau: Leuze.

Fath, A. (2016). *Rheines Wasser – 1231 Kilometer mit dem Strom*. München: Hanser.

Fath, A. (2017). Auswertung & Ergebnisse des Projekts „Rheines Wasser". Fakultätsinterne unveröffentlichte Daten, Villingen-Schwenningen.

Freundlich, H. (1906). Über die Adsorption in Lösungen. *Zeitschrift für Physikalische Chemie, 57,* 385–470.

Górecki, T., & Namieśnik, J. (2002). Passive sampling. *TrAC Trends in Analytical Chemistry, 21*(4), 276–291. https://doi.org/10.1016/s0165-9936(02)00407-7.

Guo, X., Wang, X., Zhou, X., Kong, S., Tao, S., & Xing, B. (2012). Sorption of four hydrophobic organic compounds by three chemically distinct polymers: Role of chemical and physical composition. *Environmental Science and Technology, 46*(13), 7252–7259.

Hartmann, A. (1993). *Immobilisierung von Immunoglobulin G auf ultradünnen vernetzten Langmuir-Blodgett-Filmen*. Diplomarbeit, Universität Heidelberg.

Henry, W. (1803). Experiments on the quantity of gases absorbed by water, at different temperatures, and under different pressures. *Philosophical Transactions of the Royal Society of London, 93,* 29–42 und 274–276. https://doi.org/10.1098/rstl.1803.0004 (Volltext).

Hochschule Furtwangen. (2014). „Rheines Wasser": Erstmals Ergebnisse vorgestellt. Beim Hansgrohe Wassersymposium berichtet Rheinschwimmer Andreas Fath über die Wasser-Analytik, Villingen-Schwenningen

Hummel, D. (2017). *Untersuchung der Sorption wässrig gelöster organischer Substanzen an Polymerpartikel*. Masterthesis, Hochschule Furtwangen, Studiengang NBT.

Hummel, D., Hüffer, T., Fath, A., Nestle, N., & Rueckel, M. (eingereicht). „Sorption of different 17-α-Ethinylestradiol and Estrone to micro-sized plastic particles from aqueous solution: Quantitative and qualitative sorption analysis"; Environmental Pollution.

IKSR (Internationale Kommission zum Schutz des Rheins). (2009). International koordinierter Bewirtschaftungsplan für die internationale Flussgebietseinheit. (Teil A = übergeordneter Teil) Dezember, Koblenz.

IKSR (Internationale Kommission zum Schutz des Rheins). (o. J.). Fachbericht IKSR. https://www.iksr.org/fileadmin/user_upload/DKDM/Dokumente/Fachberichte/DE/rp_De_0186.pdf.

Job, G., & Rüffler, R. (2011). *Physikalische Chemie. Eine Einführung nach neuem Konzept mit zahlreichen Experimenten*. Wiesbaden: Vieweg + Teubner/Springer Fachmedien.

Klöpffer, W. (2012). *Verhalten und Abbau von Umweltchemikalien* (2. ed.). Weinheim: Wiley-VCH.

Kolb, B. (1999). *Gaschromatographie in Bildern – Eine Einführung*. Weinheim: Wiley-VCH.

Kraus, U. R., Theobald, N., Gunold, R., & Paschke, A. (2015). Prüfung und Validierung der Einsatzmöglichkeiten neuartiger Passivsammler für die Überwachung prioritärer Schadstoffe unter der WRRL, der MSRL und im Rahmen von HELCOM und OSPAR. 01.01.2010–30.10.2012. Umweltbundesamt Forschungskennzahl 3709 22 225, UBA-FB 001938, Dessau-Roßlau.

Kunz, D. (2009). Synthesen, die gelingen Klick-Chemie. *Chemie in unserer Zeit, 43,* 224–230.

Lakowicz, J. R. (1983). *Principles of fluorescence spectroscopy*. New York: Plenum Press.

Löffler, G., & Petridas, P. E. (2014). *Biochemie und Pathobiochemie* (9. ed., p. 514). Berlin: Springer. ISBN 978–3-642-17972-3.

Matsuzawa, Y., Kimura, Z.-I., Nishimura, Y., Shibayama, M., & Hiraishi, A. (2010). Removal of hydrophobic organic contaminants from aqueous solutions by sorption onto biodegradable polyesters. *Journal of Water Resource and Protection, 2*(3), 214–221.

Meeker, J. D., Sathyanarayana, Sheela, & Swan, Shanna H. (2009). Phthalates and other additives in plastics: Human exposure and associated health outcomes. *Philosophical transactions of the Royal Society of London. Series B, Biological sciences, 1526,* 2097–2113. https://doi.org/10.1098/rstb.2008.0268.

Mills, G. A., Vrana, B., Allan, I., Alvarez, D. A., Huckins, J. N., & Greenwood, R. (2007). Trends in monitoring pharmaceuticals and personal-care products in the aquatic environment by use of passive sampling devices. *Analytical and Bioanalytical Chemistry, 387*(4), 1153–1157. https://doi.org/10.1007/s00216-006-0773-y.

Moschet, C., Vermeirssen, E. L. M., Singer, H., Stamm, C., & Hollender, J. (2015). Evaluation of in-situ calibration of chemcatcher passive samplers for 322 micropollutants in agricultural and urban affected rivers. *Water Research, 71,* 306–317. https://doi.org/10.1016/j.watres.2014.12.043.

Muhandiki, V. S., Shimizu, Y., Adou, Y. A. F., & Matsui, S. (2008). Removal of hydro-phobic micro-organic pollutants from municipal wastewater treatment plant effluents by sorption onto synthetic polymeric adsorbents: Upflow column experiments. *Environmental Technology, 29*(3), 351–361.

Reichert, W. M., Bruckner, C. J., & Joseph, J. (1987). Langmuir-Blodgett films and black lipid membranes in biospecific surface-selective sensors. *Thin Solid Films, 152*(1–2), 345–376.

Ricking, M. (2009). Umweltprobenbank des Bundes – Probenahme polarer Wasserinhaltsstoffe an Probenahmeflächen der UPB. Endbericht. Freie Universität Berlin UBA FKZ 301 02 026, Berlin.

Rüdel, H., Bester, K., Eisenträger, A., Franzaring, J., Haarich, M., Köhler, J., Körner, W., Oehlmann, J., Paschke, A., Ricking, M., Schröder, W., Schröter-Kermani, C., Schulze, T., Schwarzbauer, J., Theobald, N., Trenck, T. v. d., Wagner, G., & Wiesmüller, G. A. (2007). Positionspapier zum stoffbezogenen Umweltmonitoring. Erarbeitet vom Arbeitskreis Umweltmonitoring in der GDCh-Fachgruppe Umweltchemie und Ökotoxikologie. Fraunhofer-Institut für Molekularbiologie und Angewandte Oekologie IME, Schmallenberg.

Schmidt, R. D., & Bilitewski, U. (1992). Biosensoren. *Chemie in unserer Zeit, 26*(4), 163–174.

Schwarzenbach, R. P., Gschwend, P. M., & Imboden, D. M. (2016). *Environmental Organic Chemistry*. Hoboken: Wiley.

Tamschick, S., Rozenblut-Kościsty, B., Ogielska, M., Lehmann, A., Lymberakis, P., Hoffmann, F., et al. (2016). Impaired gonadal and somatic development corroborate vulnerability differences to the synthetic estrogen ethinylestradiol among deeply diverged anuran lineages. *Aquatic Toxicology, 177,* 503–514. https://doi.org/10.1016/j.aquatox.2016.07.001.

Vrana, B., Mills, G. A., Allan, I. J., Dominiak, E., Svensson, K., Knutsson, J., et al. (2005). Passive sampling techniques for monitoring pollutants in water. *TrAC Trends in Analytical Chemistry, 24*(10), 845–868. https://doi.org/10.1016/j.trac.2005.06.006.

Walker, C. W., & Watson, J. E. (2010). Adsorption of estrogens on laboratory materials and filters during sample preparation. *Journal of Environmental Quality, 39*(2), 744–748.

Wedler, G. (1970). *Adsorption – Einführung in Physisorption und Chemisorption.* Weinheim: Chemie.

WWF (World Wildlife Fund). (2014). Half of global wildlife lost, says new WWF Report. https://www.worldwildlife.org/press-releases/half-of-global-wildlife-lost-says-new-wwf-report.

Wypych, G. (2016). *Handbook of polymers* (2. ed.). Toronto: ChemTec.

Conclusion

4

By now, almost every sound mind understands that it is a crime against the environment to dispose of used oil in the forest or on the riverbank. When it comes to disposing of our plastic waste in nature, our conscience is still not bad enough, even though it is an equally great crime against the environment, which we ultimately pay for with our own health when we eat the resulting microplastics. If this realization that disposing of plastic waste in the environment is not a minor offense and is ultimately perceived as just as threatening to our ecosystem as used oil becomes widespread through education, then much would be achieved. My goal is to make a small contribution to this with this book.

16. Tips to Avoid Plastic Waste and its Entry into our Waters
- Prefer glass over plastic, especially for food items (milk bottle, yogurt, ketchup, dressing, etc.), there is a dense network of collection containers for glass.
- If you must use plastic bottles, then only those with a deposit!
- Only buy cosmetic products from brands that offer refill options in the store. Not just soaps but also shampoos in solid form (similar to a hockey puck, now available without a plastic container).
- When buying cosmetic products, make sure they do not contain microplastics. Indicators for this are ingredients that begin with "Poly".
- When buying fruits and vegetables, even in the supermarket, avoid the small plastic bags that you can tear off in the fresh produce department and that you can only open with a lot of finesse, as much as possible. Beef tomatoes, cucumbers, cauliflower, lettuce, bananas, etc. can also be placed on the conveyor belt as they are. You wash fruits and vegetables anyway and you don't want to eat plasticizers.
- Generally plan your shopping in advance. This means bringing your own containers to the supermarket and to all other stores that pack your goods in plastic for you (e.g., schnitzel, steaks, salads, sausage from the butcher).

A. Fath, *Microplastic*, https://doi.org/10.1007/978-3-662-69844-0_4

- Dispose of plastic waste, if it is not reusable (plastic bags can be reused, also as garbage bags), in the yellow sack. Important! Plastics can only be recycled if they are sorted by type. This means separating plastic composites (example: plastic yogurt cup with label and lid).
- At the deli counter, indicate that you want the raw ham without plastic interleaf (it's also cheaper and a hassle anyway).
- Buying larger quantities in one large container rather than many small ones reduces plastic packaging waste. Unfortunately, this often costs more. Here, the legislator must act. It cannot be that if you want to prepare 1 kg of Maultaschen, you choose three or four small packaging units just because the price per kilo is cheaper than the one-kilo pack. The saving customer is virtually forced to produce a lot of packaging waste. Incentives in the other direction must be created in this way.
- Never throw plastic remnants or expired fruit packaged into the organic waste bin. The organic waste is partly shredded and ends up as fertilizer possibly then with microplastics on our fields.
- Always have a shopping basket made of natural materials (hemp, jute, bast, willow, etc.) in the trunk for shopping (also spontaneous). No plastic bags.
- After a party at the lake or river, do not leave the trash lying around. With the next rain, the trash ends up in the water and eventually in our oceans. (The problem is known.)
- Use your own ceramic cup in the canteen or at the vending machines at work. No lids on the paper cups from coffee to go. They are made of polystyrene.
- Do not offer children drinking straws. Or only those that can be reused after the dishwasher. (In the USA, 500 million drinking straws are consumed daily). So the cocktail in the evening without a straw or from degradable renewable raw materials (real straw straws).
- Wear more textiles made of natural materials. A fleece jacket produces 1.7 g of plastic fibers during a wash cycle, which end up as microplastics in water and in fish. Or use a sealable laundry bag, from which you can remove the laundry and fibers after the washing process.
- Only put the yellow bags out the door on the day they are picked up. Rodents, wind, and rain otherwise have time to distribute the contents in the environment.
- Do not drive with winter tires longer than absolutely necessary.

The listed tips have been incorporated into the graphic of the Stuttgarter Zeitung (Fig. 4.1).

Fig. 4.1 Graphic from the Stuttgarter Zeitung. (Jana Evers, Weekend - the magazine from Sonntag aktuell)

Appendix 5

Mark Twain: A Rafting Trip on the Neckar (1878)

The river was full of logs—long, slender, barkless spruce logs—and we leaned on the bridge railing and watched as people assembled them into rafts.

These rafts had a shape and design adapted to the winding course and extraordinary narrowness of the Neckar. They were fifty to a hundred yards long and gradually tapered from nine logs wide at the stern to three logs wide at the bow. Steering is mainly done from the bow with a pole; the width of three logs there only gives room for the helmsman, for these small logs have no greater circumference than the average waist of a young lady. The connections between the different sections of the raft are slack and yielding, so that the raft can easily be bent to any curvature that the shape of the river requires.

Germany is the pinnacle of beauty in the summer, but no one has comprehended, truly perceived and enjoyed the highest extent of this gentle and peaceful beauty, who has not rafted down the Neckar.

The movement of a raft is just right; it is sluggish, gliding, gentle and noiseless; it soothes all feverish activity, lulls all nervous haste and impatience; under its calming influence, any anger, annoyance and sorrow that torment the mind fade away, and life becomes a dream, a charm, a deep and silent rapture. What a contrast it forms to the laborious wandering and the dusty, deafening railroad frenzy and the boring bumping over glaring white roads behind tired horses!

We glided silently between the green, fragrant banks, with a feeling of joy and satisfaction that kept growing. Sometimes dense masses of willows hung over the banks, completely obscuring the land behind; sometimes we had magnificent mountains on one side, densely covered with foliage to the summit, and on the other side open plains, flaming with poppies or covered with the rich blue of the cornflower; sometimes we drifted in the shadow of the forests and sometimes along the edge of long stretches of velvety grass, fresh, green and glowing grass, an eternally young charm for the eye. And the birds!—they were everywhere; they constantly crossed the river back and forth, and their jubilant song never ceased.

A. Fath, *Microplastic*, https://doi.org/10.1007/978-3-662-69844-0_5

At noon we went ashore, bought a few bottles of beer and had some chickens cooked while the raft waited; then we immediately set sail again and ate as long as the beer was cold and the chickens were hot. There is no more pleasant place for such a meal than a raft gliding down the winding Neckar, past green meadows and wooded hills, past slumbering villages and rocky heights adorned with crumbling towers and battlements.

We wanted to cover the eight-mile stretch to Heidelberg before darkness fell. In the mild glow of the sunset, we shot into the narrow passage between the dams with the rushing current. I thought I could steer under the bridge myself, so I went to the three logs at the front and took the pole and responsibility from the helmsman.

We raced along in an incredibly exciting style, and I performed the delicate duties of my office really well for a first attempt, but when I suddenly noticed that I was actually heading for the bridge itself instead of the arch underneath it, I wisely jumped ashore. The next moment my long-cherished wish was fulfilled: I saw a raft shatter. It hit the pillar right in the middle and shattered and tore like a box of matches struck by lightning.

Reference

Twain, M. (2014). Gesammelte Werke: Reise um die Welt; Reise durch Deutschland. Band 5 der Ausgewählten Werke in zwölf Bänden von Mark Twain, hrsg. von Karl-Heinz Schönfelder im Aufbau Verlag, Berlin. Aus dem Amerikanischen übersetzt von Ana Maria Brock. © Aufbau Verlag GmbH & Co. KG, Berlin 1963, 2008. Abdruck mit freundlicher Genehmigung.

MIX
Papier aus verantwortungsvollen Quellen
Paper from responsible sources
FSC® C105338